T0137807

Studies in Big Data

Volume 97

The series "Studies in Big Data" (SBD) publishes new developments and advances in the various areas of Big Data- quickly and with a high quality. The intent is to cover the theory, research, development, and applications of Big Data, as embedded in the fields of engineering, computer science, physics, economics and life sciences. The books of the series refer to the analysis and understanding of large, complex, and/or distributed data sets generated from recent digital sources coming from sensors or other physical instruments as well as simulations, crowd sourcing, social networks or other internet transactions, such as emails or video click streams and other. The series contains monographs, lecture notes and edited volumes in Big Data spanning the areas of computational intelligence including neural networks, evolutionary computation, soft computing, fuzzy systems, as well as artificial intelligence, data mining, modern statistics and Operations research, as well as self-organizing systems. Of particular value to both the contributors and the readership are the short publication timeframe and the world-wide distribution, which enable both wide and rapid dissemination of research output.

The books of this series are reviewed in a single blind peer review process.

Indexed by SCOPUS, EI Compendex, SCIMAGO and zbMATH.

All books published in the series are submitted for consideration in Web of Science.

More information about this series at https://link.springer.com/bookseries/11970

Jorge Vázquez-Herrero · Alba Silva-Rodríguez ·
María-Cruz Negreira-Rey · Carlos Toural-Bran ·
Xosé López-García
Editors

Total Journalism

Models, Techniques and Challenges

Editors
Jorge Vázquez-Herrero
Faculty of Communication Sciences
Universidade de Santiago de Compostela
Santiago de Compostela, Spain

Alba Silva-Rodríguez
Faculty of Communication Sciences
Universidade de Santiago de Compostela
Santiago de Compostela, Spain

María-Cruz Negreira-Rey
Faculty of Communication Sciences
Universidade de Santiago de Compostela
Santiago de Compostela, Spain

Carlos Toural-Bran
Faculty of Communication Sciences
Universidade de Santiago de Compostela
Santiago de Compostela, Spain

Xosé López-García
Faculty of Communication Sciences
Universidade de Santiago de Compostela
Santiago de Compostela, Spain

ISSN 2197-6503 ISSN 2197-6511 (electronic)
Studies in Big Data
ISBN 978-3-030-88030-9 ISBN 978-3-030-88028-6 (eBook)
https://doi.org/10.1007/978-3-030-88028-6

This Springer imprint is published by the registered company Springer Nature Switzerland AG
The registered company address is: Gewerbestrasse 11, 6330 Cham, Switzerland

Editorial Project

This book is part of the activities developed in:

- the research project *Digital Native Media in Spain: Storytelling Formats and Mobile Strategy* (RTI2018-093346-B-C33) funded by the Ministry of Science, Innovation and Universities (Government of Spain), Agencia Estatal de Investigación, and co-funded by the European Regional Development Fund (ERDF);
- Novos Medios research group, supported by Xunta de Galicia (ED431B 2020/20).

Editorial coordination, book layout and formatting: Jorge Vázquez-Herrero, María-Cruz Negreira-Rey and Alba Silva-Rodríguez.

Contents

An Introduction to 'Total Journalism' ... 1
Jorge Vázquez-Herrero, Alba Silva-Rodríguez,
María-Cruz Negreira-Rey, Carlos Toural-Bran, and Xosé López-García

The Innovative Nature of Journalism

Convergence and Innovation: The Conceptual and Methodological Basis of Technological Evolution and Cultural Complexity in Journalism ... 13
Ainara Larrondo Ureta, Javier Díaz Noci, and Ivar John Erdal

Journalistic Storytelling for Complexity ... 29
Jorge Vázquez-Herrero and Yael de Haan

After the Hype: How Hi-Tech Is Reshaping Journalism ... 41
Sara Pérez-Seijo and Paulo Nuno Vicente

Gamification and Newsgames as Narrative Innovations in Journalism ... 53
Jose Alberto García-Avilés, Raul Ferrer-Conill, and Alba García-Ortega

Producing Content in the Ubiquitous, Mobile Era: The Case of Digital Media ... 69
Alba Silva-Rodríguez and Juan-Miguel Aguado-Terrón

Information Visualization: Features and Challenges in the Production of Data Stories ... 83
Ana Figueiras and Ángel Vizoso

Emerging Journalisms: From Intuition to Prediction and the Constructive Approach ... 97
Xosé López-García, Carlos Toural-Bran, and Jorge Vázquez-Herrero

Big Data and Information Disorders

Misinformation Beyond the Media: 'Fake News' in the Big Data Ecosystem ... 109
Ramón Salaverría and Bienvenido León

Big Data and Disinformation: Algorithm Mapping for Fact Checking and Artificial Intelligence ... 123
David García-Marín, Carlos Elías, and Xosé Soengas-Pérez

From Misinformation to Trust: Information Habits and Perceptions About COVID-19 Vaccines ... 137
Carmen Costa-Sánchez and Carmen Peñafiel-Saiz

Models, Professionals, and Audiences

Transformation of Local Journalism: Media Landscapes and Proximity to the Public in Spain, France and Portugal ... 153
María-Cruz Negreira-Rey, Laura Amigo, and Pedro Jerónimo

Social Implications of Paywalls in a Polarized Society: Representations, Inequalities, and Effects of Citizens' Political Knowledge ... 169
Tamás Tóth, Manuel Goyanes, Márton Demeter, and Francisco Campos-Freire

The Construction of Communicative Space: The Nature of COVID-19 Information in Italy ... 181
Laura Solito and Carlo Sorrentino

Professional Profile of the Contemporary Digital Journalist ... 195
Suzana Oliveira Barbosa, Lívia de Souza Vieira, Mariana Menezes Alcântara, and Moisés Costa Pinto

Audiences First: Professional Profiles, Tools and Strategies of Digital Newsrooms to Connect with the Public ... 211
Ana-Isabel Rodríguez-Vázquez, Marius Dragomir, and Noelia Francisco-Lens

Co-creation and Curation of Contents: An Indissoluble Relationship? ... 227
José Sixto-García, Pablo Escandón Montenegro, and Lila Luchessi

The Challenges of Artificial Intelligence

Platforms in Journalism 4.0: The Impact of the Fourth Industrial Revolution on the News Industry ... 241
Josep-Lluís Micó, Andreu Casero-Ripollés, and Berta García-Orosa

Apocalypse or Redemption: How the Portuguese Media Cover Artificial Intelligence 255
João Canavilhas and Renato Essenfelder

Horizon 2030 in Journalism: A Predictable Future Starring AI? 271
Bella Palomo, Bahareh Heravi, and Pere Masip

Editors and Contributors

About the Editors

Jorge Vázquez-Herrero is Assistant Professor at the Department of Communication Sciences, Universidade de Santiago de Compostela (USC). Ph.D. in Communication and Contemporary Information (USC). He is member of Novos Medios research group and Latin American Chair of Transmedia Narratives (ICLA–UNR). He was Visiting Scholar at Universidad Nacional de Rosario, Universidade do Minho, University of Leeds and Tampere University. His research focuses on the impact of technology and platforms in digital journalism and narratives.

https://orcid.org/0000-0002-9081-3018

Alba Silva-Rodríguez is Assistant Professor of Journalism at the Department of Communication Sciences at Universidade de Santiago de Compostela. Ph.D. in Journalism and member of Novos Medios research group. She is Secretary of the RAEIC journal. As researcher, she focuses on the assessment of digital communication, especially the study of mediated conversation in social media and the evolution of media contents in mobile devices.

https://orcid.org/0000-0002-1221-5178

María-Cruz Negreira-Rey is Lecturer of Journalism at the Department of Communication Sciences at Universidade de Santiago de Compostela. Ph.D. in Communication (USC) and member of Novos Medios research group (USC). Her research focuses on local journalism and the development of local and hyperlocal media in Spain, digital journalism and social media.

https://orcid.org/0000-0001-8945-2641

Carlos Toural-Bran is Assistant Professor at the Department of Communication Sciences, Universidade de Santiago de Compostela (USC), and Ph.D. in Communication Sciences (USC). He is Chief of Staff of the Rector (USC), also President of AGACOM (Galician Association of Communication Researchers), and Secretary of Novos Medios research group. He has also been Vice Dean of the Faculty of Communication Sciences (2014–2019).

https://orcid.org/0000-0002-0961-3925

Xosé López-García is Professor of Journalism at the University of Santiago de Compostela (USC), and Ph.D. in History (USC). He coordinates the Novos Medios research group. Among his research lines, there is the study of digital and printed media, analysis of the impact of technology in mediated communication, analysis of the performance of cultural industries, and the combined strategy of printed and online products in the society of knowledge.

https://orcid.org/0000-0002-1873-8260

Contributors

Juan-Miguel Aguado-Terrón Universidad de Murcia, Murcia, Spain

Mariana Menezes Alcântara Universidade Federal da Bahia, Salvador, Brazil

Laura Amigo Université de Neuchâtel, Neuchâtel, Switzerland

Suzana Oliveira Barbosa Universidade Federal da Bahia, Salvador, Brazil

Francisco Campos-Freire Universidade de Santiago de Compostela, Santiago de Compostela, Spain

João Canavilhas Universidade da Beira Interior, Covilhã, Portugal

Andreu Casero-Ripollés Universitat Jaume I de Castelló, Castellón, Spain

Carmen Costa-Sánchez University of A Coruña, A Coruña, Spain

Yael de Haan University of Applied Sciences Utrecht, Utrecht, The Netherlands

Lívia de Souza Vieira Universidade Federal da Bahia, Salvador, Brazil

Márton Demeter University of Public Service, Budapest, Hungary

Javier Díaz Noci Pompeu Fabra University, Barcelona, Spain

Marius Dragomir Center for Media, Data and Society, Central European University, Vienna, Austria

Carlos Elías Carlos III University of Madrid, Madrid, Spain

Ivar John Erdal Volda University College, Volda, Norway

Pablo Escandón Montenegro Simón Bolívar Andean University of Ecuador, Quito, Ecuador

Renato Essenfelder Universidade Fernando Pessoa, Porto, Portugal

Raul Ferrer-Conill Karlstad University, Karlstad, Sweden;
University of Stavanger, Stavanger, Norway

Ana Figueiras iNOVA Media Lab, Universidade Nova de Lisboa, Lisboa, Portugal

Noelia Francisco-Lens Universidade de Santiago de Compostela, Santiago, Spain

Jose Alberto García-Avilés Miguel Hernández University, Elche, Spain

David García-Marín Universidad Rey Juan Carlos, Madrid, Spain

Berta García-Orosa Universidade de Santiago de Compostela, Santiago de Compostela, Spain

Alba García-Ortega Miguel Hernández University, Elche, Spain

Manuel Goyanes Carlos III University, Madrid, Spain

Bahareh Heravi School of Information and Communication Studies, University College Dublin, Dublin, Ireland

Pedro Jerónimo Universidade da Beira Interior, Covilhã, Portugal

Ainara Larrondo Ureta University of the Basque Country, Leioa, Spain

Bienvenido León Universidad de Navarra, Pamplona, Spain

Xosé López-García Universidade de Santiago de Compostela, Santiago de Compostela, Spain

Lila Luchessi National University of Río Negro, Río Negro, Argentina

Pere Masip Ramon Llull University, Barcelona, Spain

Josep-Lluís Micó Universitat Ramón Llull, Barcelona, Spain

María-Cruz Negreira-Rey Universidade de Santiago de Compostela, Santiago de Compostela, Spain

Bella Palomo University of Malaga, Malaga, Spain

Carmen Peñafiel-Saiz University of Basque Country, Bilbao, Spain

Sara Pérez-Seijo Universidade de Santiago de Compostela, Santiago de Compostela, Spain

Moisés Costa Pinto Universidade Federal da Bahia, Salvador, Brazil

Ana-Isabel Rodríguez-Vázquez Universidade de Santiago de Compostela, Santiago, Spain

Ramón Salaverría Universidad de Navarra, Pamplona, Spain

Alba Silva-Rodríguez Universidade de Santiago de Compostela, Santiago de Compostela, Spain

José Sixto-García Universidade de Santiago de Compostela, Santiago de Compostela, Spain

Xosé Soengas-Pérez Universidade de Santiago de Compostela, Santiago de Compostela, Spain

Laura Solito Dipartimento di Scienze Politiche e Sociali, Università degli Studi di Firenze, Firenze, Italy

Carlo Sorrentino Dipartimento di Scienze Politiche e Sociali, Università degli Studi di Firenze, Firenze, Italy

Tamás Tóth Corvinus University of Budapest, Budapest, Hungary

Carlos Toural-Bran Universidade de Santiago de Compostela, Santiago de Compostela, Spain

Jorge Vázquez-Herrero Universidade de Santiago de Compostela, Santiago de Compostela, Spain

Paulo Nuno Vicente Universidade Nova de Lisboa, Lisboa, Portugal

Ángel Vizoso Universidade de Santiago de Compostela, Santiago de Compostela, Spain

An Introduction to 'Total Journalism'

Jorge Vázquez-Herrero, Alba Silva-Rodríguez, María-Cruz Negreira-Rey, Carlos Toural-Bran, and Xosé López-García

Abstract We introduce the book's scope, based on the concept of 'total journalism'. The chapters are presented following the structure of the book. The contributions of the authors offer an overview of the main elements that make up total journalism today, providing an in-depth study and reflecting the trends in its development.

1 Background

Journalism has undergone an authentic metamorphosis in the first two decades of the twentieth century [30]. The popularization of the Web has led to the migration of the media industry's communication products to the Internet and the emergence of digital native media. The beginning of a new technological transition and the emergence of the social web turned platforms into major players in the communication landscape. These transformations have positioned journalism in a network society that advances under the shadow of artificial intelligence and high technology [13].

Starting the third decade of the third millennium within an agitated scenario for the communicative ecosystem because of the social consequences of the COVID-19 pandemic, journalism continues to encompass debates that have accompanied it throughout history. The debate is ongoing about its role in society, the roles of journalists, ethical challenges, sustainability models, and social involvement in its management and preservation. The essential elements of journalism, its precepts and its techniques have defined it as a social communication medium. It has a heritage that allows it to face challenges with renewed techniques and tools, which we can call the 'total journalism' needed for contemporary society. Today there is a great hybridization of practices [12] and many questions remain about the future of journalism [7, 31].

Total journalism refers to the renewed and current journalism that employs all available techniques. It preserves multimedia, hypertextuality and interactivity as

J. Vázquez-Herrero (✉) · A. Silva-Rodríguez · M.-C. Negreira-Rey · C. Toural-Bran · X. López-García
Universidade de Santiago de Compostela, Santiago de Compostela, Spain
e-mail: jorge.vazquez@usc.es

J. Vázquez-Herrero et al. (eds.), *Total Journalism*, Studies in Big Data 97,
https://doi.org/10.1007/978-3-030-88028-6_1

central elements along with the teachings harvested throughout the history of journalism. Total journalism is made for today's society and with the latest technologies, including: artificial intelligence, virtual reality, 5G technology, blockchain, big data, information visualization, transmedia storytelling, renewed verification techniques, and augmented reality. Overcoming the limitations of the physical and rigid supports of the past, today's communicative means accompany journalism to build informative pieces that provide added value and are useful to society.

The search for new paths is based on the combination of participation and the values of journalism [14], the effective articulation of the emotional turn to intervene in the new scenario [2], and the empathy that a good journalist must seek with users [11], among other emerging dimensions. The future of journalism is networked [28], although there are many questions in the current panorama, which is characterized by a hybrid system of old and new media [4]. Nevertheless, the future of journalism is a current debate to which many researchers and different sectors of society are seeking answers not only to help make it viable, but also to build better-informed democratic and pluralistic societies.

Many experiences of digital journalism show that there are ways of ensuring the future of journalism [8] because hardware and software have recently shown that, if used properly, they allow innovative formats and products that show good practices in the scenario of intelligent automation and societies that are highly sensitive to pandemics, threats, achievements, and future projects. This is the context in which we must place the challenges of the strategies of total journalism.

It is a kind of journalism marked by technologization—which works and is framed in the field of high technologies—and by a renewed humanism that feeds on the past and is oxygenated in the digital scenario. It is total journalism that creates many expectations for a society in need of information in a scenario marked by the shadow of disinformation, but which, at the same time, is accompanied by many debates that show its richness, diversity and need for social support. This twenty-first century total journalism [16] has tools, production processes, dissemination channels, narrative forms and professional profiles specific to the current technological scenario. On the other hand, the fundamental principles of journalism and its mission to tell reality based on truthfulness, loyalty, independence and citizen participation in public debate remain. Total journalism is a part of today's society and responds to today's society.

The context in which total journalism develops is complex and involves a variety of actors and technological and social factors. With an international vision, the authors of this book approach the study of total journalism from different perspectives: innovation in the journalistic field, which affects processes and products; disinformation and its relationship with big data, its effects on trust in the media or fact-checking methods; new media models, the interrelation of news actors or the relationship with the audience; as well as the challenges of the application of artificial intelligence to journalism.

2 The Innovative Nature of Journalism

In this context marked by rapid technological evolution and growing cultural complexity, the study of journalism needs to recover and update concepts as decisive as convergence or innovation. Larrondo Ureta and colleagues [15] review the main empirical studies on newsrooms convergence and address how the technological adaptation of the media is a driving force for innovation and the development of hi-tech journalism. The integration of total journalism elements and media convergence have accelerated the development of digital innovative narratives and the transformation of genres. New forms of storytelling are emerging that combine classic narrative resources with technology and interactivity. Along with a review of the relationship between journalism and narratives, Vázquez-Herrero and de Haan [29] expose the value of interactive documentary and immersive journalism in the face of complexity and with the aim of finding the best way to inform the audience. On the other hand, García-Avilés and others [9] study the application of gamification elements to the creation of innovative journalistic products. The authors analyze the newsgames The Ocean Game (2019) and The Amazon Race (2019) and point out narrative and audience participation possibilities offered by this storytelling genre.

The consolidation of new narratives and forms of interaction with the audience is possible thanks, to a large extent, to the ubiquitous connection of the user and the consumption of content at almost any time and place. Silva-Rodríguez and Aguado-Terrón [24] discuss how digital media are adapting to the mobile era and review various journalistic initiatives and products developed through mobile technology, artificial intelligence and big data, which give rise to new narrative formats and forms of participation with the public. In adapting to the multiplatform scenario, various challenges also arise in the field of information visualization. Figueiras and Vizoso [6] address this issue and select the most innovative examples from the international media scene to present an overview of the changes that visual narratives have undergone in recent decades.

The application of high technology in journalism opens up new opportunities but also generates debates. The automation of news production and user interaction processes, and the use of personal assistants and smart speakers, are based on the use of artificial intelligence and big data to generate optimized responses to user profiles. On the other hand, immersive technologies and drones bring new visual perspectives. Pérez-Seijo and Vicente [21] conclude that technology is part of news production DNA and that journalists face their role supported by it, also considering that there are risks and challenges for journalistic practice. In this scenario, the so-called emerging journalisms apply techniques and technologies in the search for quality journalism that guarantees a future in which reinvention will be a constant. López-García and colleagues [17] point to a shift from experimental intuition towards prediction and constructive and solution-based journalism, passing through various movements that have used technology as a support for journalism.

3 Big Data and Information Disorders

Among the great challenges facing journalism today is that of combating disinformation and recovering the value of the media as reliable news actors. In recent years, the public dissemination of falsehoods has focused academic attention on terms such as post-truth, fake news and alternative facts. Disinformation and misinformation cover more precisely the multitude of modalities of false content circulating in our society, according to their intentionality or unintentionality, respectively. These phenomena, however, go beyond the media. Salaverría and León [23] review types of false content and the factors that influence its dissemination, as well as providing a historical overview of this topic.

Situations such as the COVID-19 pandemic create a landscape of change and uncertainty that serves as a breeding ground for misinformation. The spread of panic and confusion in the media, coupled with the emergence of fake news, has generated a state of infodemic with negative effects on building trust in vaccines. Costa-Sánchez and Peñafiel-Saiz [5] analyze the information patterns and habits in times of pandemic in Spain and offer some keys and recommendations for good communication management on the subject of vaccines against the virus.

From a broader perspective, García-Marín and others [10] approach disinformation from an academic perspective, aware that the technological dimension of total journalism does not just affect the reality of the media and journalists. They study the assessment that journalism students make of technological tools to fight disinformation, analyze the scientific literature published in recent years, and point out the main avenues of technological development in service of fact-checking. The authors argue that the conjunction of journalism with other more technical and technological disciplines is necessary.

4 Models, Professionals and Audiences

In the context of total journalism, media models emerge or are renewed, new news actors appear, and the professional profiles of communicators are transformed. Constant innovation due to the platformization, datafication and algorithmization of the communicative sphere causes journalists to develop new profiles adapted to the contemporary digital scenario, which are studied by Barbosa and colleagues [1] in this work.

In addition to the continuing professional crisis of journalists, there is the challenge of the economic sustainability of the media, which many are trying to address with the implementation of paywalls as a measure to obtain income. Tóth and others [27] describe the social implications of such tools, which use big data in some form to adjust the limitations of news consumption. Users' refusal to pay and the inequality that purchasing power can generate both justify the need to advance in the search for new sustainable business models for news production.

As journalism adapts to global technological and communication trends and seeks solutions to their challenges, the local space is revalued as a meeting place for journalists and citizens. Local and hyperlocal digital media are experiencing significant growth at the international level and are becoming a new media model. Negreira-Rey and colleagues [19] study the local digital media maps of Spain, France and Portugal. They point out some of the keys to their development and discuss strategies for connecting with the audience and citizen participation.

Users have long since assumed an active role in the communicative processes, taking a central place in the editorial and business strategies of the media. This interest in the audience gives rise to new professional profiles aimed at designing strategies that understand, stimulate, and measure engagement with users. Rodríguez-Vázquez and others [22] summarize the experiences of 100 international digital media in the field of audience measurement, which are aimed at strengthening the connection with users. An example of the search for this connection is the integration of the dynamics of co-creation of content in the news production process. Sixto-García and colleagues [25] study these dynamics in digital native media and identify three main forms of co-creation, which are the suggestion of ideas and issues, submission of elaborated content and contribution to circulation and diffusion.

What seems evident in the current digital environment is that the voices of journalists and the public sector converge with those of other actors, creating a polyphonic and multidimensional communicative space. Solito and Sorrentino [26] approach this complex communicative space in the context of the COVID-19 pandemic. They focus their study on the main communicative actors in social networks in three regions of Italy and analyze their communicative strategies and impact on the audience.

5 The Challenges of Artificial Intelligence

The fourth industrial revolution brings with it a cultural and technological change that has a direct impact on journalism. Micó-Sanz and colleagues [18] study how big data, artificial intelligence, viral and augmented reality, and machine learning affect journalism in its relationship with the audience, business models and news production. In this hi-tech context in which total journalism is developing, the rise of artificial intelligence has made it possible to integrate it into newsrooms to facilitate routine coverage related, for example, to sports and economic information. Canavilhas and Essenfelder [3] address this issue and analyze the presence of topics related to artificial intelligence in five Portuguese media from a multidisciplinary perspective and with the aim of detecting the thematic approach, the value, the news genre or the section in which this information is published.

To close the book, Palomo and colleagues [20] take a critical look at the future of journalism and its relationship with technology. Despite the strong influence of artificial intelligence and other buzzwords, "the future of journalism is journalism". With the need to remain alert to new technological developments and the ethical debates involved in the application of technologies in the journalistic field, the 2030 horizon looks complex and promising.

6 Conclusion

The contributions of the authors of this book provide an overview of the main elements that make up total journalism today, providing an in-depth study and reflecting the trends in its development. Total journalism is determined by a highly technological and changing context, the adaptation of media and professionals to this complex environment and a communication and social scenario in which new actors, new information problems and renewed consumption habits of an increasingly active audience are emerging.

In this complex environment, hi-tech is placed at the service of the news function. The media are incorporating artificial intelligence, automation and big data management to make certain production processes more efficient or to develop new verification tools in their fight against disinformation. Advances in mobile technology lead to a scenario of ubiquitous connection, which redefines the dissemination of information through mobile devices and leads to the intensification of convergence in media organizations. Technological change induces a continuous process of innovation in which narratives, journalistic genres and formats are transformed. Content is becoming more visual and interactive, experimenting with immersive formats or introducing logics from other cultural products such as gamification. The complexity of the context affects not only the technological level, but also the social level. The new social communication platforms have opened the door to new communication actors and have fuelled problems such as disinformation.

Within this new context, the media are seeking to adapt with new organizational, production and business models. The construction of profitable and independent editorial projects is still a challenge, as is the adaptation of journalists to a market in which new professional profiles are in demand.

However, in total journalism, the essentials remain the same. The protection of truthful reporting and the battle against disinformation are becoming crucial to preserve the value of honesty in the media and to continue to build trust. At the same time, proximity between media and citizens is revalued in local journalism, which is always responsive to the needs of its community. Finally, the audience becomes central to editorial and business decisions and strategies. Passive reception by users has been transformed into active participation in news production, co-creation of content, dissemination and public debate.

References

1. Barbosa, S.O., de Souza Vieira, L., Menezes Alcântara, M., Costa Pinto, M.: Professional profile of the contemporary digital journalist. In: Vázquez-Herrero, J. et al. (eds.) Total Journalism: Models, Techniques and Challenges. Studies in Big Data (2021)
2. Beckett, C., Deuze, M.: On the role of emotion in the future of journalism. Social Media + Society **2**(3) (2016). https://doi.org/10.1177/2056305116662395
3. Canavilhas, J., Essenfelder, R.: Apocalypse or redemption: how the Portuguese media cover artificial intelligence. In: Vázquez-Herrero, J. et al. (eds.) Total Journalism: Models, Techniques and Challenges. Studies in Big Data (2021)
4. Chadwick, A.: The Hybrid Media System: Politics and Power. Oxford University Press, Oxford (2013)
5. Costa-Sánchez, C., Peñafiel-Saiz, C.: From misinformation to trust: information habits and perceptions about COVID-19 vaccines. In: Vázquez-Herrero, J. et al. (eds.) Total Journalism: Models, Techniques and Challenges. Studies in Big Data (2021)
6. Figueiras, A., Vizoso, Á.: Information visualization: features and challenges in the production of data stories. In: Vázquez-Herrero, J. et al. (eds.) Total Journalism: Models, Techniques and Challenges. Studies in Big Data (2021)
7. Franklin, B.: The future of journalism. Journal. Stud. **17**(7), 798–800 (2016). https://doi.org/10.1080/1461670X.2016.1197641
8. Deuze, M.: Considering a possible future for digital journalism. Revista Mediterránea de Comunicación **8**(1), 9–18 (2017). https://doi.org/10.14198/MEDCOM2017.8.1.1
9. García-Avilés, J.A., Ferrer-Conill, R., García-Ortega, A.: Gamification and newsgames as narrative innovations in journalism. In: Vázquez-Herrero, J. et al. (eds.) Total Journalism: Models, Techniques and Challenges. Studies in Big Data (2021)
10. García-Marín, D., Elías, C., Soengas-Pérez, X.: Big data and disinformation: algorithm mapping for fact checking and artificial intelligence. In: Vázquez-Herrero, J. et al. (eds.) Total Journalism: Models, Techniques and Challenges. Studies in Big Data (2021)
11. Glück, A.: What makes a good journalist? Journal. Stud. **17**(7), 893–903 (2016). https://doi.org/10.1080/1461670X.2016.1175315
12. Hamilton, J.F.: Hybrid news practices. In: Witschge, T. (eds.) The SAGE Handbook of Digital Journalism. Sage, London, pp. 164–178 (2016)
13. Han, S., Ye, L., Meng, W.: Artificial Intelligence for Communications and Networks. Springer, Harbin (2019)
14. Hujanen, J.: Participation and the blurring values of journalism. Journal. Stud. **17**(7), 871–880 (2016). https://doi.org/10.1080/1461670X.2016.1171164
15. Larrondo Ureta, A., Díaz Noci, J., Erdal, I.J.: Convergence and innovation: the conceptual and methodological basis of technological evolution and cultural complexity in journalism. In: Vázquez-Herrero, J. et al. (eds.) Total Journalism: Models, Techniques and Challenges. Studies in Big Data (2021)
16. López García, X., Rodríguez Vázquez, A.I.: Journalism in transition, on the verge of a 'Total Journalism' model. Intercom **39**(1), 57–68 (2016). https://doi.org/10.1590/1809-5844201614
17. López-García, X., Toural-Bran, C., Vázquez-Herrero, J.: Emerging journalisms: from intuition to prediction and the constructive approach. In: Vázquez-Herrero, J. et al. (eds.) Total Journalism: Models, Techniques and Challenges. Studies in Big Data (2021)
18. Micó-Sanz. J.-L., Casero-Ripollés, A., García-Orosa, B.: Platforms in journalism 4.0: the Impact of the Fourth Industrial Revolution on the News Industry. In: Vázquez-Herrero J et al. (eds) Total Journalism: Models, Techniques and Challenges. Studies in Big Data (2021)
19. Negreira-Rey M-C, Amigo L, Jerónimo P (2021) Transformation of local journalism: media landscapes and proximity to the public in Spain, France and Portugal. In: Vázquez-Herrero, J. et al. (eds.) Total Journalism: Models, Techniques and Challenges. Studies in Big Data (2021)
20. Palomo, B., Heravi, B., Masip, P.: Horizon 2030 in journalism: a predictable future starring AI? In: Vázquez-Herrero, J. et al. (eds.) Total Journalism: Models, Techniques and Challenges. Studies in Big Data (2021)

21. Pérez-Seijo, S., Vicente, P.N.: After the hype: how hi-tech is reshaping journalism. In: Vázquez-Herrero, J. et al. (eds.) Total Journalism: Models, Techniques and Challenges. Studies in Big Data (2021)
22. Rodríguez-Vázquez, A.-I., Dragomir, M., Francisco-Lens, N.: Audiences first: professional profiles, tools and strategies of digital newsrooms to connect with the public. In: Vázquez-Herrero, J. et al. (eds.) Total Journalism: Models, Techniques and Challenges. Studies in Big Data (2021)
23. Salaverría, R., León, B.: Misinformation beyond the media: 'Fake News' in the big data ecosystem. In: Vázquez-Herrero, J. et al. (eds.) Total Journalism: Models, Techniques and Challenges. Studies in Big Data (2021)
24. Silva-Rodríguez, A., Aguado-Terrón, J.-M.: Producing content in the ubiquitous, mobile era: the case of digital media. In: Vázquez-Herrero, J. et al. (eds.) Total Journalism: Models, Techniques and Challenges. Studies in Big Data (2021)
25. Sixto-García, J., Escandón Montenegro, P., Luchessi, L.: Co-creation and curation of contents: an indissoluble relationship? In: Vázquez-Herrero, J. et al. (eds.) Total Journalism: Models, Techniques and Challenges. Studies in Big Data (2021)
26. Solito, L., Sorrentino, C.: The construction of communicative space: the nature of COVID-19 iInformation in Italy. In: Vázquez-Herrero, J. et al. (eds.) Total Journalism: Models, Techniques and Challenges. Studies in Big Data (2021)
27. Tóth, T., Goyanes, M., Demeter, M., Campos-Freire, F.: Social implications of paywalls in a polarized society: representations, inequalities, and effects of citizens' political knowledge. In: Vázquez-Herrero, J. et al. (eds.) Total Journalism: Models, Techniques and Challenges. Studies in Big Data (2021)
28. Van der Haak, B., Parks, M., Castells, M.: The future of journalism: networked journalism. Int. J. Commun. **6**, 2923–2938 (2012)
29. Vázquez-Herrero, J., de Haan, Y.: Journalistic storytelling for complexity. In: Vázquez-Herrero, J. et al. (eds.) Total Journalism: Models, Techniques and Challenges. Studies in Big Data (2021)
30. Vázquez-Herrero, J., Direito-Rebollal, S., Silva-Rodríguez, A., López-García, X.: Journalistic Metamorphosis: Media Transformation in the Digital Age. Studies in Big Data, vol. 70. Springer, Cham (2019)
31. Wahl-Jorgensen, K., Williams,A., Sambrook, R., Harris, J., Garcia-Blanco, I., Dencik, L., Cushion, S., Carter, C., Allan, S.: The future of journalism. Journal. Stud. **17**(7), 801–807 (2016). https://doi.org/10.1080/1461670X.2016.1199486

Jorge Vázquez-Herrero Assistant Professor at the Department of Communication Sciences, Universidade de Santiago de Compostela (USC), Ph.D. in Communication and Contemporary Information (USC). He is a member of Novos Medios research group and the Latin American Chair of Transmedia Narratives (ICLA–UNR). He was visiting scholar at Universidad Nacional de Rosario, Universidade do Minho, University of Leeds and Tampere University. His research focuses on the impact of technology and platforms in digital journalism and narratives.

Alba Silva-Rodríguez Assistant Professor of Journalism at the Department of Communication Sciences at Universidade de Santiago de Compostela. She is Ph.D. in Journalism and member of Novos Medios research group. She is secretary of the RAEIC journal. As a researcher she focuses on the assessment of digital communication, specially the study of mediated conversation in social media and the evolution of media contents in mobile devices.

María-Cruz Negreira-Rey Lecturer of Journalism at the Department of Communication Sciences, Universidade de Santiago de Compostela (USC). Ph.D. in Communication and member of Novos Medios research group (USC). Her research focuses on local journalism and the development of local and hyperlocal media in Spain, digital journalism and social media.

Carlos Toural-Bran Assistant Professor at the Department of Communication Sciences, Universidade de Santiago de Compostela (USC). Ph.D. in Communication Sciences (USC). Chief of Staff of the Rector (USC) and also President of AGACOM (Galician Association of Communication Researchers) and secretary of Novos Medios research group. He has also been Vice-dean of the Faculty of Communication Sciences (2014–2019).

Xosé López-García Professor of Journalism at Universidade de Santiago de Compostela (USC), Ph.D. in History and Journalism (USC). He coordinates the Novos Medios research group. Among his research lines there is the study of digital and printed media, analysis of the impact of technology in mediated communication, analysis of the performance of cultural industries, and the combined strategy of printed and online products in the society of knowledge.

The Innovative Nature of Journalism

Convergence and Innovation: The Conceptual and Methodological Basis of Technological Evolution and Cultural Complexity in Journalism

Ainara Larrondo Ureta, Javier Díaz Noci, and Ivar John Erdal

Abstract This chapter seeks to contribute to the contextualization of current trends, based on a review of the resignification of concepts such as convergence and innovation in the field of journalism, concepts that are essential when it comes to understanding technological development and the increasing cultural complexity of Journalism. In fact, journalism today presents a hybrid scenario of old and new media, one that is beginning to be described as "hi-tech journalism", in which professionals strive to find stable employment and maintain basic principles of news journalism, such as rigor, truthfulness and quality. On the basis of this general and complete approach, the chapter is completed with an analysis of the main empirical studies on newsrooms convergence as one of the main media innovation development factors.

1 Introduction

At present journalism has major challenges and opportunities, while at the research level it is much more open to other fields, even technical ones, creating growing connections that are now beginning to be delimited [15, 16]. In this scenario, it seems necessary to tackle the matters of innovation and convergence from a refreshed academic perspective that is, nonetheless, defined by a background of over twenty years of research into digital journalism.

The increasingly technologized condition of journalism was already glimpsed in the first analyzes of the changes that digitization and the World Wide Web were bringing to the media industry at the end of the 1990s and the early 2000s.

A. Larrondo Ureta (✉)
University of the Basque Country, Leioa, Spain
e-mail: ainara.larrondo@ehu.eus

J. Díaz Noci
Pompeu Fabra University, Barcelona, Spain
e-mail: javier.diaz@upf.edu

I. J. Erdal
Volda University College, Volda, Norway
e-mail: ivar.john.erdal@hivolda.no

J. Vázquez-Herrero et al. (eds.), *Total Journalism*, Studies in Big Data 97,
https://doi.org/10.1007/978-3-030-88028-6_2

Over time, successive case studies on convergence processes in the press and Public Service Broadcasting (PSB) showed that change, more than being a mere digital adjustment of the systems of production, was systemic and of cultural importance [27, 35]. Media convergence was defined, then, as a multifaceted phenomenon facilitated by the technology that was affecting different areas of action within the media, with the goal of optimizing the creative process of journalism, in terms of planning, production and distribution. Newsroom convergence then became a key area of study [63:59, 29].

The now classic work *Convergence Culture* [31] indicated that the true scope of the convergent scenario could be felt in the internal changes within the media and culture industries and, therefore, in the social changes linked to the consumption of media. Today the media seek effective strategies in terms of engagement, given the proliferation of media available, the different usages of these media, and a role adopted by audiences that is both passive and active. It would be useful, then, to redirect attention towards dimensions that interrelate previously dissociated cultural and technological viewpoints (social Actors, technological Actants, working or productive Activities, and Audiences) [41:19].

In fact, the implementation of new technologies and their potential impacts on journalism are still a prominent theme in digital journalism research and the current research scenario provides a good opportunity to avoid the methodological bias towards newness: "The benefit of such approaches is that they make visible the importance of non-human actors like technology in how digital journalism is practiced and developed. However, Actor-Network Theory (ANT) and similar approaches can lead to an overestimation of non-human actors like technology [...] There is therefore a risk that non-human actors and actants are ascribed too much meaning and power" [71:84].

2 Web Convergence as a Driver of Change

Journalism is going through the best of times and the worst of times, although it could be that this very disruptive perception would have been shared by those who were carrying out and studying the profession decades ago. Whatever the case may be, the feeling of constant change within the media field has intensified with the rise of digital technology and its outstanding manifestation, the web, over the last three decades.

With the arrival of the first online media in the mid-1990s, press, radio and television groups and companies inaugurated a new period of continuous experimentation in the search for greater convergence among their media divisions, at all levels and in different areas. The goal was to achieve greater productive profitability based on the use of the web as a fourth major media platform. The aim was to put into practice a cross-media content strategy, that is to say, content created for one medium but distributed on others, generally the web.

In order to do this, the media companies adjusted their productive operations (content and formats) [6, 74] based on new strategies that affected the day-to-day organization of the newsroom (organization chart, coordination, physical layout, routines, professional profiles, etc.). Some companies choose to physically group together in a single room writers from different media in order to facilitate basic routines, such as sharing sources, something that created reluctance and apprehension among the more experienced professionals; others chose to go even further and fuse newsrooms in order to forge multimedia organizational structures, with the consequent staff reduction. Inevitably, these changes affected journalist profiles, and they became multi-skilled professionals with the capacity for multi-tasking and the ability to operate in multimedia environments.

To accompany these changes, the media companies took on digital technologies adapted to the different technical sub-processes (information gathering, publishing, sharing in the newsroom, etc.). Thus began a stage of greater confluence known at the research and academic level as 'journalistic convergence' or 'media convergence' [43].

The last decade has seen an intensification of this line of thought, which seeks to maximize profit through production based on multi-platform dissemination to the public, particularly on connected formats. This strategy's framework involves a growing orientation of news companies towards mobile applications [44]. This can be seen in the continuously growing use of mobile devices to access news, a growth measured in the United States as 300% since 2013, according to the Pew Research Center [53]. What is more, at the level of production and consumption, the change in paradigm is symbolized by the emergence of mobile journalism or MoJo. This is a field that has evolved, in terms of strategies, particularly quickly, as is shown by a shift from a Mobile First way of working to a Mobile Only one [23].

Audiences dedicate more time to consuming media, are more exposed to and have access to multiple sources of information—traditional offline and online media, digital natives, social media, etc. What is more, these audiences use the media with a productive approach, in the sense that they participate in the content generation process in different ways (likes, shares, comments, delivering data, etc.).

Today, consuming news in any format—television, press, radio, tablet, smartphone, etc.—is a habit that has become natural to all of us, without realizing that behind this possibility there has been an extensive process involving the restructuring of news groups and companies that has been undertaken in order to bring about an authentic convergence 'dialogue' among their different media divisions—between the printed medium and the web medium, in the case of newspaper companies, or among radio, television and the web, in the case of broadcasting companies, or among all of them, in the case of the multimedia companies. The search for new ways forward for journalism and the media thus is nowadays based, to a large extent, on a continual convergence of media that is supported by the latest digital technologies.

3 Innovation in Journalism: Responses to Technological Evolution and Expressions of Growing Cultural Complexity

As argued above, convergence is a framework of action that is behind many of the decisions regarding innovation that are happening in journalism today, even though this framework is not as influential as it was [39]. In fact, at the present time, after a 'valley stage' in terms of the intensity of convergent processes, the Internet of Things (IoT), artificial intelligence, 5G technology and even blockchain technology have laid the foundation of a renewed impulse for the convergent phenomenon, in company management, in the professional sphere and in the field of content creation [61]. This is the case of the present-day influence of the advances linked to big data and the robotization and automation of newswriting, as well as the introduction of new e-business models.

Research into innovation in journalism has become one of the main areas of study in recent years, given the interest in understanding what major changes will be introduced into media companies, above and beyond technological impact and independently of these companies' size and nature, whether public or private. Innovation has become a decided commitment of the media companies, as shown by the proliferation within them of innovation laboratories, often known as Labs or Medialabs. Google's News Initiative project has inspired many of these developments for journalism, through its News Lab laboratory, and has also sown certain doubts about possible alliances and rivalries among the major technology and media companies.

The struggle to achieve audience attention and engagement has intensified and has propelled the search for attractive and innovative formulas for offering content through multiple channels. These multi-platform stories can be simple—a single news item adapted for dissemination on different media, as happens with news and some reports—or more complex, as is the case with examples of immersive journalism and of transmedia—a single event is reported based on the use of different media and on the active role of audiences, which become prosumers, that is, both producers and consumers [23].

Media or content laboratories are a growing phenomenon, due to the journalism industry's constant need to incorporate new products making use of technological advances that arise, such as big data [61]. In the same way that, in its day, the rise of the web brought the need to develop new professional profiles adapted to the specific needs of online journalism (SEO professionals, online community managers, etc.), in recent years trends such as the automation of newswriting and so-called 'robot journalists' are becoming stronger. In a context increasingly dominated by Artificial Intelligence (AI) and big data, we see how the basic routines of journalists are being codified through algorithms. These create news products that stand out for their depth and accuracy, although they are not exempt from questions regarding their practicality and ethical suitability. However, it seems that this matter, and other even more concerning phenomena, such as fake news, will affect developments over the next few years in both the professional and research fields.

In short, media convergence-related innovation and entrepreneurship have become crucial for the advancement of journalism, based on a search for solutions to problems such as loss of audiences and income, by companies, and a lack of job opportunities, by professionals in the industry. In fact entrepreneurship is seen, in general, as one of the main routes to employment in the media industry, as shown by studies focusing on the creation of media start-ups by journalists [3].

4 Headway of Conceptual and Methodological Approaches to Convergence

What is often claimed to be the first 'significant' international example of a merger of editorial teams took place in 2000 by the US group Media General in Tampa [62:39–47]. The editorial teams of *Tampa Tribune*, the local website *Tampa Bay Online* (tbo.com) and the TV broadcasting company *WFLA-TV* were relocated in the same building [67:23].

Emerging during the early and mid-2000's, in the wake of digitization, this area of research studied news organizations focusing on the organizational and practical aspects of production.

In a seminal work on the production of regional news at the *BBC News Centre* in Bristol, Cottle and Ashton [5] integrated the study of news work and contents in the context of technological developments. Around the same time, Ursell [75] looked at how adoption of new technologies in three UK broadcasters, the *BBC*, *ITN* and *Yorkshire Television*, affected work organization and work conditions for journalists. She found increased work pressure and less time for journalism, with inevitable challenges to journalistic performance.

In an empirical study of the websites of four UK broadcasters (*BBC*, *ITV*, *Channel4* and *Channel5*), Siapera [66] concluded that television had gone online, but not changed its understandings of audiences. Across the Atlantic, Duhe et al. [12] found that nine out of ten American television newsrooms were practicing some type of convergence. However, less than half of them defined convergence as having one fully integrated newsroom.

These case studies on broadcasters were accompanied by research addressing the relationship between newspapers and online newspapers. A much-cited work is Singer's [70] study of four converged news organizations in the US: *Dallas Morning News*, *Tampa Tribune, Sarasota Herald-Tribune* and *Lawrence Journal-World*. Singer found convergence processes to be in conflict with traditional newsroom values in two major areas: the distinct culture of the medium and professional competition. Using the theoretical framework of the diffusions of innovations theory, she argued that cultural clashes block convergence, as "cultural differences have led some journalists to minimize their involvement in convergence efforts" [70:16], and that the diffusion of convergence met stumbling blocks in the form of cultural and technical differences and lack of necessary training to gain the competences needed for convergent news work.

Huang and colleagues also used the *Tampa Tribune* as their case, summarized an argument found in a lot of the literature, where opponents of convergence journalism worry that, "with less profound professional knowledge in a non-primary platform and with limited time for filing a story for multiple media platforms, reporters might not be able to produce quality journalism" [26:73].

Around the mid-2000s, there is a peak in convergence research with a flourish of case studies. In a lot of this work, the concepts of convergence and innovation go hand in hand. Steensen and Westlund [71:23] refer to an innovation bias in digital journalism studies, founded on the nature of capitalism where entities have incentives to continuously develop products and services, something that in turn are continuously changing society and markets. Alongside this, the 'need' for traditional news organizations to innovate have been very much present in digital journalism studies [51, 72], even if its status and results have been problematized [71]. Posetti [56] indeed describes the connection between an innovation discourse and technological optimism as the 'Shiny Things Syndrome', arguing that news organizations have been too occupied with innovation related to "bright, shiny things", and less on long-term strategies for sustainable innovation.

While a fair amount of early and mid-2000s case studies found media organizations dealing with convergence by adapting their existing modus operandi to digital production, a successive range of case studies indicated a maturing of convergence efforts, something that was reflected on the research side by a tendency to probe deeper into profound systemic and cultural changes. An influential work in this category was Boczkowski's [2] study of daily newspapers in the US, and their ventures into electronic publishing. He addressed the connections between technology and journalistic practice, and argued that the developments spurred on by technological change had to be seen in close relationship with the existing structures and practices of existing media. His main cases were *The New York Times*, the *Houston Chronicle* and *New Jersey Online*.

A recurring theme in this period of research was journalists' concerns about cross-media journalism, as they perceive production for more than one platform either forces them to be spread too thinly, or increases their workload without compensation. Klinenberg [36] concluded in this manner after studying the digital endeavors of the print newspaper *Metro News*, while Dupagne and Garrison [13] analyzed the integrated *Tampa News Center*.

Many of the most cited studies of this period were done in the US context. Lawson-Borders [40] compiled case studies of three "media convergence pioneers" in the US, including the Tribune Company in Chicago, Media General in Richmond, and the Belo Corporation in Dallas. Huang and colleagues [27] mapped the "top concerns in the media industry brought up by media convergence" [27:83] by conducting a survey among merged and non-merged daily newspapers and commercial TV stations in the US, more specifically a sample of 523 newspaper editors and TV news directors. Silcock and Keith [68] included two converged cases in their study, the *Tampa Tribune* and the *Arizona Republic*, and their respective TV partners. Several of these studies of the implementation of technology in newsrooms make similar conclusions to Singer [70], arguing that the professional cultures were usually not particularly open to change and innovation. Studying the newsroom of the mid-sized US news organization the *Daily Times*, Ryfe found the culture to be "remarkably resilient and resistant to change" [60:198].

In Europe, we find a cluster of cases involving ambitious editorial integration projects taking place from 2007 onward. That year, the *Daily Telegraph* launched its new integrated editorial department in London, which became a model for many other newspapers [67]. Quandt [58] analyzed the characteristics of online news stories in a wide selection of European and US news outlets, e.g., *Der Spiegel*, the *BBC*, *The Times*, *Le Monde*, *The New York Times* and *USA Today*. He found few opportunities for interaction, as well as a lack of multimedia content.

Some scholars expressed concerns about the possibility of convergence potentially leading to less diversity of voices, in an ever-escalating competition to publish the news as soon as possible [18]. Tameling and Broersma [73] found exactly this to be happening in their case study of the Dutch news outlet *De Volkskrant*. In their two-step study of integration in European newsrooms, García Avilés et al. [19, 20] covered organizations in Austria (*Österreich* and *Der Standard*), Spain (*La Verdad* and *El Mundo*) and Germany (*Die Welt* and *Hessische-Niedersächsische Allgemeine*). The authors conclude that convergence is an ongoing process, as early adopters of print-online newsroom integration were "completely remodelling their workflows, gearing them to content and section logic, following the commandments of ultrafast digital dissemination" [20:582].

There are several studies on convergence after the mid-2010's, but they are nevertheless fewer and further apart. There is a soft shift in the literature where the concept of convergence fades, giving way to terms like cross-media, and the more broad digital journalism. This is perhaps indicative of a wider scope of studies into the organization and practice of journalism, encompassing a wide variety of methodological approaches and theoretical frameworks in journalism research, as explained in more detail in the following section.

The actor-network perspective is also used by Primo and Zago [57] in a meta-study discussing different applications of ANT on digital journalism. The authors argue that while it is common to regard (digital) technology only as instruments that aid journalistic processes, "journalism would not be the same without the role played by technological artifacts" [57:43–44].

Micó and colleagues [49] looked at the Catalan public broadcaster *Corporació Catalana de Mitjans Audiovisuals* (*CCMA*) over a five year period, aiming to build bridges between two traditions of innovation research: diffusion of innovations theory and actor-network theory. They found journalism innovation to be an intricate and unpredictable matter where perception of the (dis)advantages of convergence depended on the position in the organizational structure.

The majority of studies so far have explored convergence between print and online, broadcasting and online, and combinations of all three. Less work has been done including mobile news, but this body of work now grows as mobile in many contexts is the main platform for accessing news and other forms of journalism. There are some studies of smartphones and news journalism, exploring the use of different platforms and devices [76], and mobile news consumption [50] and spatial journalism [69], to name a few examples. A recent special issue of *Digital Journalism* titled 'News: Mobilities and Mobiles' [11] contains articles exploring mobile news and innovation from a variety of approaches.

Among the advocates of a shift from a solely technological to a cultural approach, Larrondo and others [38] conducted a comparative study of newsroom convergence in European broadcasters: *BBC Scotland*, Norwegian *NRK*, Spain's *CCMA* (the Catalan Media Corporation) and *EITB* (the Basque public broadcasting company), and Flemish-Belgian *VRT*, focusing on parameters like newsrooms' physical structure, multi-skilling, professional cultures and identities, and attitudes towards convergence. Similarly, Menke and colleagues [47, 48] argue that convergence strategies are highly interrelated with newsroom cultures, basing this argument on a comparative study of newsrooms in Germany, Switzerland, the Netherlands, Austria, Spain and Portugal.

Doudaki and Spyridou [10] investigated evolutionary trends in online news presentation and delivery in the light of convergence dynamics. Based on a study of four converged newspapers in Greece, they argue that convergence ideas are still 'normalized' or toned down "due to countering forces exercised by the dominant professional culture and organizational models in the news business" [10:257].

Saridou and colleagues argue against optimistic views of convergence journalism's potential to satisfy both good journalism and good business practices, claiming that "convergence is used as a cost-effective strategy fostering low-cost and spreadable news production" [65:1006]. Based on quantitative analysis of Greek news websites, the authors provide empirical evidence for an increase in recycling of news content between 2013 and 2016, posing serious threats for content plurality and independent reporting.

Among the more recent work, we find a clear tendency to reflect on the popularity of the convergence concept in mid-2000s journalism research, arguing that the proliferation of distribution channels and easy repurposing of digital content have contributed to an emergence of a wide variety of platforms and outlets, and to increased differentiation of formats. Peil and Sparviero [52:10] find that the intense focus on convergence processes often overlook parallel processes of what they call 'de-convergence', and others call 'divergence'.

Looking back at the research on online and convergence journalism, a substantial body of work has approached the field from the perspective of technology's innovative power to transform journalism in direction of increased interactivity, participation, and, ultimately, democracy. Early research would typically either predict revolution or detect failure, in the form of comparing the potential of digital technology with the reality of media practices. One typical conclusion was that "the technological affordances of the web were sporadically used and that the internet hadn't revolutionized journalism in such a radical manner as many theorists had predicted" [10:258].

In sum, the "discourse of innovation" in journalism studies picked up speed in the early period of convergence efforts in the late nineties. In their above mentioned recent meta-study of digital journalism, Steensen and Westlund [71:82] found that one third of all scholarly publications on journalism in 2018 included the word 'innovation'. This research on innovation has, besides, a tendency to emphasize change in a positive way.

5 Innovation in Newsrooms and the Bet for Convergence: A Methodological Approach

The successive and overlapped crisis—the general economic crisis of 2008, and the particular crisis of specially printed media due to the process of digitization and the pandemic crisis of 2020—have refreshed the interest in doing some research on the working conditions and compensation of digital journalists, a topic that was mentioned a decade ago [6]. There is a clear interest in going beyond the optimistic consideration of media convergence, submitting it to the examination of social responsibility theories [30]. It seems important to interrogate practitioners about which is the opinion on convergence, using in-depth interviews [78], a method which was used by Infotendencias Group in 2010 [29, 43]. A variation of this method (semi-open interviews with journalists) was used by Bárbara Maia [4], focusing on the changes of a mutating profession—in this case, a Brazilian radio station—due to convergence.

A field which offers many chances to be investigated is to focus the different opinions and attitudes towards the implementation of the different aspects of convergence, as stated before. Actually, we are referring to the different professional cultures (of journalists and of some other professional groups, like managers) which can be observed in newsrooms. This is the approach by Menke and colleagues [47, 48], using surveys and privileging a comparative perspective. This was present in Díaz-Noci [6], as well, focusing on Hallin and Mancini's well-known typology of media systems, or by Porcu [55], using a methodological triangulation: on one hand, a strongly categorized content analysis and, on the other hand, an array of in-depth interviews to journalists and managers around two interesting concepts of learning innovation in newsroom and explorative innovation culture, even if sometimes a 'collision of convergences' may happen [55:1556].

As for methodologies used in the study of journalistic convergence, a mixed-methods approach is often privileged, so it is possible to identify several tools employed according to the different aspects of this phenomenon. An effort to measure convergence quantitatively was proposed by the Infotendencias Group in Spain [43] and applied by Díaz-Noci [6]. This tool was based in evaluating the type of collaboration amongst media (usually of the same group), the flexibility of contents and the relationship amongst newsrooms of the same group. This is a process that, from the adoption of the digital-first strategy by many media organizations around the world, is no longer the main concern of media convergence research.

Probably the methods most intensively used in research on media convergence, at least to explain changes in production and consumption routines (see, for instance,

Idiong et al. [28], an interesting insight on how media convergence may affect readership practices) and in the adoption of technological or entrepreneurial models, have been those of the ethnography. These are the methods used by Ke [42], to explain the practices of convergence (and de-convergence, which is an interesting approach in our humble opinion) in one Chinese medium, *Beijing News*, to adopt digital practices or, alternatively, keep away of the Internet. A somehow similar approach of convergence and divergence, also for the Chinese case, is adopted, in this occasion to study the changes made in copyright, by Jie [22]. It serves to a wide variety of applications of convergence at this level, including the use of new technologies (mobile devices, for instance), an aspect which has been examined by Kumar and Haneef [37].

A mixed-method approach has been also used. It is interesting, in this respect, to mention Georgeta Drulă's choice: a combined used of hierarchical cluster analysis (HCA) and of content analysis [14], and the choice preferred by Zambrano and others [77], using in this case surveys, group and individual interviews and participant observation as the main methodological tool to interpret the position of Colombian journalists in a convergence scenario. A multi-methodological design based on several qualitative methods was also employed in a European comparative research on convergence in broadcasting newsrooms [38]. Some other researchers have preferred a combination of participant observation and content analysis, for instance to analyze the changes in a specific newsroom, that of Al-Jazeera [25]. Even when very specific and quite technical aspects, such as Search Engine Optimization (SEO) tools and its use in newsrooms are investigated, such a multi-methodological approach is favored [45].

Focus groups, alongside with some other qualitative techniques, have been used as well. A good example is the research conducted by Ainara Larrondo on the public broadcasting service in the Basque Country in Spain in 2016, alongside with surveys, in-depth interviews and observation as well [38].

Multimedia (or, more properly speaking, multimodality, see [Karlsson and Holt 33]) and interactivity have been also researched, and many methods have been applied to determine to which extent are used [6]. Once again, the perception of journalists has been interpreted using in-depth interviews, mainly [34]. Some aspects, such as the modular presentation of news in television [54], have been explained using direct observation of their implementation. Even semiotics have been eventually proposed [64] or effectively applied [24], specifically to interpret the significance of multimedia (and of interactivity) affected by convergence. Sometimes multimedia means merely the implementation of video in Web media, specifically digital newspapers' editions, which is, in turn, related to business models. Since paywalls have been built, the audience metrics decline in importance in a scenario in which the goal is to attract and retain users ready to pay for news, not probably for breaking news which are everywhere, but instead willing to invest in more elaborated news items. Case studies will be increasingly used, it is to be supposed, to check the viability of such strategies [1, 17].

Content analysis (or even an intense document review, combined with some other techniques, see Jamil and Appiah-Adjei [30], Kalamar [32] has been widely used in media convergence studies as well [21, 59], both to analyze news and other transmedia products, such as contents published in social media, a trend which has been developed in the last ten years, at least. Transmedia—and also its legal related term, derivative works—is another relevant aspect researchers have started to insist on, using, once again, ethnographic methods, such as observation and in-depth interviews [46]. In some other cases, a more elaborated design is implemented, using, for instance, content analysis and a virtual ethnography approach.

It is much less common to focus on audiences and convergence. A stimulating approach, even from a methodological point of view, is Denis Kalamar's use of experimental-causal methods for data gathering, once again alongside with literature revision and in-depth interviews, the most usual tool employed in media convergence studies.

Even a legal approach can be adopted. In this respect, it is important to consider that multimedia means, in legal terms, a complex work, which can be categorized as a collaborative, a derivative or a collective work. It is important also to underline, and to follow through observation and legal analysis (we favor a transnational, comparative functionalist analysis of the rules) the adaptation of the law to new necessities of media convergence. The insistence and the lobbying activity of many media corporations, singularly the greatest newspaper groups in Europe, and in some other parts of the world such as Australia, towards legal reforms to confront giants like Google (and make them pay, see Díaz-Noci [7–9], which attract all attention to the collective work—and put aside collaborative and derivative works—is another aspect to be examined.

6 Final Remarks

This chapter has looked at convergence, innovation and entrepreneurship as a strategic opportunity for journalism to move forward, considering that at present this field is evolving in a context of massive data which hyperconnects it, like never before, to its sources and public and, therefore, with its goals and functions within society. Convergence and innovation and, by extension, entrepreneurship have become key concepts in the professional and academic spheres, meaning that journalism is moving forward in step with technological developments. Today these developments are becoming more evident than ever, which begs for a re-examination of the meaning and implications of the aforementioned concepts, both for the practice and the study of journalism.

As was mentioned in this chapter, media convergence opened up a pathway of constant experimentation that has kept up until today; this experimentation has been in the areas of technologies, tools, content, formats, relationships with audiences and funding models. In this regard, innovation can be seen as a philosophy that accompanies and complements convergence, one that is useful for conveying the essence of the changes and implications that convergence has brought and will continue to bring, along with technology and creativity. For this reason, although convergence has lost momentum as a fashionable concept in both the study and profession of social communication, much of the research in this field continues to be linked to this macro-phenomenon.

There was a moment in which convergence was a fashionable term, often presented as an unavoidable sign of progress and "conceived as a gradual, evolutionary, and multidimensional process" [29:24]. Many changes have happened from then onwards, not being the least of them the fact that the media industry has suffered an enormous crisis starting in 2008, which has led to the adoption, accelerated by the COVID-19 health situation in 2020, of different paywalls (freemium or metered, for instance), or both subscription or membership models, and to the compulsory adoption of remote work and to the redefinition of what is a newsroom, no longer a physical space exclusively, so the polyvalent aspects of journalists have been strengthened.

The 'discourse of innovation' emphasizes change over continuity, digital over analogue, something that is especially important to have in mind when reviewing the role of concepts like convergence and innovation in journalism studies. In the current context there is a tendency to overestimate the newness of the things that stand out in research findings, even if they do not necessarily play a more important role in news production [71:84]. Therefore, being a fact that technological perspectives are currently totally accompanied by more culturally oriented ones, it is also necessary to expand the scope from looking merely at how mobile devices and algorithms produce and distribute differently news contents. Convergence strategies are still highly interrelated with cultural dynamics that are decisive for innovation in contents and in professional organizational patterns which genuinely imply an advance for the journalism's role and function.

Funding Gureiker Consolidated Research Group (Basque University System, Basque Government, ref. IT1112-16) and DigiDoc Research Group (included in the Journalism and Digital Documentation Research Unit, Catalan University Agency, Ref. AGAUR 2017-SGR-1103). This paper is one of the results of the research project *News, networks, and users in the hybrid media system. Transformation of media industries and the news in the post-industrial era* (RTI2018-095775-B-C43) funded by the Ministry of Science, Innovation and Universities (Government of Spain) and the ERDF structural fund (2019–2021).

References

1. Baranova, E.A., Zheltukhina, M.R., Shnaider, A., Zelenskaya, L., Shestak, L., Redkozubova, E., Zdanovskaya, L.: New media business philosophy in conditions of mass media convergence. Online J. Commun. Media Technol. **10**(4), 1–9 (2020)
2. Boczkowski, P.J.: Digitizing the News: Innovation in Online Newspapers. MIT Press, Cambridge (2005)
3. Bruno, N., Nielsen, R.K.: Survival is Success: Journalistic Online start-ups in Western Europe. Reuters Institute for the Study of Journalism, Oxford (2012)
4. Cerqueira, B.M.: The journalists and a profession in mutation: tThe voices of the professionals of a Curitiba Rrdio. Braz. Journal. Res. **14**(2), 506–523 (2018). https://doi.org/10.25200/BJR.v14n2.2018.1093

5. Cottle, S., Ashton, M.: From BBC newsroom to BBC newscentre: on changing technology and journalist practices. Convergence **5**(3), 22–43 (1999)
6. Díaz-Noci, J.: Newsroom convergence. A comparative research. In: Larrondo, A., Meso, K., Tous, A. (eds.), Shaping the News Online. A Comparative Research on International Quality Media. LabCom, Covilhã, pp. 301–341 (2014)
7. Díaz Noci, J.: Copyright and news reporting. A comparative legal study of companies', journalists' and users' rights. In: Meso, K., et al. (eds.) Active Audiences and Journalism: Analysis of the Quality and Regulation of the User Generated Contents, pp. 183–210. Servicio Editorial de la Universidad del País Vasco, Bilbao (2015)
8. Díaz Noci, J.: Copyright and law tendencies: a critical approach of press publishers' right or link tax and of upload filtering for user generated contents. In: #6OBCIBER Proceedings, pp. 381–395. Universidade do Porto, Oporto (2018)
9. Díaz Noci, J.: Authors' rights and the media. In: Pérez-Montoro, M. (ed.) Interaction in Digital New Media, pp. 147–173. Palgrave-McMillan, London (2018)
10. Doudaki, V., Spyridou, L.P.: News content online: Patterns and norms under convergence dynamics. Journalism **16**(2), 257–277 (2015)
11. Duffy, A., Ling, R., Kim, N., Tandoc, E., Westlund, O.: News: mobiles, mobilities and their meeting points. Digit. Journal. **8**(1), 1–14 (2020)
12. Duhe, S.F., Mortimer, M.M., Chow, S.S.: Convergence in North American TV newsrooms: a nationwide look. Convergence **10**(2), 81–104 (2004)
13. Dupagne, M., Garrison, B.: The meaning and influence of convergence: aA qualitative case study of newsroom work at the Tampa News Center. Journal. Stud. **7**(2), 237–255 (2006)
14. Drulă, G.: Formas de la convergencia de medios y contenidos multimedia: Una perspectiva rumana. Comunicar **22**(44), 131–140 (2015)
15. Eldridge, S.A., Hess, K., Tandoc, J.E., Westlund, O.: Editorial: digital journalism (Studies)–defining the field. Digit. Journal. **7**(3), 315–319 (2019)
16. Erdal, I.J.: Convergence in/of Journalism. In: Oxford Research Encyclopedia, pp. 1–17. Oxford University Press, Oxford (2019)
17. Friedrichsen, M., Kamalipour, Y.: Digital Transformation in Journalism and News Media: Media Management, Media Convergence and Globalization. Springer, London (2017)
18. Fenton, N.: New Media, Old News: Journalism and Democracy in the Digital Age. Sage, London (2010)
19. García-Avilés, J.A., Meier, K., Kaltenbrunner, A., Carvajal, M., Kraus, D.: Newsroom integration in Austria, Spain and Germany. Journal. Pract. **3**(3), 285–303 (2009)
20. García-Avilés, J.A., Kaltenbrunner, A., Meier, K.: Media convergence revisited: lessons learned on newsroom integration in Austria, Germany and Spain. Journal. Pract. **8**(5), 573–584 (2014)
21. Garifullin, V.Z., Khasanova, K.S.: Specifics of media convergence in Tatarstan. Turk. Online J. Des. Art Commun. **7**, 840–845 (2017)
22. Gu, J.: From divergence to convergence: institutionalization of copyright and the decline of online video piracy in China. Int. Commun. Gaz. **80**(1), 60–86 (2018)
23. Gutiérrez-Caneda, B., Pérez-Seijo, S., López-García, X.: Analysing VR and 360-degree video apps and sections. A case study of seven European news media outlets. Rev. Lat. Commun. Soc. **75**, 149–167 (2020)
24. Haase, F.A.: Presentation' and 'representation' of contents as principles of media convergence: a model of rhetorical narrativity of interactive multimedia design in mass communication with a case study of the digital edition of the New York Times. Semiotica **226**, 89–106 (2017)
25. Hassan, M.M., Elmasry, M.H.: Convergence between platforms in the newsroom. Journal. Pract. **13**(4), 476–492 (2019)
26. Huang, E., Rademakers, L., Fayemiwo, M.A., Dunlap, L.: Converged journalism and quality: a case study of the Tampa Tribune news stories. Convergence **10**(4), 73–91 (2004)
27. Huang, E., Davison, K., Shreve, S., Davis, T., Bettendorf, E., Nair, A.: Facing the challenges of media convergence. Media professionals' concerns of working across media platforms. Convergence **12**(1), 83–98 (2006)

28. Idiong, A., Idiong, N.S., Udoakah, N.: Influence of media convergence on newspaper readership in Akwa Ibom State, Nigeria. J. Stud. Humanit. Soc. Sci. **7**(2), 42–63 (2018)
29. Infotendencias Group: Media convergence. In: Siaperas, E. (ed.) The Handbook of Global Online Journalism, pp. 21–38. Wiley, London (2012)
30. Jamil, S., Appiah-Adjey, G.: Journalism in the era of mobile technology: the changing pattern of news production and the thriving culture of fake news in Pakistan and Ghana. World Media J. Russ. Media Journal. Stud. **1**(3), 42–64 (2019)
31. Jenkins, H.: Convergence Culture. Where Old New Media Collide. New York University Press, New York (2008)
32. Kalamar, D.: Convergence of media and transformation of audience. Informatologia **49**(3–4), 190–202 (2016)
33. Karlsson, M., Holt, F.: Journalism on the Web. In: Nussbaum, J. (ed.), Oxford Encyclopedia of Communication. Oxford University Press, Oxford (2016). https://doi.org/10.1093/acrefore/9780190228613.013.113
34. Kartveit, K.: How do they do it? Multimedia journalism and perceptions of the practice. Journalism **21**(10), 1468–1485 (2020)
35. Killebrew, K.C.: Managing Media Convergence. Blackwell Publishing, Iowa (2005)
36. Klinenberg, E.: Convergence: news production in a digital age. Ann. Am. Acad. Pol. Soc. Sci. **597**(1), 48–64 (2005)
37. Kumar, A., Haneef, M.S.: Convergence of technologies and journalists. Translation of journalistic practices through ANT perspectives. Estudos em Comunicação **22**, 105–122 (2016)
38. Larrondo, A., Domingo, D., Erdal, I.J., Masip, P., Van der Bulck, H.: Opportunities and limitations of newsroom convergence. A comparative study on European public service broadcasting organizations. Journal. Stud. (17)**3**, 277–300 (2016)
39. Larrondo, A., López-García, X.: Media convergence' resignification in Hi-Tech Journalism. In: Pedrero, E., Pérez, A (eds.), Post Digital Communication Cartography: Media & Audiences in COVID-19 Society. Thomson Reuters Aranzadi, Madrid (2020)
40. Lawson-Borders, G.: Media Organizations and Convergence. Case Studies of Media Convergence Pioneers. Lawrence Erlbau, New Jersey (2006)
41. Lewis, S.C., Westlund, O.: Actors, actants, audiences, and activities in cross-media news work. Digit. Journal. **3**(1), 19–37 (2015)
42. Li, K.: Convergence and deconvergence of Chinese journalistic practice in the digital age. Journalism **19**(9–10), 1380–1396 (2018). https://doi.org/10.1177/1464884918769463
43. López-García, X., Pereira, X.: Convergencia digital. Reconfiguración de los medios de comunicación en España. Servicio de Publicaciones de la Universidad de Santiago de Compostela, Santiago de Compostela (2010)
44. López-García, X., Silva-Rodríguez, A., Vizoso-García, A.A., Westlund, O., Canavilhas, J.: Mobile journalism: systematic literature review. Comunicar **59**, 9–18 (2019)
45. Lopezosa, C., Codina, L., Díaz-Noci, J., Ontalba-Ruipéz, J.A.: SEO and the digital news media: from the workplace to the classroom. Comunicar **63**, 65–75 (2020)
46. Martins, E.: Convergence and transmedia storytelling in journalism: transformations in professional practices and profiles. Braz. Journal. Res. **2**(2), 168–187 (2015)
47. Menke, M., Kinnebrock, S., Kretzschmar, S., Aichberger, I., Broersma, M., Hummel, R., Kirchhoff, S., Prandner, D., Ribeiro, N., Salaverría, R.: Convergence culture in European newsrooms. Comparing editorial strategies for cross-media news production in six countries. Journal. Stud. **6**(19), 881–904 (2018)
48. Menke, M., Kinnebrock, S., Kretzschma, S., Aichberger, I., Broersma, M., Hummel, R., Kirchhoff, S., Prandner, D., Ribeiro, N., Salaverría, R.: Insights from a comparative study into convergence culture in European newsrooms. Journal. Pract. **13**(8), 946–950 (2019)
49. Micó, J.L., Masip, P., Domingo, D.: To wish impossible things. Convergence as a process of diffusion of innovations in an actor-network. Int. Commun. Gazette **75**(1), 118–137 (2013)
50. Molyneux, L.: Mobile news consumption: a habit of snacking. Digit. Journal. **6**(5), 634–650 (2018)
51. Pavlik, J.V.: Innovation and the future of journalism. Digit. Journal. **1**(2), 181–193 (2013)

52. Peil, C., Sparviero, S.: Media convergence meets deconvergence. In: Sparviero, S., Peil, C., Balbi, G. (eds.), Media Convergence and Deconvergence. Global Transformations in Media and Communication Research. Palgrave Macmillan, Cham, pp. 3–30 (2017)
53. Pew Research Center.: Americans favor mobile devices over desktops and laptops for getting news (2018). Available at https://pewrsr.ch/39LX04T. Accessed 01 Jan 2021
54. Pjesivac, I., Cantrell-Bickley, Y., Hazinski, D.: Digital convergence in the newsroom: experimenting with modular production of television news in grady newsource. Journal. Mass Commun. Educator **73**(3), 346–357 (2018)
55. Porcu, O.: Exploring innovative learning culture in the newsroom. Journalism **21**(10), 1556–1572 (2020)
56. Posetti, J.: Time to Step Away from the 'Bright, Shiny Things'? Towards a Sustainable Model of Journalism Innovation in an Era of Perpetual Change. Reuters Institute for the Study of Journalism, Oxford (2018)
57. Primo, A., Zago, G.: Who and what do journalism? An actor-network perspective. Digit. Journal. **3**(1), 38–52 (2015)
58. Quandt, T.: News on the World Wide Web? A comparative content analysis of online news in Europe and the United States. Journal. Stud. **9**(5), 717–738 (2008)
59. Rizzo, G., Gruszynski, A.: Shaping information at digital native Nexo in the scenario of journalistic convergence. Braz. Journal. Rev. **1**(2), 394–421 (2020). https://doi.org/10.25200/BJR.v16n2.2020.1257
60. Ryfe, D.M.: Broader and deeper: a study of newsroom culture in a time of change. Journalism **10**(2), 197–216 (2009). https://doi.org/10.1177/1464884908100601
61. Salaverría, R.: Periodismo digital: 25 años de investigación. Artículo de revisión. El Profesional de la Información **28**(1), 1–27 (2019)
62. Salaverría, R., Negredo, S.: Integrated journalism: media convergence and newsroom organization. Sol 90, Madrid (2009)
63. Salaverría, R., García-Avilés, J.A., Masip, P.: Concepto de convergencia periodística. In: López-García, X., Pereira, X. (eds.) Convergencia digital. Reconfiguración de los medios de comunicación en España, pp. 41–64. Servicio de Publicaciones de la Universidad de Santiago de Compostela, Santiago de Compostela (2010)
64. Sánchez-García, P., Salaverría, R.: Multimedia news storytelling: semiotic-narratological foundations. El Profesional de la Información **28**(3) (2019). https://doi.org/10.3145/epi.2019.may.03
65. Saridou, T., Spyridou, L.P., Veglis, A.: Churnalism on the rise? Assessing convergence effects on editorial practices. Digit. Journal. **5**(8), 1006–1024 (2017)
66. Siapera, E.: From couch potatoes to cybernauts? The expanding notion of the audience on TV channels' websites. New Media Soc. **6**(2), 155–172 (2004)
67. Siapera, E., Veglis, A.: The Handbook of Global Online Journalism. Wiley, London (2012)
68. Silcock, B.W., Keith, S.: Translating the tower of Babel? Issues of definition, language, and culture in converged newsrooms. Journal. Stud. **7**(4), 610–627 (2006)
69. Schmitz Weiss, A.: Place-based knowledge in the twenty-first century: the creation of spatial journalism. Digit. Journal. **3**(1), 116–131 (2015). https://doi.org/10.1080/21670811.2014.928107
70. Singer, J.B.: Strange bedfellows? The diffusion of convergence in four news organizations. Journal. Stud. **5**(1), 3–18 (2004)
71. Steensen, S., Westlund, O.: What is Digital Journalism Studies? Routledge, London (2020)
72. Storsul, T., Krumsvik, A.H.: Media Innovations. A Multidisciplinary Study of Change. Nordicom, Gothenburg (2013)
73. Tameling, K., Broersma, M.: De-converging the newsroom: strategies for newsroom change and their influence on journalism practice. Int. Commun. Gaz. **75**(1), 19–34 (2013)
74. Thurman, N., Lupton, B.: Convergence calls: multimedia storytelling in British news websites. Convergence **14**(4), 439–455 (2008)
75. Ursell, G.: Dumbing down or shaping up? New technologies, new media, new journalism. Journalism **2**(2), 175–196 (2001)

76. Wolf, C., Schnauber, A.: News consumption in the mobile era. Digit. Journal. **3**(5), 759–776 (2015)
77. Zambrano, W.R., García, D., Barrios, A.: El periodista frente a los nuevos retos y escenarios de la convergencia mediática colombiana. Estudios sobre el Mensaje Periodístico **25**(1), 587–607 (2018)
78. Xiong, H., Zhang, J.: How local journalists interpret and evaluate media convergence: an empirical study of journalists from four press groups in Fujian. Int. Commun. Gaz. **80**(1), 87–115 (2018)

Ainara Larrondo Ureta Senior Lecturer on Journalism at the University of the Basque Country (Leioa), Ph.D. in Journalism and Master in Contemporary History. Visiting Researcher at the CCPR (University of Glasgow). Main director of 'Gureiker' Research Group (Basque University System) and KZBerri (Innovation Education Group in Online Journalism Teaching). Her research lines are online journalism, innovation in media and communication, media and gender and organizational and political communication.

Javier Díaz Noci Full Professor on Communication at the Pompeu Fabra University (Barcelona), Ph.D. in both History (1992) and Law (2016). During the decades of 1980 and 1990 he worked as a journalist in, among some other media, the National Radio of Spain (RNE). His research lines are online journalism, digital inequality and the news, media history and copyright applied to digital news.

Ivar John Erdal Associate Professor at Volda University College. His research interest lies mainly in digital journalism, mobile media, cross-media production and organization studies, and his work has appeared in journals such as *Journalism StudiesConvergence*, and *Journalism Practice*. Erdal earned his Ph.D. in media studies from the University of Oslo in 2008, on a study of digital news production and cross-media journalism at the Norwegian public service broadcaster NRK.

Journalistic Storytelling for Complexity

Jorge Vázquez-Herrero and Yael de Haan

Abstract The evolution of digital media has fostered the emergence of new and adapted forms of storytelling. Journalism combines classic narrative resources with technology and interactive strategies to inform, explain and represent reality. After an initial stage of adaptation and experimentation, it is necessary to reflect on the contribution of the new formats to contemporary, quality journalism for new active audiences. This chapter summarizes the evolution of digital journalistic narrative to identify the differentiating elements that have come from the intersection of documentary and journalism, as well as the new approaches that have arisen from immersive technologies. A panoramic vision will allow us to establish a roadmap to advance the production and research of storytelling for a complex world.

1 Introduction

Narrative is inherent to the human being. It is a practice present "in all times, all places and all societies" [4:1]. From the most ancient manifestations, the oral tradition, painting, chants or dances, humans searched for a way out of the need to tell stories, to register and build memory. Journalism is nothing more than storytelling—*Journalism is a story* is the title of a book that compiles articles by journalist Manuel Rivas—and it has thus constantly sought ways to represent reality and transmit it to society.

Our global society is a so-called Network Society [8], marked by the irruption of information and communication technologies. Many changes have occurred in this society, placing journalism in a moment of rapid and constant change [16] and facilitating a renewal in formats and narrative models driven by technology. Nevertheless, we still need stories as a survival manual [47]. Storytelling allows us to organize

J. Vázquez-Herrero (✉)
Universidade de Santiago de Compostela, Santiago de Compostela, Spain
e-mail: jorge.vazquez@usc.es

Y. de Haan
University of Applied Sciences Utrecht, Utrecht, The Netherlands

J. Vázquez-Herrero et al. (eds.), *Total Journalism*, Studies in Big Data 97,
https://doi.org/10.1007/978-3-030-88028-6_3

experiences in memory, to transfer meaning [5] and to build and structure visions of the world we aspire to know [21].

Despite the transformation of media, technologies and platforms, the constant in storytelling remains the deepening of human understanding and the widening of connectedness:

> We need this creative practice for its own sake, but more than that, we need the process of continuously expanding our means of storytelling, because it allows us to expand our ability to know who we are and to collectively reimagine who we might become. [31:362]

In this chapter, we focus on new storytelling and particularly consider non-fiction genres that try to involve the public through different technologies that seek user agency. While social media and emerging platforms have created the possibility for more interaction, here we consider more long-form genres that seek to involve the user, including digital long-forms, interactive documentaries and immersive journalism.

1.1 Digital Media and Digital Narratives

The digital environment was, from the very first moment, a new canvas for storytelling. At the beginning, in the 1980s, we encountered experimental projects in multimedia laboratories such as the Apple Multimedia Lab and the MIT Media Lab. The popularization of multimedia devices for creation and consumption, as well as the global irruption of the Internet as a new channel, allowed a progressive expansion of the forms of digital communication.

However, nothing is born without a connection to the past. Different theories such as remediation [6], mediamorphosis [15] and media ecology [37] reflect the coexistence of old and new models, with constant exchanges between them. With the arrival of a new technology, the first step is usually the imitation of a previous instance, simultaneously with more experimental and innovative methods. Later, consolidation became the desired step that not all technologies and forms of expression can achieve. Decadence, rejection or being surpassed by something new are frequent outcomes in the history of formats, platforms and technologies. It is a phenomenon we have increasingly become accustomed to in the hurried evolution of the field of communications. Although the threat of going down in history as something irrelevant, ephemeral or passing exists and persists, there is always an opportunity to learn.

The digital arena provides the narrative with multimedia, hypertextual and interactive support to develop diverse qualities. In addition, the rise of new platforms opens up new possibilities. Think, for example, of the ephemeral content of Snapchat and Instagram Stories, the micro-videos of TikTok or the interactive productions of Netflix. Each of these formats brings its own language, aesthetics and logic. Producers and media take part in them with the purpose of reaching an audience by alternative means, transmitting stories through several channels simultaneously and seeking the participation of users. The struggle for attention is increasingly fierce and, once again, storytelling and technology are allied to look for more persuasive, experiential and efficient strategies.

Digital storytelling, as a widely used term, encompasses narration using computational media that combines still and moving images, sound, animations, infographics and other multimedia elements. Beyond the technological parameters, digital storytelling has the potential to turn the audience into participants. Active users, prosumers, fans and co-creators are offered mechanisms to get involved—alternatively, they seek them out—which gives them the ability to have an impact on the story.

Needless to say, not all stories transmitted by digital media can be described as digital storytelling [1]. Once the initial phase of imitation of previous media has been overcome, digital narratives must have their own differentiating attributes. Similarly, not all stories seek high audience engagement and the experience is not always the same. The diversity of formats, technologies and platforms also influences the configuration of the relationship between the creator, the content and the user.

In the last two decades, different techniques and strategies of digital narratives have been put into practice that emphasize specific aspects, such as participation, experience and immersion. We address these points later from different perspectives.

Contemporary digital narratives are situated within complex narratives [7] that are carried out in the computer as a programmable digital medium and consider the creator and the broadcaster to exist on the same level. We find complexity in our environment, in society, in the relationship between the local and the global. The world is becoming more complex, so we need stories to explain this complexity, demanding new ways of storytelling that are more impactful and accessible at the same time. Thus, to tell such stories, journalists have to understand and interact with new genres.

1.2 Journalism: Evolution, Innovation and Narratives

From the journalistic field, the evolution of the digital media resulted in important changes that derived from convergence processes [25]. Technology is not a factor that is indifferent to journalism: accelerated technological changes have generated disruptions in the media ecosystem [35]. Since the invention of the printing press, journalism has been linked to the development of industrial society. Later, the computerization of newsrooms had a great impact, such as the development of precision techniques [30]. Changes also reached and impacted on the profession, content and relationships between professionals, organizations and users [34]. Today, discussions are still open regarding the implementation of automated processes in newsrooms.

With the arrival of the news media on the Internet around 1994, hypertext and the use of multimedia became progressively incorporated into journalistic work. At the end of the first decade of the twenty-first century, there was increasing talk of hypertext formats for the coverage of special events [26] and so-called multimedia journalism [17]. However, interactivity—which is the most singular characteristic of the digital medium and the one that adds a new conception of user participation [44]—experienced slower development and is still, to a large extent, an utopia. Although there is evidence of the search for interactive ways to tell a story, current developments still have important limitations. Content selection or participation through social networks are common features, but the reach of the user agency is limited to a few cases.

The exploration of renewed forms of storytelling in journalism came with the processes of innovation [33], especially in the line of narrative and formats. The creation of media labs [36] also contributed in this sense, but resource limitations in media that were diminished by the 2008 crisis reduced the possibilities of innovation. Reference media such as *The New York Times*, *The Washington Post* and *The Guardian* have taken the lead and demonstrated this with initiatives and projects in recent years around interactive non-fiction, 360° video, gamification, virtual and augmented reality, and interactive visualizations.

The interest in exploring new ways of telling stories in journalism has been reflected in different trends and modalities. Narrative journalism—a genre that utilizes techniques and resources from other disciplines such as literature [22]—is being credited with the ability to create a meaningful context and engage audiences [42]. The use of digital media resources, especially multimedia [24], promotes narrative journalism as an opportunity for the future [32]. Slow journalism and long-form journalism [27, 28] are also emerging in the face of an atmosphere of over-information and accelerated consumption in which reflection, pause and quality must make a difference.

We have seen the development of these modalities in an outstanding way in reportages, documentaries and special coverages. However, the complexity of today's media ecosystem puts more elements on the table. From the combination of platforms and media, transmedia projects emerge, seeking the expansion of the narrative universe and the complicity of the audience, above all by relying on social networks and mobile devices. The irruption of immersive technologies, which are gradually becoming more accessible, has allowed the emergence of immersive journalism, while the standardization of technologies and devices along with multidisciplinary collaboration has favored the production of interactive and experimental non-fiction works.

2 Intersections of Documentary and Journalism: Towards an Interactive and Participatory Dimension

The representation of reality has roots in documentary and journalism. With different origins and historical trajectories, cinematography and the printing press influenced the development of both forms of expression, as technologies and differentiated supports. Because of their methods of consumption, their resources and different points of view about this reality—more authorial and artistic perspectives are developed in the documentary. However, the evolution of the media has led to convergence [25] and hybridization. In the same way that narrative journalism has taken on features of literature, the documentary has developed in the interactive medium, offering approaches to reality that move away from some of its fundamental characteristics, while reinforcing others.

The confluence of documentary cinematography with the interactive medium, starting in the 1980s and passing through numerous formats and media [43], led to the decline of the traditional author-text-user relationship [18]. An interactive documentary is "any project that starts with an intention to document the 'real' and that uses digital interactive technology to realize this intention" [3:125]. Stories in this format are not linear anymore: there can be multiple structures—parallel, alternative or optional—including linear but also free navigation. Non-linearity is a clear example of the transfer of control from the author to the user, who is no longer a spectator but becomes an element of the story. The multimedia and interactive nature of stories have facilitated this new relationship, through decision making, freedom of navigation, personalization and participation.

News media organizations have developed interactive documentary projects, especially between 2010 and 2017, after which production decreased and stabilized. According to William Uricchio [40], journalism is going through a process of redefinition in which experience becomes a priority. It is a stage of innovation and a search for new narrative solutions, but also a period marked by the technology of each moment. Social networks become indispensable platforms and, moreover, constantly present new features. Successive technologies take away the importance of those already explored: the interactive documentary was relatively consolidated as a webdoc, using the Web as the main support, but the rise of virtual reality took away its importance in a substantial way. The technical sophistication, the specialization of the production teams and the high costs of interactive documentary diminish the impact of this format, which is now reserved for special coverage.

The differentiating potential of the interactive documentary is in reaching a point of agency: a significant interactivity in which the user really does act and thus becomes immersed. Participation through user contribution and co-creation is the most developed state of interaction, however it is not common. Full interactivity is still a promise, not a reality [45]. Paradigmatic examples can be found in *Quipu Project*, a transmedia documentary that implemented a telephone system and message recording replicated on the web, or more discreetly in the collaborative extension of *A Short History of the Highrise* or the *Hollow* community.

The interactive documentary in the journalistic context offers new ways to be informed, with an elaborate and deep context. Even the journalistic approach can play an important role in the autonomy vs. activism binomial—closer to the cinematographic documentary, with a reinforced meaning when interactivity is increasing. Interactivity becomes a tool to reach the engagement of the audience, in a moment of maximum complexity and competition for attention. The search for commitment and emotional effect forces us to redefine the author-text-user relationship in the interactive documentary, where tensions between development, purpose and effect are evident [44]. Interest in reception analysis increases among professionals and academics, with the aim of providing answers about engagement in interactive stories. The desired audience involvement becomes more evident when immersive technologies are applied to representing reality.

3 New Approaches to Journalistic Storytelling Through Immersive Technologies

Global news outlets are experimenting with new kinds of storytelling in 360° videos, virtual reality, augmented reality and interactive websites [2, 39, 46]. It is now possible to not only watch, listen to or read a story, but also to feel, experience and relive stories from a first-person perspective. In 2010, Nonny de la Peña introduced the term 'immersive journalism' as an innovative storytelling technique that transports the user to a virtual world [12]. The idea is that immersive journalism increases the emotional engagement of the user [10].

Since 2015, this new form of journalistic storytelling has gained popularity due to the increased availability of immersive technologies for consumers, such as high-end consumer headsets, smartphone-based headsets and the introduction of 360° cameras, to the 360 YouTube platform and Facebook 360 [14, 46]. News organizations have since produced a range of immersive stories. But this idea of immersion is not new to journalism. In the 1960s, journalists used narrative techniques to engage audiences in stories [32], for example investigative journalist Günther Walraff immersed himself in an immigrant life and adopted the identity of a Turkish guest worker in Germany. Meanwhile, award-winning journalist Barbara Ehrenreich spent months working as a cashier for Walmart, trying to survive on minimum wage in the United States. Both journalists immersed themselves in an issue and wrote their stories in a narrative form with personal and emotional styles, which were aimed at involving the audience (see also [19, 29]).

New immersive technologies have now made it possible for journalists to engage the user more intensely. Today it is not so much about the journalist immersing themselves in the story, but rather they employ various techniques to immerse the audience in the story. What exactly does immersive journalism comprise and how do journalists use it in practice? These questions will be answered below, concluding with how the user actually experiences this new form of journalistic storytelling.

3.1 Immersive Journalism: A Mix of Technology, Interaction and Narrative

According to researchers of the Tow Center of Digital Journalism, "immersive technology promises to bring audiences closer to a story than any previous platform"

[2:4]. While much focus has been on the immersive technologies that can create this engagement, more recently scholars have introduced a conceptual model for immersive journalism, which comprises three elements: immersive technology, interaction; and the narrative of a story [10].

Immersive technology is that which enables the user to be included in a virtual world while isolated from the real world [38]. In that virtual world, the technology allows the user to look around the surroundings, by for example having a 360° view. The user is also able to experience the story in different sensory modalities, including different forms of video and audio [23]. These immersive technologies can range from 360° camera to complete VR cave systems.

The second element of immersive journalism is the level of interaction. The user can, for example, look around in a story, walk around in a virtual world, or grab objects in a virtual environment. The user is sometimes also able to influence the pace of the story or the progression of the storyline.

The third element is the narrative of the story. This is not so much about the topic of the story, but more about which role the user plays in it. In journalistic stories, the reader, viewer or listener is usually an observer of a story, but in immersive stories it is possible for the user to be a story participant. The user can, for example, see their own virtual body, join in with other characters, and be directly spoken to [9, 13].

The three elements of immersive journalism may lead to a sense of presence—a psychological state of mind that arises when the awareness of the mediating device (e.g. a VR-headset) decreases and when the user feels present in the virtual world [38]. This presence, can in turn increase the emotional engagement of the user. Immersive journalism is thus defined by technological features (i.e. the inclusiveness of the device), narrative features (i.e. first- or third-person storytelling) and interactive features (i.e. determining the level of autonomy and agency for the user). The interactive possibilities can be limited—such as merely providing a 360° view—or extensive—allowing a user to influence the pace of the production, have direct interaction with the characters in the story or to independently navigate through a virtual space. While the user is an observer in the story in some productions, they are an (active) participant in others.

3.2 The Practice of Immersive Journalism

A content analysis of 200 productions that journalists labelled as immersive journalism shows that while immersive journalism can indeed range in variety, journalists mostly focused on the technology, either using 360° cameras or virtual reality to transport users to a virtual world. Most productions were no more than a 360° view of a specific location: observing the life of Antarctic penguins, being present at a soccer match, or observing public mourning at the site of the Bataclan terrorist attack in Paris [10]. Although renowned news outlets such as *The Guardian* and *The New York Times* experimented with more advanced and inclusive VR technologies, interaction possibilities have proven to be scarce.

More than five years since the excitement of VR reached the journalistic field, journalists seem to have moved beyond the experimental to a more strategic phase of producing immersive stories. The initial phase was mostly focused on trying out innovative technologies, and less focused on telling a journalistic story. We see now that it has moved beyond the fascination of technologies to strategize how to tell a journalistic story that engages and involves the user through a mix of inclusive technologies and interactive options, allowing the user to participate in a story [20]. Providing more interactive options is felt to create more emotional impact as the user can have more agency in the story.

At the same time, these choices also create new journalistic dilemmas. With the journalist providing more interactive options in a story, a tension is created between the autonomy of the journalist and the agency of the user. In some productions, the user can choose between different storylines, choosing their own path through the production they are interested in while skipping others. In such productions, the author gives the user more control of the structure of the story and the information they acquire. But how much agency should be granted to the user? And what does that mean for the story being told? Journalists struggle with how to tell a journalistic story and maintain their autonomy while involving and engaging the user in the story. A related dilemma is that journalists question what role to take when creating immersive productions. Often immersive productions are made with interdisciplinary teams, both with developers and people with a film or artistic background. In this team, the journalist needs to make sure that productions are based on journalistic principles, which means not only keeping to the facts but also not taking an activist or a more artistic role.

3.3 The Effect of Immersive Journalism on the Public

Journalists have been experimenting with innovative productions that have challenged their skills, their role in the production process and their relationship with the audience. But do these productions lead to more emotional engagement among users? Several studies have shown that the more inclusive the technology, the higher the level of presence [39, 41]. These studies compared immersive stories in VR or 360° camera with more traditional formats such as newspaper or video. Another study looked at the effect of the different elements of immersion, revealing that adding interaction possibilities leads to a higher level of presence and emotional engagement. Having the user be a participant and not merely an observer in a story increases the user's emotional engagement [11]. Where journalists struggle to find the right balance between control and agency, public effects studies show that working on some kind of controlled agency is worth the effort.

Concluding, immersive journalism has moved beyond the experimental phase, where the focus was merely on bringing the inclusive technology, to a new genre in which a balance needs to be found between emotionally engaging the user in a story while maintaining journalistic principles of neutrality and facticity.

4 Conclusions

Journalism is constantly facing the challenge of preserving its role in society, against the threat of new voices and platforms. Through multimedia, interactive and immersive narratives, journalism is constantly seeking the best way to inform the public.

Narrative renewal is based on techniques that allow the user to be at the center of the narrative, to make decisions and to participate in it. Consequently, this impacts how stories are narrated, which technology is used and what form interaction with users should take.

After a first stage of exploration, we see that while a constant renewal of innovative technologies will advance in the coming years, underlying this shift is a discussion around what role users should have in the story and how this relates to the maker or journalist. Soon, we will see how some practices will be consolidated while others will be overtaken by yet-to-be-invented technology. There is no doubt that technology-supported narrative modalities bring positive effects for the representation of reality. However, for this effect to resonate, journalists and news organizations will have to rethink their working processes. Firstly, these new types of interactive stories demand new journalistic routines. While journalists are used to working autonomously and with a top-down approach towards the audience, these new stories ask to involve the user and take a more user-centered approach. Secondly, these productions demand skills from other disciplines, such as IT, design and documentary, and therefore a more interdisciplinary approach is required. Lastly, there is a challenge between the agency of the user and the autonomy of the journalist. How much autonomy should the journalist hold and how much should be granted to the user? Should the user be able to shape their own experience and decide for themselves which information they will receive and in what order? Or should the journalist remain in charge?

What do interactive and immersive narratives contribute to the representation of complex reality? Increasing involvement and audience engagement are favored by the interaction enabled by immersive and interactive technologies. Consequently, there is an effect on emotion and empathy reinforced by such techniques. There is no guarantee of this result, although there are indications of the potential for interactive and immersive storytelling to favor this effect. In fact, the most salient condition of immersive journalism is to give the user the sense of being there. Regarding other interactive formats, participation and co-creation stand out. In short, a greater involvement of the active audience is begin sought.

Why is it important to understand the reception of the audience? In this chapter we discussed the significance of the experience, which is a challenge and a goal in itself for any product and is at the heart of design and product improvement. Experience is the core element of these interactive and immersive projects.

However, despite the benefits for news and reporting described above, these narrative modalities still have challenges ahead in terms of production: high costs, the need for specialized and multidisciplinary professionals, technological sophistication, great competition of online entertainment, and economical restrictions in media companies. Underlying these practical considerations, such interactive journalistic storytelling asks for a reconsideration of the relationship between the journalist and their audiences.

Research in this field must continue to address important questions, not only regarding business models, profitability and audience research, but also fundamental ethical issues and changing journalistic role patterns. The expansion of the described practices in journalism will depend on the future steps of production and research.

Funding This research has been developed within the research project *Digital Native Media in Spain: Storytelling Formats and Mobile Strategy* (RTI2018-093346-B-C33), funded by the Ministry of Science, Innovation and Universities (Government of Spain) and the ERDF structural fund.

References

1. Alexander, B.: The New Digital Storytelling. Praeger, California (2011)
2. Aronson-Rath, R., Milward, J., Owen, T.: Virtual Reality Journalism. Columbia Journalism School, New York (2016)
3. Aston, J., Gaudenzi, S.: Interactive documentary: setting the field. Stud. Documentary Film **6**, 125–139 (2012). https://doi.org/10.1386/sdf.6.2.125_1
4. Barthes, R.: Introduction à l'analyse structurale des récits. Communications **8**, 1–27 (1966)
5. Behmer, S.: Digital Storytelling: Examining the Process with Middle School Students. Iowa State University (2005)
6. Bolter, J.D., Grusin, R.: Remediation: Understanding New Media. The MIT Press, Massachusetts (1999)
7. Bonilla, D.: Retos para la creación (y la investigación) de narrativas complejas. In: Longhi, R.R., Lovato, A., Gifreu, A. (eds.) Narrativas complexas. Ria Editorial, Aveiro, pp. 11–34 (2020)
8. Castells, M.: The Internet Galaxy. Oxford University Press, Oxford (2001)
9. Cummings, J., Bailenson, J.: How immersive is enough? A meta-analysis of the effect of immersive technology on user presence. Media Psychol. **19**, 272–309 (2016)

de Bruin, K., de Haan, Y., Kruikemeier, S., Lecheler, S., Goutier, N.: A first-person promise? A content-analysis of immersive journalistic productions. Journalism, Online First (2020). https://doi.org/10.1177/1464884920922006

11. de Haan, Y., de Bruin, K., Goutier, N., Kruikemeier, S., Lecheler, S.: Immersive journalism and the engaged audience? A multi-method study on immersive journalism and its effects on the public. In: Future of Journalism Conference, 12–13 Sep, Cardiff (2019)
12. De la Peña, N., Weil, P., Llobera, J., et al.: Immersive journalism: immersive virtual reality for the first-person experience of news. Presence Teleop. Virt. **19**(4), 291–301 (2010)
13. Domínguez, E.: Going beyond the classic news narrative convention: the background to and challenges of immersion in journalism. Frontiers Digit. Humanit. **4**(10), 1–10 (2017)
14. Doyle, P., Gelman, M., Gill, S.: Viewing the Future? Virtual Reality in Journalism (2016). Available at https://kng.ht/3jK2Jfp. Accessed 01 Dec 2020
15. Fidler, R.: Mediamorphosis. Pine Forge Press, Thousand Oaks, Understanding New Media (1997)
16. Franklin, B.: The future of journalism. Journal. Stud. **17**, 798–800 (2016). https://doi.org/10.1080/1461670X.2016.1197641

17. George-Palilonis, J.: The Multimedia Journalist. Storytelling for Today's Media Landscape. Oxford University Press, Oxford (2012)
18. Gifreu, A.: El documental interactivo. Evolución, caracterización y perspectivas de desarrollo. Editorial UOC, Barcelona (2013)
19. Glück, A.: What makes a good journalist? Empathy as central resource in journalistic work practice. Journal. Stud. **17**(6), 893–903 (2016). https://doi.org/10.1080/1461670X.2016.1175315
20. Goutier, N., de Haan, Y., de Bruin, K., Kruikemeier, S., Lecheler, S.: From "cool observer" to "emotional participant": The practice of immersive journalism (in review)
21. Herrera, R.: Érase unas veces: Filiaciones narrativas en el arte digital. Editorial UOC, Barcelona (2014)
22. Herrscher, R.: Periodismo narrative. Cómo contra la realidad con armas de la literatura. Universitat de Barcelona, Barcelona (2012)
23. Higuera-Trujillo, J.L., Maldonado, J.L.T., Millán, C.L.: Psychological and physiological human responses to simulated and real environments: a comparison between photographs, 360 panoramas, and virtual reality. Appl. Ergon. **65**, 398–409 (2017)
24. Jacobson, S., Marino, J., Gutsche, R.E.: The digital animation of literary journalism. Journalism **17**, 527–546 (2016). https://doi.org/10.1177/1464884914568079
25. Jenkins, H.: Convergence Culture: Where Old and New Media Collide. New York University Press, New York (2006)
26. Larrondo, A.: La metamorfosis del reportaje en el ciberperiodismo: concepto y caracterización de un nuevo modelo narrativo. Commun. Soc. **22**, 59–88 (2009)
27. Lassila-Merisalo, M.: Story first—publishing narrative long-form journalism in digital environments. J. Mag. New Media Res. **15**(2), 1–15 (2014)
28. Le Masurier, M.: Slow journalism. Journal. Pract. **10**, 439–447 (2016). https://doi.org/10.1080/17512786.2016.1139902
29. Lecheler, S.: The emotional turn in journalism needs to be about audience perceptions. Digit. Journal. **8**(2), 287–291 (2020). https://doi.org/10.1080/21670811.2019.1708766
30. Meyer, P.: Precision journalism. A reporters guide do social science methods. Indiana University Press, Bloomington (1993)
31. Murray, J.: Hamlet on the Holodeck: The Future of Narrative in Cyberspace. The MIT Press, Massachusetts (2017)
32. Neveu, E.: Revisiting narrative journalism as one of the futures of journalism. Journal. Stud. **15**, 533–542 (2014). https://doi.org/10.1080/1461670X.2014.885683
33. Paulussen, S.: Innovation in the newsroom. In: Witschge, T., Anderson, C.W., Domingo, D., Hermida, A. (eds.) The SAGE Handbook of Digital Journalism. Sage, London (2016)
34. Pavlik, J.: The impact of technology on journalism. Journal. Stud. **1**, 229–237 (2000). https://doi.org/10.1080/14616700050028226
35. Pavlik, J.V., Dennis, E.E., Mersey, R.D., Gengler, J.: Conducting research on the world's changing mediascape: Principles and practices. Media Commun. **7**, 189–192 (2019). https://doi.org/10.17645/mac.v7i1.1982
36. Salaverría, R.: Los labs como fórmula de innovación en los medios. El Profesional de la Información **24**(4), 397–404 (2015). https://doi.org/10.3145/epi.2015.jul.06
37. Scolari, C.: Los ecos de McLuhan: ecología de los medios, semiótica e interfaces. Palabra Clave **18**, 1025–1056 (2015). https://doi.org/10.5294/pacla.2015.18.4.4
38. Slater, M., Wilbur, S.: A framework for immersive virtual environments (FIVE): speculations on the role of presence in virtual environments. Presence **6**(6), 603–616 (1997)
39. Sundar, S.S., Kang, J., Oprean, D.: Being there in the midst of the story: how immersive journalism affects our perceptions and cognitions. Cyberpsychol. Behav. Soc. Netw. **20**(11), 672–682 (2017)
40. Uricchio, W.: Mapping the Intersection of Two Cultures: Interactive Documentary and Digital Journalism. MIT Open Documentary Lab, Massachusetts (2015)
41. van Damme, K., All, A., De Marez, L., Van Leuven, S.: 360° video journalism: experimental study on the effect of immersion on news experience and distant suffering. Journal. Stud. **20**(14), 2053–2076 (2019). https://doi.org/10.1080/1461670X.2018.1561208

42. van Krieken, K., Sanders, J.: Framing narrative journalism as a new genre: a case study of the Netherlands. Journalism **18**, 1364–1380 (2017). https://doi.org/10.1177/1464884916671156
43. Vázquez-Herrero, J., Gifreu-Castells, A.: Interactive and transmedia documentary: Production, interface, content and representation. In: Túñez-López, M., et al. (eds.) Communication: Innovation & Quality. Studies in Systems, Decision and Control vol. 154, pp. 113–128 (2019)
44. Vázquez-Herrero, J., López-García, X.: When media allow the user to interact, play and share: recent perspectives on interactive documentary. New Rev. Hypermedia Multimedia **25**, 245–267 (2019). https://doi.org/10.1080/13614568.2019.1670270
45. Vázquez-Herrero, J., López-García, X., Irigaray, F.: The technology-led narrative turn. In: Vázquez-Herrero, J., et al. (eds.) Journalistic Metamorphosis. Studies in Big Data, vol. 70, pp. 29–40 (2020)
46. Watson, Z.: VR for News: The New Reality? (2017). Available at https://bit.ly/3rJCfgR
47. Wilson, E.O.: The power of story. Am. Educ. **26**, 8–11 (2002)

Jorge Vázquez-Herrero Assistant Professor at the Department of Communication Sciences, Universidade de Santiago de Compostela (USC). Ph.D. in Communication and Contemporary Information (USC). He is a member of Novos Medios research group and the Latin American Chair of Transmedia Narratives (ICLA–UNR). He was visiting scholar at Universidad Nacional de Rosario, Universidade do Minho, University of Leeds and Tampere University. His research focuses on the impact of technology and platforms in digital journalism and narratives.

Yael de Haan Professor of Applied Sciences for Journalism in Digital Transition at the University of Applied Sciences Utrecht in the Netherlands. She obtained her Ph.D. at the Amsterdam School of Communication Research (ASCoR) at the University of Amsterdam. Her research focuses on the consequences of digitalization in journalism productions, consumption and ethics.

After the Hype: How Hi-Tech Is Reshaping Journalism

Sara Pérez-Seijo and Paulo Nuno Vicente

Abstract Digital journalism has experienced disparate and immeasurable transformations over the last decade. The impact of the Internet of Things (IoT) and high technologies was such that led to a whole reconfiguration of the processes of news content production, distribution and consumption. A shift which has been named hi-tech journalism. Media and audience have witnessed a paradigm change in which journalism becomes automated (robot or algorithmic journalism), personalized (virtual assistants, smart speakers, chatbots) and more immersive (immersive journalism). Even it takes visual journalism into the sky (drone journalism). Through this chapter it will be explored and explained the potential opportunities of the latest technologies in journalism practices, but also the threats, risks and challenges posed by these new devices and software.

1 A Technology-Driven Shift

Technology sets the pace of change and innovation in the digital sphere. Hardware and software have become dogmas of this sort of doctrine called digital innovation, key for both online native and migrated news outlets to compete and achieve differentiation, as if were actually mandatory in the current scenario [27]. An investment that represents, at the same time, a potential boost for engaging new audiences, attract digital native users and reconnect with the old, fragmented ones.

During the last 20–25 years technology has greatly evolved. The technology-driven change came first with what was commonly known as new technologies, but, once software took command [46], an even major transformation began with the Internet of Things (IoT) and a whole set of high technologies (e.g., augmented

S. Pérez-Seijo (✉)
Universidade de Santiago de Compostela, Santiago de Compostela, Spain
e-mail: s.perez.seijo@usc.es

P. N. Vicente
Universidade Nova de Lisboa, Lisboa, Portugal

J. Vázquez-Herrero et al. (eds.), *Total Journalism*, Studies in Big Data 97,
https://doi.org/10.1007/978-3-030-88028-6_4

reality, virtual reality, drones, artificial intelligence, 5G, blockchain). And actually this is the shape the journalism that will tell the future is adopting [44].

High tech journalism, also known as hi-tech journalism, refers to the use and application of sophisticated, emerging technologies in different areas of journalism with the aim of improving, to some extent—way, cost, time, understanding, perspective, etc.—a process, a model, or even a user experience [56].

Since 2014, this high-tech journalism has gained more and more strength [59], leading to a ubiquitous paradigm shift [60] in which journalism becomes automated (robot journalism), intelligent (chatbots, voice assistants and smart speakers) and more immersive (virtual reality, augmented reality and 360° video storytelling), as well as personalized (artificial intelligence systems). Even takes visual journalism where journalists could not reach before (drone journalism).

The impact has been such that also novel multimedia news storytelling forms have emerged [60], aimed to be more visual, interactive, immersive and gamified [43]. Novel ways of telling the news stories that, often, turns the consumption into a true experience in which users acquire a more active role.

In short, these technological transformations have not only impacted on the processes of news production, distribution, consumption and reception, but also have affected the business models and the relationships between audience and journalist/newsroom. However, even though the future hand-in-hand with hi-tech seems full of potential novel opportunities for journalism in general and newsrooms in particular, as the following sections will explain, these emerging technologies can, at the same time, compromise and threaten the quality of news, the user's privacy, the right to be accurately informed, the role of journalists as storytellers and, among others, even the working conditions [50, 60].

2 High-Technologies Applied to Journalism

2.1 Automated Journalism

Automated journalism, also known as algorithmic journalism, computational journalism and robot journalism, refers to the application of computer programs—namely, algorithms—to news work, with the aim to organize, interpret and present news pieces from structured data sets [17, 41]. An algorithm is a set of specific instructions programmed to perform a given task and solve a precise problem.

Within the vast research field of Data Science, machine learning is the branch of artificial intelligence dedicated to the computational ability of automating data analysis and model building. In simple terms, data scientists develop methodologies and techniques that allow computer programs to learn from data, hence the name, which have the potential to be applied to highly time and labor-intensive tasks for humans. In this precise sense, automated journalism corresponds to the application of algorithmic news judgment and the automation of specific journalistic tasks related to news reporting, writing, curation, data analysis and online social platforms' dissemination.

In the last decade, research and development in the field of automated journalism has been carried away by companies like Automated Insights, Narrative Science, Yseop and Arria, encountering an early adoption by major news media outlets like Associated Press, Forbes, *BBC*, ProPublica and *Los Angeles Times*, among others [22]. News organizations have been experimenting with the application of algorithms on topics such as sports, finance and economics, weather forecasting, geological surveys (e.g. earthquakes) and crime. The numerical representation and database organization of this data facilitates the application of automated data extraction and analysis tools, as well as the generation of news stories whose narrative structures are previously defined [69]. Automating news production is, thus, also closely associated with developments in natural language generation (NLG), the computerized production of human natural language from a digital representation.

For the task of analyzing large quantities of documents, consequently accelerating the process of journalistic discovery and reducing its costs, the algorithmic turn in news production has also been tested in the scope of investigative journalism, namely in public affairs reporting, towards data-rich storytelling [11, 62]. In this context, algorithms have been used for data mining in the public interest, envisioning the use of computation and artificial intelligence to find hidden patterns in large volumes of data.

Although scarce, existing studies on the perceived readability and credibility by readers of autonomous production of journalistic content show ambiguous results. In previous experimental research, subjects rated computer-written articles as more credible and higher in journalistic expertise but less readable [30]. On the other hand, credibility perceptions of human, automated and combined content and sources are assumed equal, just varying on topic-specific factors (e.g. sports), suggesting that currently the outcome of algorithmic journalistic production it is largely indiscernible from its human counterpart [71] or only discernible when clearly disclosed the role of automation in the news article [70].

Considering the contemporary sourcing and reception of news stories, algorithmic curation is also becoming a crucial line both in professional practice, as well as in research for communication and journalism studies. As the use of social media and web platforms pose new challenges for the journalistic verification process, algorithms are being developed for the task of supporting news producers evaluating the credibility of social media contributors [23]. At the reception level, due to their high relevance in personal information consumption, online search engines recommendation policies (e.g., Google News) and social media platforms' feeds (e.g. Facebook, Twitter, YouTube) mimic story selection mechanisms that parallel traditional editorial news organizations.

However, very little is still known about which values translate into the automation mechanisms that define how the news feed is structured—friend relationships, user engagement and preferences, sociodemographic data, media affordances, among other factors—stressing the importance of algorithmic fairness, transparency and accountability by non-journalistic companies [49, 73]. Although computerization of newsrooms has decades, the specific use of algorithms and machine learning in journalism poses new opportunities, as well as challenges in the ethical and legal domains [36, 40, 72].

2.2 Chatbots

In online environments, the use of chatbots, i.e. computer programs also known as software agents which are conceived, designed and developed to simulate a human-to-human conversation, has been proliferating in recent years, both in its text and text-to-speech renditions. These conversational systems are often anthropomorphized into virtual assistants and given a name by which they are called and activated, a humanized voice they respond to, and they are assigned a virtual gender. This programmed performativity as if they were human revamps the recognition that computers are contemporarily social actors reflecting the rules of human interaction.

The application of chatbots in the context of news media organizations translates into a trend for a more conversational journalism, where the direct interaction between the audiences and news workers is mediated by a virtual assistant, namely in online social media [42]. The botification of news [7] has been materialized by media organizations such as *NBC*, which launched a chatbot inside Kik, a chat app, launching news stories based on keywords sent by users, and such as *BBC*, which adopted Telegram's open source software (Bot API) in order to implement a chatbot placed at the end of selected online articles that followed a multiple-choice question and answer format using pre-scripted material [37].

The growth of bots has been regarded as impacting on news production and distribution towards the personalization of news content [39], as well as on the communication patterns between news organizations' representatives and their publics (e.g. [24, 29]), representing a possible avenue for the development of sustainable news business strategies [35].

2.3 Voice Assistants and Smart Speakers

The association of developments in the field of natural language generation (NLG) with audio synthesis technologies, particularly those applied to the artificial production of human speech, find in voice assistants and smart speakers a cutting-edge frontier for digital journalism. Advances in NLG already translate into exploratory nonlinear story structures for messaging bots and artificial intelligence speakers (e.g. Amazon's Alexa, Google's Home, Microsoft's Cortana, Apple's HomePod). Users can interact in a conversational way with these devices, making clear that the Internet of Things (IoT) is paving its way into journalism.

These communicative robots are more and more relevant for media and communication as they embody a new frontier in human–machine communication. As meaning making machines, they stimulate emergent ways of interacting with humans and progressively fill in social roles. More than just channels or gadgets, these conversational software agents are designed to become digital interlocutors. They are rapidly evolving and changing our relationship with the media, our representation as subjects, and the structure of our culture [32, 33, 68].

Intelligent virtual assistants' (IVAs) basic tasks currently include information provision (e.g. news bulletins, weather, traffic, historical facts), music and radio playing, audiobooks' reading, video streaming and e-commerce. IVAs are already in our homes (e.g. Samsung Family Hub 2.0 refrigerators and TVs), in our vehicles (e.g. Mercedes-Benz's Ask Mercedes), on our wrists (e.g. Amazfit smartwatch) and in our pockets (smartphones, tablets).

IVAs depend on our social data. At the base of their design is the technological realization that every human behavior is passible of interpretation as a piece of information and thus have a computational representation. IVAs are also, thus, profiling machines: they recognize patterns and learn from our daily routines. This data, on a first instance individual and disconnected, is compiled on a data set [20]. From there, it is known that IVAs datafy social interaction through measurable types: human behavioral patterns technologically interpreted and constructed [12]. In communicative terms, computer code has thus a performativity in which data circulation plays a primary role [45]. Although, there is currently no sense of public accountability for IVAs. Virtually nothing is known about how they perform (if at all) informed consent to the collection, use, and transfer to third parties of users' personal information.

As data-driven media, IVAs constitute a turn in human communication that has far-reaching societal implications. They rest on the interplay of computation power and algorithmic accuracy, pattern inference from data sets and the belief in data objective epistemic superiority [10]. Given that algorithmic mediation is founded in making humans detectable, traceable and recordable, fundamental concerns arise about the advance of a network-distributed surveillance society, algorithmic bias, insecurity and gender divides [13, 15, 25, 66].

2.4 360° Video, VR and AR Storytelling

Immersive journalism was for the first time described as "the production of news in a form in which people can gain first-person experiences of the events or situation described in news stories" [14:291]. Ten years later and passed the immersive race that started around the end of 2015 [18, 57], 360° video and virtual reality are being used in newsrooms as a multimedia resource to produce news stories and offer a more immersive news experience.

Major news outlets such as *The New York Times*, *El País*, *Publico*, *ZDF*, *Euronews*, *Corriere della Sera*, *Chosun Ilbo* or *The Guardian* have created their own 360° video news stories, often related to social issues. Even digital native media have tested the potential possibilities of VR and spherical videos, as in the case of *Vice News* and *elDiario.es*. From this experimentation around the globe, novel genres and narrative forms have emerged in an attempt for adapting the storytelling to the new immersive possibilities [38, 53].

Regarding user experience, 360° videos and CGI (Computer Generated Imagery) pieces have introduced changes in news consumption. The viewing experience with a VR headset (e.g. Google Cardboard, Oculus Go, Samsung Gear) moves away the traditional way in which viewers just see, hear or read information. Instead, immersive journalism enables users to choose the angle of view in a surrounding scene [16] since the traditional frame is gone [8], and thus news turns into a 'storyliving' experience [48]. Borders disappear and the distant realities and remote locations of news become virtually closer when wearing a VR headset [47, 61]. That is in fact the innovation compared to prior forms of producing immersive nonfiction content, from literary journalism to webdocs and newsgames.

The potential of 360° video reports and VR experience is understood in terms of gratifications and experience [52]. According to research, immersive journalism has a positive impact on engagement, presence or place illusion, empathy, authenticity, enjoyment, credibility and even story-sharing intention compared to non-immersive formats [5, 52, 63, 64, 67]. Nevertheless, it faces also ethical challenges related to objectivity, transparency, image manipulation, witnessing and, among others, the role of the journalist [3, 9, 31, 51, 58].

Augmented reality is included in the scope of immersive storytelling, along with 360° video and virtual reality [4, 55]. However, augmented reality does not immerse users into a new reality as they were there, which is the essence of VR and 360° storytelling, but consists of overlaying objects and information into a real-time space. Introduced in journalism as potential tool for enhancing user engagement [54], augmented reality had led to a novel storytelling form in which, sometimes, audience assume a leading role while interacting [65].

This technology attracted the attention of major news outlets like *Esquire*, *The New York Times*, *USA Today* or *The Guardian*. Even though augmented reality news contents are usually created for a mobile consumption, thus as an individual experience, there are also examples of this technology in TV programs [6]. In these cases, the audience is able to see extra visual information, related to a particular topic being presented in real time, directly superposed on the set.

2.5 Drone Journalism

Drone journalism consists of the use of unmanned aerial vehicles (UVAs), often called drones, in news gathering. Drones allow journalists to obtain high-definition aerial images that provide users with a bird's eye view [19] and a bigger picture [55] of the places in the news, which in fact lead to novel forms of witnessing the stories [60]. Drones take visual journalism into the sky, which thus brings not only new ways to provide news coverage [28], but also new perspectives and forms of seeing the covered stories—e.g. underline the news content, add meaning, provide surprise or uniqueness—[1].

Developed outside the newsrooms as a technology with multiple applications, drones have entered the journalism practice as an innovation [19], a high-technology tool [56] that offer journalists new opportunities for reporting in general and producing visual content in particular. In fact, it has been introduced in journalism as an emergent form of producing visual content that replaces news helicopters—or other manned aircraft—traditional footage as a cheaper, easier-to-use and faster alternative to get aerial images [28]. Moreover, the remote piloting or algorithmic control of drones constitute safer options for journalists and camera operators while covering violent or dangerous situations.

But what makes drone journalism different from other forms is the possibility of accessing to remote, difficult or unsafe locations that, otherwise, journalists could not reach [21]. The range of potential uses is vast, from live streaming and breaking news to investigative uses and war reporting, even for producing immersive news content in 360° video [55], just as the events where a drone can add value to the news coverage: sports competitions, protest marches, natural disasters, armed conflicts, etc. [44].

During the last decade, legacy media (e.g. *CNN*, *CBS*, *RTVE*, *The New York Times*, *BBC*, *Clarín*) and even digital native media (e.g. *El Español*) have been exploring and using drones with journalistic purposes—by buying it or by subcontracting services. Several benefits have been pointed out; however, drone journalism also faces legal challenges [2, 26, 34], which in some countries hold back its further development and integration into the newsrooms, and also it raises privacy and ethical questions [1, 19].

3 Will Hi-Tech Be Actually the Future of Journalism?

In its third decade, digital journalism has the responsibility of consolidating the emerging technologies, developments and storytelling forms that newsrooms from all over the globe have explored, practiced and in some cases even almost integrated during the past years. Transformations that came hand-in-hand with the above-mentioned high technologies and whose impact virtually reaches all the processes of journalism: production, distribution and consumption. This hi-tech journalism, primarily led by artificial intelligence and virtual and augmented reality, has also introduced significant changes in the way audiences and journalists relate to each other, i.e., in the relationship between user and media.

The technological-led turn brought new models, new ways of disseminating the information, and also managing it, new options for content personalization, new forms of multimedia and visual storytelling, including storyliving experiences, and a whole new frontier in this human–machine communication currently being developed. But beyond this, hi-tech journalism proved to be useful, as a potential contributor to public interest when used for verification, investigation, data analysis and also immersion.

All the transformations that digital journalism experienced since the beginning of the century are actually immeasurable since hardware and software are already part of the news production DNA. These technologies have improved digital journalism in disparate ways and have demonstrated that journalists are no longer alone doing their job, which is in essence producing information, as they now have the support of the machines: algorithms and bots that can automate particular tasks, collect and manage information for them, and even sometimes assume their role as writers and communicators. Of course, not without challenges and risks for the journalistic practice.

Despite all the possibilities and opportunities posed by technologies, there are also threats that reveal that hi-tech journalism is not always synonym of a high-quality journalism. These technologies and devices have brought legal and ethical risks related to privacy, manipulation, bias, accountability and copyright, among others. But also challenges related to audience perception of credibility and objectivity when the news content is the result of a technological, and thus non-human, mediation. So, after the hi-tech hype, reality begins.

Funding This article has been developed within the project *Digital native media in Spain: storytelling formats and mobile strategy* (RTI2018-093346-B-C33), from the Ministry of Science, Innovation and Universities (Government of Spain) and co-funded by the ERDF structural fund. On the other hand, the author Sara Pérez-Seijo is beneficiary of the Training University Lecturers' (FPU) Program funded by Spanish Ministry of Science, Innovation and Universities (Government of Spain).

References

1. Adams, C.: Tinker, tailor, soldier, thief: an investigation into the role of drones in journalism. Digit. Journal. **7**(5), 658–677 (2019). https://doi.org/10.1080/21670811.2018.1533789
2. Aguado, G.: Repercusión en el ejercicio del periodismo de la regulación del uso de drones en Europa. In: Flores Vivar, J. (coord) Tecnologías del ecosistema periodístico. Realidad inmersiva, drones y otras tecnologías disruptivas en la nueva ecología de los medios. Comunicación Social, Salamanca (2019)
3. Aitamurto, T.: Normative paradoxes in 360° journalism: contested accuracy and objectivity. New Media Soc. **21**(1), 3–19 (2019). https://doi.org/10.1177/1461444818785153
4. Aitamurto, T., Aymerich-Franch, L., Saldivar, J., Kircos, C., Sadeghi, Y., Sakshuwong, S.: Examining augmented reality in journalism: presence, knowledge gain, and perceived visual authenticity. New Media Soc. (Online First) (2020). https://doi.org/10.1177/1461444820951925
5. Archer, D., Finger, K.: Walking in another's virtual shoes: Do 360–degree video news stories generate empathy in viewers? Tow Center Digit. Journal. Columbia (2018)
6. Azkunaga, L., Gaztaka, I., Eguskiza, L.: Nuevas narrativas en televisión: la realidad aumentada en los telediarios de Antena 3. Revista de Comunicación **18**(2), 25–50 (2019). https://doi.org/10.26441/rc18.2-2019-a2
7. Barot, T.: The botification of news. Nieman Lab (2015). Available at https://bit.ly/2OLdaUQ. Accessed 21 December 2020

8. Benítez, M.J., Herrera, S.: El reportaje inmersivo a través de vídeo en 360°: caracterización de una nueva modalidad de un género periodístico clásico. In: de Lara, A., Arias, F. (eds.) Medi-amorfosis. Perspectivas sobre la innovación en periodismo. Sociedad Española de Periodística, Elche, pp. 196–212 (2017)
9. Benítez, M.J., Herrera, S.: Periodismo inmersivo 360°. Telos: Cuadernos de comunicación e innovación **111**, 22–25 (2019)
10. Boyd, D., Crawford, K.: Critical questions for big data: provocations for a cultural, technological, and scholarly phenomenon. Inf. Commun. Soc. **15**(5), 662–679 (2012). https://doi.org/10.1080/1369118X.2012.678878
11. Broussard, M.: Artificial intelligence for investigative reporting: using an expert system to enhance journalists' ability to discover original public affairs stories. Digit. Journal. **3**(6), 814–831 (2015). https://doi.org/10.1080/21670811.2014.985497
12. Cheney-Lippold, J.: We Are Data: Algorithms and the Making of Our Digital Selves. New York University Press, New York (2017)
13. Chung, H., Iorga, M., Voas, J., Lee, S.: Alexa, can I trust you? Computer **50**(9), 100–104 (2017). https://doi.org/10.1109/MC.2017.3571053
14. De la Peña, N., Weil, P., Llobera, J., et al.: Immersive Journalism: immersive virtual reality for the first–person experience of news. Presence **19**(4), 291–301 (2010). https://doi.org/10.1162/pres_a_00005
15. Deleuze, G.: Postscript on the societies of control. October **59**, 3–7 (1992)
16. Domínguez, E.: Going beyond the classic news narrative convention: the background to and challenges of immersion in journalism. Frontiers Digit. Humanit. **4**, 1–11 (2017). https://doi.org/10.3389/fdigh.2017.00010
17. Dörr, K.N.: Mapping the field of algorithmic journalism. Digit. Journal. **4**(6), 700–722 (2016). https://doi.org/10.1080/21670811.2015.1096748
18. Doyle, P., Gelman, M., Gill, S.: Viewing the future? Virtual reality in journalism. Knight Foundation (2016). Available at https://kng.ht/3vfX2LQ. Accessed 15 Dec 2020
19. Duncan, M., Culver, K.B.: Technologies, ethics and journalism's relationship with the public. Media Commun. **8**(3), 101–111 (2020). https://doi.org/10.17645/mac.v8i3.3039
20. Elmer, G.: Profiling Machines: Mapping the Personal Information Economy. The MIT Press, Cambridge (2004)
21. Fischer, D.A.: Dron't stop me now: prioritizing drone journalism in commercial drone regulation. Columbia J. Law Arts **43**(1), 107–146 (2019). https://doi.org/10.7916/jla.v43i1.4127
22. Fırat, F.: Robot journalism. In: Vos, T.P., Hanusch, F. (eds.) The International Encyclopedia of Journalism Studies. Wiley-Blackwell, Hoboken. https://doi.org/10.1002/9781118841570.iejs0243
23. Fletcher, R., Schifferes, S., Thurman, N.: Building the 'Truthmeter': training algorithms to help journalists assess the credibility of social media sources. Convergence **26**(1), 19–34 (2017). https://doi.org/10.1177/1354856517714955
24. Ford, H., Hutchinson, J.: Newsbots that mediate journalist and audience relationships. Digit. Journal. **7**(8), 1013–1031 (2019). https://doi.org/10.1080/21670811.2019.1626752
25. Foucault, M.: Discipline and Punish. Vintage, New York (1979)
26. Gallardo-Camacho, J., Breijo, V.R.: Relationships between law enforcement authorities and drone journalists in Spain. Media Commun. **8**(3), 112–122 (2020). https://doi.org/10.17645/mac.v8i3.3097
27. García-Avilés, J.A., Carvajal, M., Arias, F.: Implementation of innovation in Spanish digital media: analysis of journalists' perceptions. Rev. Lat. Comun. Soc. **73**, 369–384 (2018). https://doi.org/10.4185/RLCS-2018-1260
28. Goldberg, D., Corcoran, M., Picard, R.G.: Remotely Piloted Aircraft Systems and Journalism: Opportunities and Challenges of Drones in News Gathering. Reuters Institute for the Study of Journalism, Oxford (2013)
29. Gómez-Zará, D., Diakopoulos, N.: Characterizing communication patterns between audiences and newsbots. Digit. Journal. **8**(9), 1093–1113 (2020). https://doi.org/10.1080/21670811.2020.1816485

30. Graefe, A., Haim, M., Haarmann, B., Brosius, H.-B.: Readers' perception of computer–generated news: credibility, expertise, and readability. Journalism **19**(5), 595–610 (2016). https://doi.org/10.1177/1464884916641269
31. Gregory, S.: Immersive witnessing: From empathy and outrage to action. Witness (2016). Available at https://bit.ly/3veBG1i. Accessed 15 Dec 2020
32. Guzman, A.L.: Human-Machine Communication: Rethinking Communication, Technology, and Ourselves. Peter Lang, New York (2018)
33. Hepp, A.: Artificial companions, social bots and work bots: communicative robots as research objects of media and communication studies. Media Cult. Soc. **42**(7–8), 1410–1426 (2020). https://doi.org/10.1177/0163443720916412
34. Holton, A.E., Lawson, S., Love, C.: Unmanned aerial vehicles: opportunities, barriers, and the future of "drone journalism." Journal. Pract. **9**(5), 634–650 (2015). https://doi.org/10.1080/17512786.2014.980596
35. Hong, H., Oh, H.J.: Utilizing bots for sustainable news business: understanding users' perspectives of news bots in the age of social media. Sustainability **12**(16), 1–16 (2020). https://doi.org/10.3390/su12166515
36. Jamil, S.: Artificial intelligence and journalistic practice: the crossroads of obstacles and opportunities for the Pakistani journalists. Journal. Pract. (Online First) (2020). https://doi.org/10.1080/17512786.2020.1788412
37. Jones, B., Jones, R.: Public service chatbots: automating conversation with BBC News. Digit. Journal. **7**(8), 1032–1053 (2019). https://doi.org/10.1080/21670811.2019.1609371
38. Jones, S.: Disrupting the narrative: immersive journalism in virtual reality. J. Media Pract. **18**(2–3), 171–185 (2017). https://doi.org/10.1080/14682753.2017.1374677
39. Kunert, J., Thurman, N.: The form of content personalisation at mainstream, transatlantic news outlets: 2010–2016. Journal. Pract. **13**(7), 759–780 (2019). https://doi.org/10.1080/17512786.2019.1567271
40. Lewis, S.C., Sanders, A.K., Carmody, C.: Libel by algorithm? Automated journalism and the threat of legal liability. Journal. Mass Commun. Q. **96**(1), 60–81 (2018). https://doi.org/10.1177/1077699018755983
41. Linden, C.-G.: Decades of automation in the newsroom. Digit. Journal. **5**(2), 123–140 (2017). https://doi.org/10.1080/21670811.2016.1160791
42. Lokot, T., Diakopoulos, N.: News bots. Digit. Journal. **4**(6), 682–699 (2016). https://doi.org/10.1080/21670811.2015.1081822
43. López-García, X., Pérez-Seijo, S., Vázquez-Herrero, J., García-Ortega, A.: New narratives in the age of visualization. In: Toural-Bran, C., Vizoso, Á., Pérez-Seijo, S., Rodríguez-Castro, M., Negreira-Rey, M.-C. (eds.) Information Visualization in the Era of Innovative Journalism, pp. 51–63. Routledge, New York (2020)
44. López-Hidalgo, A.: El periodismo que contará el futuro. Chasqui **131**, 239–256 (2016). https://doi.org/10.16921/chasqui.v0i131.2733
45. Mackenzie, A.: The performativity of code: software and cultures of circulation. Theory Cult. Soc. **22**(1), 71–92 (2005). https://doi.org/10.1177/0263276405048436
46. Manovich, L.: Software Takes Command. Bloomsbury, New York (2013)
47. Marconi, F., Nakagawa, T.: The Age of Dynamic Storytelling. A Guide for Journalists in a World of Immersive 3–D Content. Associated Press, New York (2017)
48. Maschio, T.: Storyliving: An ethnographic study of how audiences experience VR and what that means for journalists. Google News Lab (2017). Available at https://bit.ly/3rHlzY0. Accessed 15 Dec 2020
49. Milosavljević, M., Vobič, I.: Human still in the loop. Digit. Journal. **7**(8), 1098–1116 (2019). https://doi.org/10.1080/21670811.2019.1601576
50. Murcia, F.J., Ufarte, M.J.: Mapa de riesgos del periodismo hi-tech. Hipertext.net **18**, 47–55 (2019). https://doi.org/10.31009/hipertext.net.2019.i18.05
51. Nash, K.: Virtual reality witness: exploring the ethics of mediated presence. Stud. Documentary Film **12**(2), 119–131 (2017). https://doi.org/10.1080/17503280.2017.1340796

52. Nielsen, S.L., Sheets, P.: Virtual hype meets reality: users' perception of immersive journalism. Journalism (Online First) (2019). https://doi.org/10.1177/1464884919869399
53. Paíno, A., Rodríguez, M.I.: Proposal for a new communicative model in immersive journalism. Journalism (Online First) (2019). https://doi.org/10.1177/1464884919869710
54. Pavlik, J.V.: Journalism in the Age of Virtual Reality: How Experiential Media Are Transforming News. Columbia University Press, New York (2019)
55. Pavlik, J.V.: Drones, augmented reality and virtual reality journalism: Mapping their role in immersive news content. Media Commun. **8**(3), 137–146 (2020). https://doi.org/10.17645/mac.v8i3.3031
56. Pérez-Seijo, S., Gutiérrez-Caneda, B., López-García, X.: Periodismo digital y alta tecnología: de la consolidación a los renovados desafíos. index.comunicación **10**(3), 129–151 (2020). https://doi.org/10.33732/ixc/10/03Period
57. Pérez-Seijo, S., López-García, X.: Las dos caras del periodismo inmersivo: el desafío de la participación y los problemas éticos. In: López Paredes, M. (ed.) Nuevos escenarios en la comunicación: retos y convergencias, pp. 279–305. Centro de Publicaciones PUCE, Quito (2018)
58. Pérez-Seijo, S., López-García, X.: Five ethical challenges of immersive journalism: a proposal of good practices' indicators. In: Rocha, Á., Ferrás, C., Paredes, M. (eds.) Information Technology and Systems. ICITS 2019. Advances in Intelligent Systems and Computing 918. Springer, Cham, pp. 954–964. https://doi.org/10.1007/978-3-030-11890-7_89
59. Salaverría, R.: Periodismo en 2014: balance y tendencias. Cuadernos de periodistas **29**, 9–22 (2015). https://cutt.ly/ygqQksB
60. Salaverría, R., de-Lima-Santos, M.F.: Towards ubiquitous journalism: impacts of IoT on news. In: Vázquez-Herrero, J., Direito-Rebollal, S., Silva-Rodríguez, A., López-García, X. (eds.) Journalistic Metamorphosis. Studies in Big Data 70. Springer, Cham, pp. 1–15 (2020). https://doi.org/10.1007/978-3-030-36315-4_1
61. Sánchez Laws, A.L., Utne, T.: Ethics guidelines for immersive journalism. Frontiers Robot. AI **6**(28), 1–13 (2019). https://doi.org/10.3389/frobt.2019.00028
62. Stray, J.: Making artificial intelligence work for investigative journalism. Digit. Journal. **7**(8), 1076–1097 (2019). https://doi.org/10.1080/21670811.2019.1630289
63. Suh, A., Wang, G., Gu, W., Wagner, C.: Enhancing audience engagement through immersive 360-degree videos: an experimental study. In: Schmorrow, D., Fidopiastis, C. (eds.) Augmented Cognition: Intelligent Technologies, pp. 425–443. Springer, Cham (2018)
64. Sundar, S.S., Kang, J., Oprean, D.: Being there in the midst of the story: how immersive journalism affects our perceptions and cognitions. Cyberpsychol. Behav. Soc. Netw. **20**(11), 672–682 (2017). https://doi.org/10.1089/cyber.2017.0271
65. Tejedor-Calvo, S., Romero-Rodríguez, L.M., Moncada-Moncada, A.J., Alencar-Dornelles, M.: Journalism that tells the future: possibilities and journalistic scenarios for augmented reality. El Profesional de la Información **29**(6), 1–14 (2020). https://doi.org/10.3145/epi.2020.nov.02
66. UNESCO.: I'd blush if I could: Closing gender divides in digital skills through education. UNESCO, Paris (2019). Available at https://bit.ly/3qDooaW. Accessed 21 December 2020
67. Van Damme, K., All, A., De Marez, L., Van Leuven, S.: 360 video journalism: experimental study on the effect of immersion on news experience and distant suffering. Journal. Stud. **20**(14), 2053–2076 (2019). https://doi.org/10.1080/1461670X.2018.1561208
68. Vicente, P.N.: From narrative machines to practice–based research: making the case for a digital Renaissance. Estudos em Comunicação **27**(2), 67–78 (2018a). https://doi.org/10.20287/ec.n27.v2.a05
69. Vicente, P.N.: Ledes and story structures. In: Vos, T.P., Hanusch, F. (eds.) The International Encyclopedia of Journalism Studies. Wiley-Blackwell, Hoboken (2018b)
70. Waddell, T.F.: Attribution practices for the man–machine marriage: how perceived human intervention, automation metaphors, and byline location affect the perceived bias and credibility of purportedly automated content. Journal. Pract. **13**(10), 1255–1272 (2019). https://doi.org/10.1080/17512786.2019.1585197

71. Wölker, A., Powell, T.E.: Algorithms in the newsroom? News readers' perceived credibility and selection of automated journalism. Journalism **22**(1), 86–103 (2018). https://doi.org/10.1177/1464884918757072
72. Wu, S., Tandoc, E.C., Salmon, C.T.: A field analysis of journalism in the automation age: understanding journalistic transformations and struggles through structure and agency. Digit. Journal. **7**(4), 428–446 (2019). https://doi.org/10.1080/21670811.2019.1620112
73. Wu, S., Tandoc, E.C., Salmon, C.T.: When journalism and automation intersect: assessing the influence of the technological field on contemporary newsrooms. Journal. Pract. **13**(10), 1238–1254 (2019). https://doi.org/10.1080/17512786.2019.1585198

Sara Pérez-Seijo Ph.D. student in Communication and Contemporary Information at Universidade de Santiago de Compostela and member of the Novos Medios research group. Her research focuses on digital native media, new digital narratives, such as the case of 360 video storytelling or Immersive Journalism, and the journalism ethics in the current digital environment. She is a beneficiary of the Faculty Training Program (FPU) funded by the Spanish Ministry of Science, Innovation and Universities.

Paulo Nuno Vicente Assistant Professor at Universidade Nova de Lisboa. He is the founder and coordinator of iNOVA Media Lab, a digital innovation laboratory dedicated to immersive and interactive narrative, human-machine communication, new media literacies, science communication, social media and information visualization. He holds a Ph.D. in Digital Media (UT Austin Portugal) and he is an honored recipient of the prestigious German Marshall Fund of the United States Fellowship (2016) and Calouste Gulbenkian Prize—Knowledge (2019).

Gamification and Newsgames as Narrative Innovations in Journalism

Jose Alberto García-Avilés, Raul Ferrer-Conill, and Alba García-Ortega

Abstract This chapter explores how innovative narratives, supported by a combination of playful approaches and technological convergence, provide a reconfiguration of digital news storytelling. Newsgames integrate two opposing logics: the culture of journalism, based on truthfulness and credibility, and the culture of games, characterized by the creation of imaginary worlds, persuasion and mechanics. We analyse *The Ocean Game* (2019) and *The Amazon Race* (2019), which use different procedural strategies. We examine the development of gamification in journalistic storytelling, which uses game elements to enhance the user experience. The relationship that gamified news products establish with the audience illuminates changes in the rhetorical and structural dynamics between the news organization, the media workers, and the users. Thus, innovation in journalistic narrative through gamification and newsgames might translate into effective ways of producing content that combines rigour in substance with attractiveness in form, while preserving journalistic quality and incorporating the playful elements of games.

1 Introduction: Innovation in Online News Narrative Genres

This chapter explores how innovative narratives, supported by a combination of playful approaches and technological convergence, provide a reconfiguration of digital news storytelling. The use of playful approaches exemplifies what has been called 'Total Journalism' [39] as it integrates game elements into journalistic processes by using a wide range of social knowledge, technologies, and automation models. Media convergence and newsroom integration of print, broadcast and online

J. A. García-Avilés (✉) · A. García-Ortega
Miguel Hernández University, Elche, Spain
e-mail: jose.garciaa@umh.es

R. Ferrer-Conill
Karlstad University, Karlstad, Sweden

University of Stavanger, Stavanger, Norway

J. Vázquez-Herrero et al. (eds.), *Total Journalism*, Studies in Big Data 97,
https://doi.org/10.1007/978-3-030-88028-6_5

platforms facilitated the emergence of genres with high levels of interactivity and hyper-textuality, enhancing multimedia narratives and innovative content production [28:455–456, 53]. In the digital space, news genres are modified, enriched, and renewed, while new ones emerge that exploit its audio-visual, immersive and transmedia potential in innovative ways [36].

Journalism is a narrative art that borrows its writing techniques from literature and its investigative and interpretative methodologies from social sciences. Genre studies offer an approach for organizing storytelling conventions into a structure of representation that reveals the ways in which news content works as a sort of standardized truth [52]. News narratives can be understood as multimedia artifacts that reveal these structures in action [5]. We regard a narrative genre as the expression of content through the specific forms and structures with which a journalistic team seeks to connect with the recipients of the information. Therefore, news genres are the specific ways in which any story is articulated according to shared journalistic conventions, such as immersive reportages, podcasts, or webdocs [26].

Online narrative genres display a hybrid nature, as they combine a great variety of resources and interactive elements and share similar characteristics. These genres usually have a collective authorship embodied in multidisciplinary teams integrated by developers, data journalists, designers, video and audio editors, computer graphics, etc. [45]. Through interactive narratives, a team of professionals seeks to make users understand the information through the resources of text, image, audio, video, animation, computer graphics, data, etc. [37].

Users configure the structure of these genres through their own navigation, being able to choose among the available options and make certain decisions. In some instances, users can introduce changes in the points of view of the story or in the characters' actions, feeling that they enter the virtual world where the story takes place and are able to control the point of view, choosing even the angle of vision and camera movements [35]. Narratives can incorporate aesthetics and mechanics based on games, and playful thinking to promote learning experiences and solve problems through the narrative itself [29:327]. In 2003, Henry Jenkins announced the beginning of "a new era of media convergence that makes the flow of content through multiple channels inevitable" and coined the term 'transmedia narrative', which encourages the participation of users in the construction of the story through different platforms, expressly emphasizing their active role "in the process of expansion and construction of the narrative" [34:5].

Journalism innovation is defined as "the capacity to react to changes in products, processes and services through the use of creative skills that allow a problem or need to be solved, in such a way that it allows the introduction of something new that adds value to consumers and promotes the viability of the journalistic organization" [27:31]. Innovations in news genre and content have not received as much scholarly attention as innovations in other areas such as technology, production, or commercialization.

We believe that news narratives are innovative because they provide an effective way of designing content that is adapted to the story, achieving success in terms of an increase in audience, user engagement, advertising revenue, brand image or prestige for the news outlet. Although it is very difficult to catalogue the diversity of hybrid narrative genres that can be found on the Internet, there is a growing interest in experimenting with news genres [53].

Media organizations that implement innovative genres can differentiate themselves from their competitors and attract new audiences. To cite just a few international examples, outlets such as *The Guardian*, *Al Jazeera*, *The New York Times*, *The Washington Post*, or *Folha de São Paulo* stand out for their narrative innovation [46]. Longform journalism projects require advance planning, multidisciplinary work, and execution with wide production margins, far from the immediacy of everyday news coverage. The process includes several stages of design, experimentation, testing, development, dissemination, and learning. Of course, technology plays a fundamental role in innovation, but other relevant aspects must also be considered, such as management capacity, professional culture, creativity in storytelling and the ability to produce quality journalism.

Gamification has proven to be useful in attracting younger audiences and boosting news consumption [19], while newsgames provide "more options of informing, sense-making, storytelling and persuasion than simply remediating 'old' forms of news production" [56:12]. Their ability to show information in an experiential way, as well as the possibility of placing the user at the centre of the narrative, allows reaching new audiences accustomed to virtual and participatory environments [44].

In the next section, we examine how newsgames innovate as they seek to integrate two opposing logics: the culture of journalism, based on truthfulness and credibility, and the culture of games, characterized by the creation of imaginary worlds, persuasion and mechanics that provide a playful experience and user engagement [20]. These hybrid genres potentially might increase storytelling quality: the stories can be enhanced with interactive and playful elements, and its share-ability brings about new possibilities for reaching younger users and creating conversations. We argue that, more than novelty itself, true narrative innovation in newsgames is closely related to journalistic standards and translates into effective ways of designing useful content that combines rigor in substance with attractiveness in form, in a playful way.

2 The Concept and Development of Newsgames

The term 'newsgame' is attributed to video game designer Gonzalo Frasca [4, 25], who considers it a form of video game equivalent to political cartoons: "simulation meets political cartoon" (cit. in [25:3]). However, the industry's confusing definition [40] and the lack of a formal academic definition [56] make it difficult to establish their limits as a narrative genre. 'Newsgame' is often used as an umbrella-term, "a 'label' attached to a wide and rather heterogeneous spectrum of digital artefacts falling in between journalism, play, simulation and participatory action" [56:2]. It is therefore not surprising that similar formats are frequently labelled 'newsgames', 'political games', 'quizzes', 'web documentaries' or 'multimedia stories' [40:431]. Nevertheless, two characteristics make it possible to differentiate newsgames from other narratives: (1) they use procedural rhetoric to transmit information; and (2) the user experiences game dynamics, gaining experiential insight into interdependencies beyond the events or stories depicted [56].

Unlike other journalistic formats based on verbal or visual rhetoric, newsgames use the expressive and persuasive potential of processes to create complex arguments. Bogost's term 'procedural rhetoric' defines "the practice of persuading through processes in general and computational processes in particular" [3:3]. Newsgames' procedural rhetoric allows showing complex concepts in a didactic, experiential way [30, 44].

The first newsgames focused on their ability to convey political ideas. According to Frasca, creator of some of the first newsgames such as *Kabul Kaboom* (2001), *September 12th* (2002) and *Madrid* (2004), newsgames can be useful for analysing, debating, and editorializing important news stories. Treanor and Mateas [51] argue that newsgames use rules and game mechanics that act as an additional rhetorical outlet, offering a deeper commentary than the related political cartoons. This approach would mostly define production and academic research between 2003 and 2009, a period in which the main studies argued that newsgames were not designed for the neutral presentation of the facts, but as a form of ideological expression [40].

In the mid-2000s, several news outlets such as *The New York Times* and *CNN* signed collaboration agreements with producers of 'serious games' to create and publish political opinion newsgames. Among them, *Food Imported Fooly* (The New York Times 2007), a newsgame about the difficulties faced by Food and Drug Administration (FDA) inspectors published in the Times Select opinion section, and *CNN*'s *Campaign Rush* and *Presidential Pong* (CNN 2008), focused on the 2008 presidential elections. However, both agreements only lasted a couple of years, relegating newsgames to an occasionally used genre.

Two events changed both the industry and academia perspective of this genre: the launching of *Cutthroat Capitalism* (Wired 2009) and the publication of *Newsgames: Journalism at play* [4]. *Cutthroat Capitalism,* published by Wired magazine, focuses on a report on the economic system of piracy in Somalia. Unlike most newsgames, *Cutthroat Capitalism* did not convey opinions, but instead showed the negotiation processes between Somali pirates and merchant ships in an interactive way. Through a first-person narrative, users must put themselves in the shoes of a Somali leader and they kidnap and negotiate the rescue of the crews. *Cutthroat Capitalism* thus became one of the first newsgames to provide an objective presentation of a news event.

In 2010, *Newsgames: Journalism at play* [4] was published, presenting an exhaustive analysis of the newsgames produced until then. According to the approach and the narrative strategy used, the authors distinguish between current events—including opinion newsgames, infographics, documentary, puzzles, literacy, community, and platforms. However, if Frasca's proposal—video games with editorial purposes—is too specific, that of Bogost and colleagues seems too general; newsgames are considered "any intersection of journalism and games" [4:13]. This is too broad a classification that includes narratives and genres that are often considered independent from mainstream newsgaming. Nevertheless, the book became a turning point in the study of newsgames.

According to Wiehl [56], newsgames can be used "for expository, exploratory and persuasive matters as well as for making differentiated comment" [56:2]. Ferrari [17] distinguishes between video games that respect the theories and traditions of journalism—newsgames—and those that manifestly show an editorial line, which he refers to as "editorial games". For their part, Plewe and Fürsich [44] state that, despite including rules and mechanics oriented along the lines of the designers' opinions, newsgames should not be limited to the role of political cartoon or editorial comment, because they "can also provide contextual information in the way journalistic documentaries or features could" [44:3]. In this regard, Sicart provides one of the most complete definitions to date: "newsgames are computer games used to participate in the public sphere with the intention of explaining or commenting on current news" [47:27].

Three main perspectives are involved in the design of newsgames: to encourage debate around current events, to transmit ideas or viewpoints, and to interact with news stories [29]. Nevertheless, the genre has shown a surprising ability to engage their audience and has achieved prestigious awards in the fields of design and journalism. *How Y'all, Youse and You Guys Talk* (2013) was *The New York Times*' most popular piece of 2013. The *Financial Times*' *Uber Game* (2018) won gold at the annual Serious Play Awards. For its part, ProPublica's *The Waiting Game* (2018) was a finalist in the Online Journalism Awards' Excellence and Innovation in Visual Digital Storytelling.

Despite public and critical recognition, some academics and journalists hold reservations about newsgames because they consider them prone to misinterpretation or lack of interpretation at all [56:10]. According to Plewe and Fürsich [44], the architecture of a video game tends to favour conceptual simplification, a quick outcome, and Manichaeism in terms of good or bad characters, winners, or losers, favouring excessive plot trivialization. Poorly designed newsgames may thus favour a simplistic view of problems, oblivious to the complexity of real life [23]. According to Meier [40], many newsgames simplify the content and omit relevant aspects to understand a news story, so that they often require complementary contextual information.

Several studies show that newsgames can transmit information in a manner consistent with journalistic quality standards [30]. Wiehl [56] states that newsgames have the potential "to serve as a mode for exploring intricate interdependencies, to adequately present complex facts, to make qualified arguments and to stimulate critical thoughts" because they provide "more options of informing, sense-making, storytelling and persuasion than simply remediating 'old' forms of news production" [56:2]. Newsgames allow users to experience for themselves the events that are narrated [57], providing the possibility of showing different paths or perspectives. The genre can therefore help young people become aware of serious issues such as corruption (*The good, the bad and the accountant*, El Confidencial 2018), the economy (*Objectif Budget*, Le Figaro 2017) or immigration (*The Refugee Challenge*, The Guardian 2014).

Newsgames face two major challenges: the need for technical knowledge and resources on the part of newsrooms—often only accessible to large corporations—, and a cultural change that regards games an appropriate genre for the dissemination of news [22]. According to Newman [41], video games were considered from the beginning as a distraction for the youngest, lacking interest from an academic point of view. As some critics argue, the connection between youth culture and video games reduces them to "mere trifles—low art—carrying none of the weight or credibility of more traditional media" [41:14].

Older generations tend to hold reservations about this new form of media coverage because newsgames represent a new 'information-driven media' [4] in which news content is constructed through interaction with a game system. Several studies based on interviews with newsgame players show that users less accustomed to virtual environments often prefer traditional genres [40] or believe that the genre does not provide quality journalism [57]. Users' active role represents a significant change in the way information is consumed. Therefore, both the industry and academia need to work together in a process of media literacy that demonstrates the narrative and expressive potential of newsgames.

3 Analysis of Newsgame Design: The Cases of *the Ocean Game* and *the Amazon Race*

Newsgames with similar aesthetics and structure often offer very different results in terms of information processing. Beyond usability, other factors such as user immersion, emotion and motivation are involved [31] and the choice of game mechanics also determines how the user interacts with the information [30]. In the field of video game design, the mechanics and dynamics have a huge subjective load to convey opinions about the story being told [47]. For this reason, besides the verbal and visual elements involved in the creation of arguments, to understand a newsgame in its entirety it is useful to analyse the procedural logics used in its construction.

This difference in information processing can be seen in *The Ocean Game* (2019) and *The Amazon Race* (2019), which have a very different procedural strategy. Both newsgames are inspired by an investigative report, use a game system based on first-person decision-making and provide complementary information through links to the main investigations. However, the processes that trigger each of the mechanics differ significantly. While *The Amazon Race* uses the decision-making system exclusively to advance the narrative—the storylines are supported by visual and verbal components—in *The Ocean Game* the main storyline of the story is built through decision-making.

The Ocean Game (2019) is a newsgame produced by *Los Angeles Times* about the rise in sea level off the coast of California. This interactive format, based on an investigative report on climate change and its consequences on the western coast of the United States, challenges users to save a coastal city in eight turns. To do so, they must make a series of decisions such as building a stone wall, adding more sand to the beach, or launching a plan to buy the houses built on the waterfront and move them away from the coast. For its part, *The Amazon Race* (2019), produced by the Australian public broadcaster's *ABC News* Story Lab, focuses on an investigative report on the working conditions of employees at the Amazon Melbourne warehouse.

The Ocean Game's playful strategy is characterized by its simplicity. The interface is a fixed screen with a looping animation of the California coast that varies according to the player's decisions and their consequences. The characters are limited to animated icons that give their opinion on the decisions based on their preferences. Four main characters represent the sectors involved: a yellow character for the business sector, a blue one for surfers, a green one for environmentalists and a red one for the coastguard.

The objective of the game is straightforward: users must save the houses on the beachfront in less than eight turns. Each turn follows the same structure: users are presented with a problem and offered three different options. Except for the first turn, when the user has the possibility of consulting an expert to gather more information, all the decisions offer the same three options: build a rock wall, add more sand to the beach or propose a plan to move the houses away from the coast. Users must decide which option is the most appropriate, considering two elements of pressure: it must be kept within budget—represented by a progress bar at the top of the interface—and each measure has a different cost.

Despite the simplicity of the playful strategy, the choice of game mechanics is loaded with meaning. Although players are free to decide when to implement each measure, they always face the same result. Both the rock wall and the purchase of sand for the beach are temporary measures that disappear after a few turns. The only effective measure is the purchase of the houses. In this way, the designers construct the newsgame's main argument: the only possible solution is to move the houses away from the coast. Furthermore, although the option of buying houses is available from the outset, families are more receptive and accept less money as the game advances and the sea level rises. Thus, the argument is built that the more inevitable the rise in sea levels seems, the easier it is to convince families to leave their homes and move away from the coast.

On the other hand, *The Amazon Race* (2019) develops a different play strategy. Although the dynamics of the game are more complex, the mechanics are not so subjectively charged. The narrative is built on the statements of the workers and former workers of the Amazon Melbourne warehouse, so that all the events and situations are based on the experiences of the interviewees and provide information through windows and help messages. The game mechanics are limited to providing interactivity and facilitating navigation between the different parts of the game. In addition, the playful strategies favour the immersion of the player, such as the use of certain objects—smartphone, work console, etc.—, interaction with other people and personal decision-making—spending time with colleagues or carrying out surveys on job satisfaction.

The entire narrative is based on the player's decisions. It uses a branched first-person narrative structure in which there are different ways to advance. After each decision, the player receives feedback through a message inspired by real situations experienced by the interviewees. In addition, the newsgame includes a carefully designed mini-game in which the user must go through the warehouse and pick up the object indicated on the work console before time runs out. At the beginning, the user is shown a screen with the kind of object to be picked up, instructions on the controls, a countdown and a marker with the pick rate.

Unlike the other newsgame, *The Amazon Race* offers most of the information through verbal and visual rhetoric. The narrative is structured around a series of messages based on the decisions made by the player. In this case, the game mechanics are used to make the content more playable—through mini-games—and to advance the narrative. They are purely instrumental because they are not designed to build arguments, but to portrait the hard conditions of the workers according to the interviews carried out.

4 Gamification in News Media

A decade after the popularization of newsgames, a new, playful journalistic innovation found its way into the newsroom. Gamification proposes a different approach by infusing game elements in journalistic storytelling [19]. The main difference between newsgames and gamification is that while the former carry their own procedural rhetoric and a self-contained narrative to transmit editorial information [51], gamification adopts a more traditional journalistic storytelling technique and inserts game elements to enhance the user experience [21]. In other words, newsgames are digital games that invite the public to participate and interact with current affairs content, while gamification integrates game elements within normal journalistic processes without the goal of creating a full-fledged game [47].

The use of gamification in news media in the last decade responds to a surge of implementation of gamified approaches across online platforms with the aim of enhancing online user engagement [55]. During the early 2010s, a multitude of industries and fields implemented gamified strategies, amped by the hype of the term [6]. The growth of gamification is an incarnation of two major developments. On the one hand, gamification promised to entice younger generations under the wing of newfound popularity of digital gaming culture to new technological advances [58]. On the other hand, an increasingly optimistic view on the potential outcomes of gamification, such as increased user engagement and audience loyalty [42, 59]. To temper what seemed over-exaggerated expectations, a wave of scholarship outlines the potential dangers of gamifying. These dangers stem from the instances of exploitation, surveillance and transparency issues to which gamification can lead [9, 13, 24].

While the use of playful approaches as a problem-solving solution is nothing new [7, 33], journalism's adoption of gamification is the "product of cultural, economic, political and technological forces in different times and spaces" [2:257]. After decades in which news organizations tried to keep news and entertainment separate, the irruption of gamification is most innovative in that it blends news and games within the usual journalistic format. 'Real news' has been a predominantly serious endeavor that has tended to shy away from entertainment and a connection with the audience [10]. More popular topics or formats were considered by critics and other journalists a sign of bad journalism [16]. Historically, games such as crosswords, puzzles and quizzes have been present in newspapers, but in different sections, clearly demarcated as news and entertainment. Even with newsgames, these are standalone artifacts, that while combining editorial content and games, they do so as a separate journalistic product [45].

In fact, Vos and Perreault [54] propose that since the late 2000s, journalists have discursively constructed the gamification of journalism as an institutional renegotiation of the journalist–audience relationship. This implies that gamification can be understood both as the use of game mechanics in non-gaming contexts [12], and as a process by which an institution slowly adopts the tenets of play and game thinking. One of the first proponents of this discursive turn toward playful and ludic approaches was first championed by Stephenson [50], who called for rethinking the seriousness of news and for accommodating notions of games and play in the production and consumption of news. For journalists to re-discover the audience, they "could intentionally incorporate elements into their articles as a means of improving the traditional labor of media: informing the public" [15:8]. Such an overarching institutional change is what Dowling [14] identifies as journalism's "ludic turn" and includes all forms of game-oriented journalism artifacts, from newsgames to immersive VR applications, all encompassed as the "ultimate empathy machine" [14:164].

We understand gamification in news media as distinct instances in which game elements and game thinking have been embedded in the story as a narrative strategy [18]. These instances help us visualize how gamification can be understood as an innovative genre in its own right, one that combines and integrates different social aspirations, technological artifacts and storytelling techniques within journalism. Moreover, gamification pushes news organizations to rely on boundary work [8] and the collaboration between journalists and actors that usually reside outside the newsroom, such as technologists and game designers [45]. For example, one of the most well-known cases of gamification, *Pirate Fishing*, published in 2014, is the outcome of a tight collaboration between *Al Jazeera*'s senior reporter Juliana Ruhfus and Ivan Giordano, head of digital media at Altera Studio. In this case, the journalist leads the piece, which uses a storytelling technique where users begin the story as junior reporters and must consume content as the story progresses. Users collect evidence and earn points and badges, which allows them to advance to subsequent levels of the story. While the degree of immersion and playfulness is high (a trait carried by Altera Media), the story prioritizes a new way to present editorial content that entices and attracts the audience in new ways.

Not all instances of gamified journalism, however, share the same priorities. According to Ferrer-Conill [19], when journalists gamify the news, they set in motion an interplay of the professional and commercial logics of journalism with the utilitarian and the hedonic logics of gamification. While the professional logic of journalism provides the democratic norms and values of traditional journalism, the commercial logic establishes the imperative need of journalism to be economically viable [43, 48]. At the same time, while the hedonic logic of gamification focuses on the participative and empowering nature of games, the utilitarian logic tries to push behavior to be enacted through the game mechanics [32]. When these logics interact, combined with the symbolic and material practices of journalists and game designers, explain why and how gamification is implemented within journalism and how it adheres to any of these logics.

The first case of gamification implemented by a major news organization is a perfect aid to exemplify these tensions. In 2010, *The Guardian* created *Investigate your MP's expenses*, a crowdsourcing microsite that used game-mechanics to engage users. With the help of a Yahoo! former engineer, the journalists at the Guardian used progress bars, points, and leaderboards to motivate users to access, review and classify about half a million documents regarding the Members of Parliament expenses scandal [11]. While the game elements were kept to the minimum, the narrative was successful at incentivizing behavior, and thus the gamified system reflected the utilitarian logic of gamification. Similarly, the microsite aimed to empower the audience in an act of watchdog journalism, which reflects the professional logic of journalism as the main driver of the case.

Many other examples of gamification in journalism show the incredible flexibility of gamification as a narrative strategy. From *MTV*'s 2012 *Fantasy Football Election Coverage*, in which users predicted the US elections as if they were football players to the *Bleacher Report* Writer's Ranking system, in which the sports news website gamified the user-generated-scheme. From *The Times of India* Times Points gamified loyalty program to *Al Jazeera*'s News Quiz Passport, a gamified audience research strategy. Gamification is poised to serve many purposes with game-like experiences. This is important because the relationship these gamified news products try to establish with the audience illuminates changes in the rhetorical and structural dynamics between the news organization, the different types of journalists, media workers and the audience.

Just like with newsgames, gamified news experiences face significant challenges. The negotiation between technical and journalistic knowledge is a hurdle difficult to overcome. An unbalanced agreement can lead to stress on logics that pushes the story further away from its journalistic nature. Furthermore, while most of the research is done on the producers, little has been studied on how different frames communicated by news in the gamified context are experienced by the audience [1]. If gamified news can capture youth's civic agency and empower their role as active participants in news ecosystems [38], news organizations must understand how audiences understand their hedonic and utilitarian narratives. Finally, news editors and managers need to overcome the steep economic investment of gamified stories, often facing unclear return on investment. In a period of a commercial crisis, strong news organizations rushed to test this innovation to reach out and engage with younger audiences. It is in a moment of uncertainty when news organizations innovate and invest in technology while searching for solutions [49], but it is not an investment that is at the hand of small or struggling organizations.

5 Conclusion

Compared to traditional reporting, newsgames might be an innovative way to attract the users' attention and involve them in dealing with complex realities and current events, such as climate change or precarious employment. The genre shows the implications of a topic, encourages critical thinking, and helps to generate debate among the users. In this way, newsgames can integrate both the mechanics of the game and journalistic standards, encouraging young people to inform themselves about issues that do not usually interest them.

The analysis of two newsgames, *The Amazon Race* and *The Ocean Game,* shows how they use a different procedural strategy, although both share a similar purpose and the same decision-making-based game system. *The Amazon Race* builds its arguments mainly through classic rhetorical formulas based on visual and textual components; although it also includes procedural rhetorical formulas, these tend to be objectively depicted. On the contrary, *The Ocean Game* constructs its main arguments through a procedural form, in which the visual and textual elements are limited to defining the different response options and the consequences of each one of them. These two different strategies intend to show the causes and consequences of a current event in a playful and interactive way.

At the same time, gamification is tied to the material manifestations of gamified news systems and whether news organizations can grasp the empowering narratives that gamification can deliver, or whether a simple deployment of game mechanics would render storytelling mere technical experimentation. The complex interplay between the professional and commercial logics of journalism and the hedonic and utilitarian logics of gamification shapes how news organizations and journalists implement gamified systems. The friction between professionalism and the market, and the increasing transgression of institutional journalistic boundaries will determine the next phases of the gamification process.

Both gamification and newsgames share the goal of placing users at the centre of the experience. In this way, the news content could be integrated into the narrative and allows players to get a deeper and more direct understanding of the information, expanding the creative possibilities of an explanatory, gamified, personalized, and immersive journalism. Another important aspect of these strategies is the richness of the hybridization of different codes, languages, and mechanisms in the same discursive unit.

Thus, the adoption of innovative narrative genres enhances the expressive possibilities of journalism through gamification, game-playing and immersion. The combination of these properties offers valuable potential for news organizations in a market of overabundance of supply, with a limited attention economy and an increasingly fragmented and selective audience, especially among the youth.

In short, innovation in journalistic narrative through gamification and newsgames might translate into effective ways of producing news content that combines rigour in substance with attractiveness in form, while preserving journalistic quality and using the playful elements of games.

References

1. Arafat, R.K.: Rethinking framing and news values in gamified journalistic contexts: a comparative case study of Al Jazeera's interactive games. Convergence **26**(3), 550–571 (2020). https://doi.org/10.1177/1354856520918085
2. Belair-Gagnon, V., Revers, M.: The sociology of journalism. In: Vos, T.P. (ed.) Journalism, pp. 257–280. Walter de Gruyter, Boston (2018)
3. Bogost, I.: Persuasive Games: The Expressive Power of Videogames. MIT Press, Cambridge (2007)
4. Bogost, I., Ferrari, S., Schweizer, B.: Newsgames: Journalism at Play. MIT Press, Cambridge (2010)
5. Buozis, M., Creech, B.: Reading news as narrative: A genre approach to journalism studies. Journal. Stud. **19**(10), 1430–1446 (2018). https://doi.org/10.1080/1461670X.2017.1279030
6. Burke, B.: What's next: The gamification of everything. Gartner (2011). Available at https://gtnr.it/2Ntrnp2. Accessed 15 Dec 2020
7. Caillois, R.: Man, Play and Games. University of Illinois Press, Chicago (1961)
8. Carlson, M.: Boundary work. In: Vos, T.P., Hanusch, F. (eds.) The International Encyclopedia of Journalism Studies. Wiley, Hoboken (2019)
9. Conway, S.: Zombification? Gamification, Motivation, and the User. J. Gaming Virtual Worlds **6**(2), 143–157 (2014)
10. Costera Meijer, I.: The paradox of popularity: How young people experience the news. Journal. Stud. **8**(1), 96–116 (2007). https://doi.org/10.1080/14616700601056874
11. Daniel, A., Flew, T.: The Guardian reportage of the UK MP expenses scandal: a case study of computational journalism. In: Papandrea, F., Armstrong, M. (eds.) Record of the Communications Policy and Research Forum 2010, pp. 186–194. Network Insight, Sydney (2010)
12. Deterding, S., Khaled, R., Nacke, L., Dixon, D.: From game design elements to gamefulness: defining 'gamification'. In: Proceedings of the 15th International Academic MindTrek Conference: Envisioning Future Media Environments, pp. 9–15. ACM, Tampere (2011)
13. DeWinter, J., Kocurek, C.A., Nichols, R.: Taylorism 2.0: Gamification, scientific management and the capitalist appropriation of play. J. Gaming Virtual Worlds **6**(2), 109–127 (2014)
14. Dowling, D.: The Gamification of Digital Journalism: Innovation in Journalistic Storytelling. Routledge, New York (2021)
15. Dozier, D.M.: A test of the Ludenic news reading theory using R-Factor analysis. In: Annual Meeting of the Association for Education of Journalism, Ottawa, 1975. Available at http://eric.ed.gov/?id=ED122278. Accessed 12 Dec 2020
16. Dubied, A., Hanitzsch, T.: Studying celebrity news. Journalism **15**(2), 137–143 (2014). https://doi.org/10.1177/1464884913488717
17. Ferrari, S.: Newsgame, or editorial game? News games (2009). Available at https://b.gatech.edu/3vw232O. Accessed 12 Dec 2020
18. Ferrer-Conill, R.: Points, badges, and news. A study of the introduction of gamification into journalism practice. Comunicació **33**(2), 45–63 (2016)
19. Ferrer-Conill, R.: Gamifying the News. Exploring the Introduction of Game Elements into Digital Journalism. Karlstad University Press, Karlstad (2018)
20. Ferrer-Conill, R., Foxman, M., Jones, J., et al.: Playful approaches to news engagement. Convergence **26**(3), 457–469 (2020). https://doi.org/10.1177/1354856520923964

21. Ferrer-Conill, R., Karlsson, M.: The gamification of journalism. In: Gangadharbatla, H., Davis, D.Z. (eds.) Emerging Research and Trends in Gamification, pp. 356–383. IGI Global, Hershey (2015)
22. Foxman, M.: Play the news: fun and games in digital journalism. Columbia Journalism School, Tow Center for Digital Journalism (2015). Available at https://bit.ly/3r1DXtu. Accessed 12 Nov 2020
23. Frasca, G.: Ludologists love stories, too: notes from a debate that never took place. DiGRA Conference: Level Up, 4–6 Nov, Utrecht (2003). Available at https://bit.ly/3qVUHSO. Accessed 15 Nov 2020
24. Fuchs, M.: Gamification as twenty-first-century ideology. J. Gaming Virtual Worlds **6**(2), 143–157 (2014)
25. Galloway, A.: Social realism in gaming. Game Stud. **4**(1) (2004). Available at https://bit.ly/3bRX70r. Accessed 20 Nov 2020
26. García-Avilés, J.A.: Reinventing television news: innovative formats in a social media environment. In: Vázquez-Herrero, J., Direito-Rebollal, S., Silva-Rodríguez, A., López-García, X. (eds.) Journalistic Metamorphosis, pp. 143–155. Springer, Cham (2020)
27. García-Avilés, J.A., Carvajal-Prieto, M., De-Lara-González, A., Arias-Robles, F.: Developing an index of media innovation in a national market: the case of Spain. Journal. Stud. **19**(1), 25–42 (2018). https://doi.org/10.1080/1461670X.2016.1161496
28. García-Avilés, J.A., Meier, K., Kaltenbrunner, A.: Converged media content: reshaping the 'legacy' of legacy media in the online scenario. In: Franklin, B., Eldridge, S.A. (eds.) The Routledge Companion to Digital Journalism Studies, pp. 449–458. Routledge, London (2016)
29. García-Ortega, A., García-Avilés, J.A.: Los newsgames como estrategia narrativa en el periodismo transmedia: propuesta de un modelo de análisis. Revista Mediterránea de Comunicación **9**(1), 327–346 (2018). https://doi.org/10.14198/MEDCOM2018.9.1.19
30. García-Ortega, A., García-Avilés, J.A.: When journalism and games intersect: examining news quality, design and mechanics of political newsgames. Convergence **26**(3), 517–536 (2020). https://doi.org/10.1177/1354856520918081
31. González-Sánchez, J.L., Gutiérrez-Vela, F.L.: Jugabilidad como medida de calidad en el desarrollo de videojuegos. CoSECivi (2014). Available at https://bit.ly/3ttS1xf. Accessed 20 Nov 2020
32. Hamari, J. Gamification. Motivations & Effects (PhD Thesis). Aalto University, Aalto (2015)
33. Huizinga, J.: Homo Ludens: A Study of the Play-Element in Culture. Routledge, New York (1949)
34. Jenkins, H.: Transmedia storytelling. Moving characters from books to films to video games can make them stronger and more compelling. Technol. Rev. **2**, 1–14 (2003). Available at https://bit.ly/3lo3Yln. Accessed 20 Nov 2020
35. Jones, S.: Disrupting the narrative: immersive journalism in virtual reality. J. Media Pract. **18**(2–3), 171–185 (2017). https://doi.org/10.1080/14682753.2017.1374677
36. Larrondo Ureta, A.: El relato transmedia y su significación en el periodismo. Una aproximación conceptual y práctica. Trípodos **38**, 31–47 (2016)
37. Lassila-Merisalo, M.: Story first. Publishing narrative long-form journalism in digital environments. J. Mag. New Media Res. **15**(2), 1–15 (2014)
38. Literat, I., Chang, Y.K., Hsu, S.Y.: Gamifying fake news: engaging youth in the participatory design of news literacy games. Convergence **26**(3), 503–516 (2020). https://doi.org/10.1177/1354856520925732
39. López García, X., Rodríguez Vázquez, A.I.: Journalism in transition, on the verge of a 'Total Journalism' model. Intercom **39**(1), 57–72 (2016)
40. Meier, K.: Journalism meets games: newsgames as a new digital genre. Theory, boundaries, utilization. J. Appl. Journal. Media Stud. **7**(2), 429–444 (2018). https://doi.org/10.1386/ajms.7.2.429_1
41. Newman, J.: Videogames. Routledge, London (2004)
42. Paharia, R.: Loyalty 3.0: How to Revolutionize Customer and Employee Engagement with Big Data and Gamification. McGraw-Hill, New York (2013)

43. Peters, C., Broersma, M.: Rethinking Journalism: Trust and Participation in a Transformed News Landscape. Routledge, New York (2013)
44. Plewe, C., Fürsich, E.: Are newsgames better journalism? Empathy, information and representation in games on refugees and migrants. Journal. Stud. **19**(16), 2470–2487 (2018). https://doi.org/10.1080/1461670X.2017.1351884
45. Plewe, C., Fürsich, E.: Producing newsgames beyond boundaries: Journalists, game developers, and the news business. Convergence **26**(3), 486–502 (2020). https://doi.org/10.1177/1354856520918076
46. Ritter-Longhi, R., Flores, A.M.: Web journalist narratives as an element of innovation: cases of Al Jazeera, Folha de S. Paulo, The Guardian, The New York Times and Washington Post. Intercom **40**(1), 21–40 (2017)
47. Sicart, M.: Newsgames: Theory and design. In: Stevens, S.M., Saldamarco, S.J. (eds.) International Conference on Entertainment Computing, pp. 27–33. Springer, Berlin (2008)
48. Shoemaker, P.J., Reese, S.D.: Mediating the Message in the 21st Century: A Media Sociology Perspective. Routledge, New York (2014)
49. Steensen, S.: Online journalism and the promises of new technology: a critical review and look ahead. Journal. Stud. **12**(3), 311–327 (2011). https://doi.org/10.1080/1461670X.2010.501151
50. Stephenson, W.: The ludenic theory of news reading. Journal. Mass Commun. Q. **41**(3), 367–374 (1964). https://doi.org/10.1177/107769906404100306
51. Treanor, M., Mateas, M.: Newsgames. Procedural rhetoric meets political cartoons. In: Krzywinska. T., Kennedy, H.W., Atkins, B. (eds.) Proceedings of the DiGRA International Conference (2009)
52. Van Leeuwen, T.: News genres. In: Wodak, R., Koller, V. (eds.) Handbook of Communication in the Public Sphere, pp. 345–362. De Gruyter Mouton, Berlin (2010)
53. Vázquez-Herrero, J., Direito-Rebollal, S., Silva-Rodríguez, A., López-García, X.: Journalistic Metamorphosis: Media Transformation in the Digital Age. Studies in Big Data 70. Springer Nature, Cham (2020)
54. Vos, T.P., Perreault, G.P.: The discursive construction of the gamification of journalism. Convergence **26**(3), 470–485 (2020). https://doi.org/10.1177/1354856520909542
55. Werbach, K., Hunter, D.: For the Win: How Game Thinking Can Revolutionize Your Business. Wharton Digital Press, Philadelphia (2012)
56. Wiehl, A.: Newsgames—Typological approach, re-contextualization and potential of an underestimated emerging genre. IFLA Lyon (2014). Available at https://bit.ly/2OEufji. Accessed 08 Nov 2020
57. Wolf, C., Godulla, A.: Newsgames im Journalismus. Haben sie Potenzial? Was sagen die Nutzer? Journalistik 2 (2018). Available at https://bit.ly/3qZMEnL. Accessed 12 Nov 2020
58. Zackariasson, P.: Old things—new names. In: Dymek, M., Zackariasson, P. (eds.) The Business of Gamification, pp. 219–226. Routledge, London (2016)
59. Zichermann, G., Linder, J.: The Gamification Revolution: How Leaders Leverage Game Mechanics to Crush the Competition. McGraw-Hill, New York (2013)

Jose Alberto García-Avilés Full Professor of Journalism at Miguel Hernández University (Elche), where he lectures at the Master Program in Journalism Innovation. His research interests are news quality and storytelling. He currently participates in the research project "Journalism innovation in democratic societies: Index, impact and prerequisites in international comparison (JoIn-DemoS)" funded by the DFG German Research Foundation.

Raul Ferrer-Conill Assistant Professor of Media and Communication Studies at Karlstad University and an Associate Professor of Journalism at the University of Stavanger (Norway). His research investigates digital journalism, audience engagement and gamified processes, and the structural changes of the datafied society. His work has been published in journals such as *Digital JournalismJournalism Studies*, and *Television & New Media*.

Alba García-Ortega Doctoral fellow at the Communication Research Group (GICOV) of the Miguel Hernández University (UMH). She holds a B.A. Degree in Journalism and a M.A. Degree in Journalism Innovation. She did her doctoral dissertation on newsgames production and collaborates as Graduate Teaching Assistant at the UMH in Reporting and journalistic research, Multimedia narratives and Journalism on social media. Her research focuses on media innovation, journalism gamification and newsgames.

Producing Content in the Ubiquitous, Mobile Era: The Case of Digital Media

Alba Silva-Rodríguez and Juan-Miguel Aguado-Terrón

Abstract Mobile media transform traditional paradigms by providing new synchronous, localized and individualized formats and by modifying social contexts and the methods of production, emission and the reception of information. This chapter will analyze the effects of ubiquity on the journalistic environment and describe the affordances offered by mobile devices in this context. Outlined here are the main contents and narratives that arise in a scenario in which the participation and interaction of users is prevalent, and where the ubiquitous guarantees narrative micro-sequences that are mainly consumed in the interstices of people's everyday lives.

1 Introduction

Journalism is an industry in a constant state of flux. In the last decade, mobile news consumption has grown in importance all over the world [34, 54]. Adapting to an audience that consumes content through mobile devices [35] is a priority for media companies. The technological advances of recent years have made mobile devices key to three defining trends for the media ecosystem. Firstly, they are a platform through which all kinds of content are channeled. Secondly, they play a crucial role in the collection of a huge amount of data on user behavior and lastly and concurrently, they are the central node of social relations and the user's digital identity [2]. Aspects such as ubiquity, personalization, location, security, connectivity and accessibility are some of most important ways that mobility has impacted the journalistic field.

The consumption of news from smartphones is a modern-day reality. Six out of ten adults receive information on their mobiles and 37% of users connect to the Internet solely from these devices—meaning they do not subscribe to high-speed

A. Silva-Rodríguez (✉)
Universidade de Santiago de Compostela, Santiago de Compostela, Spain
e-mail: alba.silva@usc.es

J.-M. Aguado-Terrón
Universidad de Murcia, Murcia, Spain
e-mail: jmaguado@um.es

J. Vázquez-Herrero et al. (eds.), *Total Journalism*, Studies in Big Data 97,
https://doi.org/10.1007/978-3-030-88028-6_6

broadband connection at home [8]. Furthermore, given that mobile devices bring together social, professional and media consumption activity, the omnipresence of news in this flow of interactions introduces new aspects for consideration.

For Salaverría, mobile journalism is an anticipation of ubiquitous journalism that is closely related to the concept of citizen journalism [22, 42]. It provides personalized and uninterrupted information. In this context, the consumption and production of journalistic content will be carried out through an interconnected system of devices that will allow an increasingly corporeal communication (through voice, gestures, eye movements or even the mind) [43], and in which the anchoring element between devices and access environments will be the user's digital identity [2]. Baccin et al. [9] argue that mobile journalism benefits from wearable devices because they facilitate the production, circulation and consumption of content in a ubiquitous way. Everyday ubiquitous portability [3] is at the forefront of a news greening process that imposes the condition of continuous, on-demand and personalized service onto journalism [44]. In the era of "mass self-communication" [14], the importance of social networks is crucial as new patterns of news consumption and participation are established that allow a constant and ubiquitous connection with information [25]. Mobile technologies also favor and prioritize shorter and more dispersed consumption patterns [32] as well as situational consumption patterns (related to the user's situation) and relational consumption patterns (framed in social dynamics) [2].

From a product and professional process perspective, Carlan-da-Silveira [11] defines ubiquitous journalism as one that has thirteen properties: hybrid space, empowerment, digital database journalism, multi-media, contextuality, integration, continuity, automation, ubiquitous format, redistribution and reproduction, bi-directional personalization, mobile first copywriting and multi-competence.

2 The Implications of Mobile Technologies on Social Interactions: Journalistic Opportunities

The Canadian sociologist McLuhan [31] stated that the media needed to be understood as forms of technology. He conceived these technologies as extensions of the organs, the senses and the functions of the human being. Mobile devices, also known as the fourth screen [4, 15] have been recognized as the remote control of our lives [39], and as an extension of the human body. In Finland, the very concept 'mobile' refers to *kannykka* which means 'extension of the hand' [37]. This idea is relevant, because in addition to being a tool for accessing content (that can be seen, listened to or read), the mobile device is also a tool for interacting with the environment [6]. This allows for the content to be acted on, and also for others, or the situation that we find ourselves in, to be acted upon.

Over the last few years, there have been many authors who have delved into the strengths and potential of mobile communication. Geser [21] spoke of the powerful influence of the smartphone in social interactions [24], arguing that it could help to build a 'nomadic intimacy', creating a personal, individual and private sphere. It can therefore be argued that there is a psychosocial benefit, as this device provides the impression that the user is accompanied instead of alone. Fidalgo contends that the mobile phone is the best antidote against loneliness, that it is the umbilical cord to our 'happy place', which fosters peace and harmony [19]. In addition, one of its most salient features is that of customization. With this technology, communication is privatized for the first time [45]. Added to this are other equally compelling attributes, such as security, control and accessibility [53]. Geser also refers to mobiles as a tool that connects real world emergencies with the environment by allowing someone in a dangerous situation to access help through the use of their device [21].

The mobile must be understood as a metamedium [29] and from a dual perspective: intended for logistical purposes, such as in tracking and being used in cases of emergency, and from an expressive point of view, together with emotions and social relationships [28]. This idea is shared by Campbell [10], who distinguishes two functions of social coordination according to the purposes of this technology. The first is micro-coordination, associated with instrumental purposes. The second is the hyper-coordination model, linked to the expressive and relational dimension of the mobile communication, such as contacting family or messaging with friends.

The smartphone is much more than a technological device. It is also a cultural tool. The space experienced through the mobile screen is a 'synesthetic' space [1] in which visual representations interact with sound and tactile forms in a hybrid and complex manner. Even beyond the expressive space of a synesthetic nature, the mobile superimposes the symbolic and the actantial levels. It brings together a network of meanings that make up the content with the network of meanings that constitute our social experience of everyday situations. For this reason, the symbolic practices around the mobile device have been defined by Pink and Hjorth [38] as 'emplaced visualities' that intertwine the physical and sensory experiences of real space and the social relations of interaction mediated by technology.

The spatio-temporal conception is modified by the arrival of mobile technologies. Gergen [20] delves into the creation of absent spaces with the extension and popularization of these devices. Though physically absent, the individual who speaks with another through a telephone is present in the space of mobile communication. From a distance, the listener on the other end assists physically, if involuntarily. It is in this coming together of spaces where the dilution of barriers between the public and the private has special relevance [47]. On the other hand, the logic of being 'always on' (always connected) reshapes the traditional contexts in which technology was seen as a kind of mechanism to replace more traditional ideas and pursuits. This is the case, for example, with travel, means of transport, the street, bars and establishments, and the place of work or education.

It can be said that we are facing a new form of social organization [13] and technology can be considered the icon of twenty-first century globalization, one that is both capable of interfering with, and maintaining social ties [45]. We live in a disruptive environment [17] that modifies interpersonal relationships and other aspects of social and cultural life [46]. In effect, it has arisen as a form of innovation aimed at a marginal niche of customers in the telecommunications market. With the progressive improvement of benefits and with the reduction of prices, this technology becomes affordable for the majority, offering new services that were not possible before with landlines [30]. The mobile device prefigures the total interface [2] as it constantly and ubiquitously intermediates the symbolic level, one enhanced by digital technology, and the everyday level of our interaction with our immediate surroundings.

These transformations in turn activate a plethora of changes in the processes (of production, distribution and consumption) and in the formats (expressive resources, the access interface and the relationship with the user) of journalism in the era of ubiquitous connectivity. The following two sections will delve into this double transformation, of processes and formats.

3 Journalism at the Crossroads of AI, Big Data and Mobile Technology

The second decade of the twenty-first century consolidated the relationship between exhaustive data processing (big data), artificial intelligence (algorithms that allow to automate complex tasks in a flexible way) and mobile technology. It allowed for the ubiquitous, permanent and massive collection of data on the behavior of the user, and the located application of algorithms in the interaction with the environment [2]. The hyper-connected society is defined precisely by the feedback loop between three vectors: the technologies of ubiquity (from mobile technology and 5G to the Internet of Things) that enhance the capacities of collection, the processing and massive exploitation of data (big data) that in turn allow the development, training and application of intelligent algorithms (the third vector) for decision-making, process automation and the selection of alternatives. The increased capacities and the acceleration of operations originated from the application of AI is, once again, rooted in a greater demand for ubiquitous connectivity, and other similar motives. Note, for example, the exponential development of the power and coverage of mobile networks in less than a decade. This is the result of a demand that feeds itself.

Journalism is not and cannot be detached from this confluence of factors. Being a part of it, mobile journalism becomes ubiquitous. This is type of journalism that, thanks to the intensive use of algorithmic systems and artificial intelligence, disseminates news produced by journalists, users and robots on multiple digital devices so that it can be consumed from anywhere, at any time, through a constant flow of personalized and multisensory information [44]. 'Environmental' journalism, as defined by Hermida [23], is characterized by offering a broad, asynchronous, light and constantly active communication system that fosters new types of interactions around the news. However, this context alters the traditional models of consumption and how journalistic content is received. Dependence on non-proprietary platforms and media disintermediation are some of its most disruptive consequences in the media ecosystem [55]. Furthermore, ubiquitous and multiplatform mobile connectivity has created a context of hypermediation—of acceleration and multiplication of mediations—in which the loss of quality and reliability are obvious drawbacks [26]. The problem of disinformation dynamics is an impact factor in this context. Nonetheless, technological developments also seem to offer possible solution strategies, such as the application of AI tools in verification techniques for the identification of falsified information.

Ubiquitous journalism refers to the flow of information in real time and the ability to produce and consume content anywhere and at any time through the Internet [51]. This concept is often related to citizen journalism [22, 42]. Ubiquitous communication promotes the deregulation of agendas and lack of coordination of social roles. In other words, a scenario arises in which the limits between the world of work and leisure, and between the timetables associated with work and with family, are increasingly blurred [19]. Technology has created a context in which information experiences a process of disintermediation. The technological impact of recent years has created a new communicative paradigm focused on ubiquitous journalism. This has transformed the public's relations with information and changed the concept of news itself [44].

Based on the omnipresence of ubiquitous connectivity, the development of data processing capacity and intelligent algorithms also introduces changes in the processes and instrumental possibilities of digital media activity. This starts with the qualitative leap in natural language processing. The prospective and analysis reports of consumption indicate, for example, that the services of transcription, automatic translation and audio/text or text/audio will be adopted little by little. The same reports project a horizon of new opportunities and changes in the information production processes [12, 36].

Artificial intelligence and big data directly affect the three key processes of journalistic activity: obtaining information (relevant, for example, in investigative journalism and other applications of data journalism), production (including different kinds of automated writing) and distribution (recommendation systems based on comprehensive metrics) [36]. The extraction of information involves the use of algorithms for processing and detecting patterns in massive data flows. News automation consists of the application of artificial intelligence (AI) to newsmaking and involves the identification of repeated routines that can be encoded into algorithms that generate products similar to those obtained from the same task performed by humans [50].

One of the benefits of the application of artificial intelligence in journalism lies in the possibility of knowing how information has been prepared and identifying the origin and the different stages of the production process. In other words, AI bringing about greater transparency. In Spain there are examples that illustrate this. The newspaper *Público* has the TJ Tool, one created to combat misinformation. It automatically tracks eight indicators (sources, supporting documents, context, date, editorial location, author, editorial reasons and transparency policy) offering complete traceability of the information.

Leading digital media are experimenting with this technology. *The Times* has launched James, a recommendation engine powered by artificial intelligence. The Scandinavian company *Schibsted* already publishes semi-automated covers for media outlets such as *Aftenposten* and *VG*. Automatic translations, subtitles and technological improvements for synthesized voices allow for the development of content in a faster, more instantaneous way. Many news agencies are aware of the possibilities that are generated when the audience is demanding updated content. Media outlets like the *BBC* take advantage of these technologies to broadcast last minute events. The British corporation used this technique to report on the election night of December 2019. 689 semi-automated local stories were published in just a few hours. In 2020, *RTVE*, in alliance with the EFE news agency, developed a news writing service through an Artificial Intelligence system that narrated information related to the Second Division B matches of La Liga.

On the other hand, the creation of multimedia formats from text is also being tested. Converting a textual story into a visual story adapted to mobile phones with quotes and animations is now a reality.

Artificial intelligence makes it possible to offer more personalized and relevant services and more efficient ways of presenting and distributing content. Technological improvements and the evolution of mobile networks such as 5G will allow the media to broadcast high-definition video through mobile phones, making it easier to work and wrap content from any location. In addition to the smartphone, new items equipped with sensors and connected to the Internet have emerged, the so-called Internet of Things (IoT) [44]. These pose new challenges and demands for the journalistic sector. The expansion of mobile technologies favors the experimentation of narratives in a new environment. This world is virtual, robotic and highly sensory in nature, thanks to the confluence of the real and virtual worlds and the convergence of the Internet with everyday objects.

Finally, the impact of artificial intelligence on ubiquitous journalism signifies a new chapter in the development of metrics. These extend beyond the classic access/reading to get an idea of different aspects of user behavior in relation to the media outlet, a specific story or the interaction between different pieces or products [16]. The recommendation systems and the conversion or loyalty strategies—key in the consolidation of revenue models not based on advertising—increasingly rely on the use of intelligent algorithms for the profiling of the audience or the exploration of the approach of stories [40]. This has transformed the production timeline in a de facto way. If before the journalist's activity began with the idea of the story and ended with publication, AI and the interrelation of media fostered by mobile technology have opened up another horizon for the monitoring of stories and reports and their interrelation with other products related to information.

4 Ubiquitous Narratives and Digital Media

Based on these parameters, journalistic narratives have evolved from the adaptation of formats focused on the content of the preceding media, to consolidation, focused on fully exploiting the specific uses of mobile technology (motion sensors, location, etc.), as well as the changes caused by it in social dynamics and content consumption. In terms of content, narratives emerge that seek to satisfy the needs of a segmented audience that demands personalized and specialized information. Unsurprisingly, the latest Reuters trend prediction report confirms the shift towards the revaluation of formats such as email, mobile alerts and podcasts in order to establish direct connections with consumers and fight against the power of platforms [36].

From the year 2000 the media began to experiment with products adapted for mobile consumption. *The New York Times* published its first free mobile version in 2006. The idea was to provide access to news from a list of headlines that could be sent by email. It was very similar to the service of informational alerts or notifications. This type of content marks the beginnings of mobile journalism. Two years later, that same American newspaper launched a native iPhone application. In Spain, the first newspapers to experiment with mobile content were *El País*, in January 2009, and the radio stations *40 Principales* and *Cadena Ser*.

In 2008, Ahonen [7] publicized seven benefits of mobile technologies that could be adapted to the media context. The first was that the mobile could be a personal news source, for private consumption, which accompanies the user permanently, always connected, with a fixed payment channel. Also that it was a device that drives creativity, which enables users to create content, and the first media to know exactly what the real audience data are. Also that only the mobile could capture the social context of media consumption. However, the reality is that digital media are not prioritizing the adaptation of specific content on mobile platforms. In fact, they are treated less as autonomous means of communication, with a particular language and with unique characteristics, and more as mere distribution channels. Previous research corroborates this statement, having already illustrated the trend of reusing content and the duplication of traditional business models [33].

Where the evolution and adaptation of mobile content is most evident is in native applications. Many newspapers have experimented with the development of apps, alerts and notifications as a solution to dependence on third-party platforms, and as a way to exert greater editorial and commercial control. Applications are created for different purposes that evolve over time. To browse and search for information; to edit, create and view documents; to distribute content (Instagram, Facebook, Flipboard); to send text messages and share multimedia messages (WeChat, Snapchat, Telegram, Signal, WhatsApp); for blogging and sharing news (Tumblr, Blogger, WordPress); to manage newsroom workload (Slack); to analyze audience data (Hootsuite, TweetDeck); and use geolocation to identify and estimate geographic locations (weather forecast applications, for example) [49]. The mobile as a multifunction device generates endless opportunities for the digital content industry.

There are multiple examples related to content innovation on mobile screens. The most interesting products are created in the laboratories that the media create specifically to produce and distribute mobile content. For example, the *Guardian Mobile Innovation Lab* is one of a number of British labs that have experimented with various narratives such as interactive podcast players, smarticles and offline news applications.

Interactive videos edited with still images and/or animations and subtitles, for the most part, have become very popular in digital media. *Al Jazeera* and *NowThis* have pioneered this fully visual format aimed at short and instant consumption. For example, videos by *The Guardian* that are located on its YouTube channel with the title *Animations and Explainers*. Among its most prominent videos is one summarizing of the Cambridge Analytica scandal. In the last four years, the proportion of users who consume newsletters and information alerts from mobile devices has increased considerably. Publishers have sought to generate more direct traffic to their websites to retain subscribers and attract an audience [36]. The vast majority of digital media have newsletters specialized in certain topics such as the Daily Briefing from *The Guardian*, or Daily Brief of *Quartz*. In doing so, they have reinvented this format and, using a very simple structure, turned it into a viable editorial product.

However, the media are also experimenting with long-form narratives, going so far as to adapt traditional genres, such as documentaries, to the small screen. Such is the case of the *BBC*'s *Your phone is now a refugee's phone*, an immersive journalistic piece that tracks the journey of a refugee through his phone. Designed vertically for mobile consumption, it places the viewer in the shoes of the protagonist, helping them understand the confusion and fear that refugees experience. On the other hand, long-form storytelling, popularized by *The New York Times*' interactive multimedia report *Snow Fall*, fits to screens following the same parallax structure to create smooth transitions between different parts of the story. The *BBC*'s *Modern women in the land of Genghis Khan* is another example.

Mobile devices have also played an important role in popularizing the adoption of immersive formats (virtual reality and 360° video). Faced with the obstacles posed by the demands of computing power or dedicated devices, the design of Google Cardboard inaugurated the possibility in 2015 of incorporating elementary optics (literally mounted on cardboard cards) to the smartphone. With a double spherical viewing software adapted to the use of the accelerometer and gyroscope, it allowed for an immersive video experience of reasonable quality at a minimum cost. Converted into an accessible format, immersive videos were quickly adopted by the news media as a tool to place the user within the story. This provided a more emotional experience and connected the media outlet and the news event with the users' environmental space, something especially relevant in the case of information and news reports focused on human interest, color or spectacular events. Some digital media, such as *The New York Times* and *The Guardian*, published specific virtual reality (VR) applications that compiled news stories and reports in spherical format on very different topics. Stories included those on candidate rallies in the 2016 US presidential elections, the situation of the refugees from Syria, the women's marches in New York, and life in an isolation cell. Research on the journalistic use of immersive formats confirms that emphasizing locations in stories encourages exploration, builds audience loyalty, reinforces contextual understanding, and furthers emotional engagement [27].

Though the potential usefulness of big data and adaptive algorithms is clear, stories that are segmented and adapted to the interests of the audience are not fully established in digital media. This comes despite the fact that personalization is one of the main characteristics of mobile journalism. There are, however, examples related to the use of filters to offer content linked to user profiles. The *BBC*'s *Life expectancy calculator* story is an example of a personalized mobile narrative.

Beyond these experiences, artificial intelligence and big data are of great importance to mobile journalistic narratives. The automation of processes that favours the development of AI has a specific chapter on segmentation and personalization in user interaction interfaces. In particular, for those interfaces that involve natural language processing, that is, conversational dynamics. The consumption of news on mobile devices in short, immediate formats, adapted to the user's situation or interests, constituted an ideal breeding ground for experimentation with chatbots in the mid-2010s [52]. The use of conversational formats, similar to the mechanics of instant messaging applications, facilitated the appearance of a news channel that also provided intensive metrics on the informative interests of users [5]. So, starting in 2016, news organizations began to develop conversational robots or chatbots that functioned as interactive services for man–machine conversation through text, image or video. Some of them are extensions of non-proprietary applications (for example, in Facebook Messenger) and others belong to the media themselves (*Politibot* or *Quartz*). Instant messaging companies such as Line, Facebook Messenger, Skype and Viber have designed bots that integrate into social media networks. The example of *Politibot* is particularly noteworthy. In June 2016, the company launched a chatbot experiment on Telegram to report on the North American electoral campaign and that received updated information on the results by town.

In fact, the beginning of the 2020s and the context of the pandemic have led to a rebirth of audio formats in mobile environments. The podcast is the clear forerunner [18], though group audio chat applications, like Clubhouse, aren't far behind. Beyond chatbots, something that AI brings to the table that adds much depth in terms of formats is the expansion of voice assistants to multiple devices, both to surrounding environments or multi-device ambiences: for smartphones, smartwatches, smart speakers and vehicle-integrated systems. Digital media's access to news stories resembles that of a brand-new channel. This new channel to access the news also contains elements of radio and podcasting, and certain aspects of the mechanics of music content platforms, meaning that digital media are increasingly integrating skins or audio applications for smart speakers and voice assistants on mobile devices, reaching a greater audience [35]. In Spain, *El País* and *elDiario.es* already have applications for voice assistants. However, it is radio stations that seem to envision the new niche as a natural space for expansion. The same is true for smart watches. Starting in 2015, the media launched their first applications for a device that, for now, works as an accessory in a multiplatform scenario where each one fulfils a specific purpose [48].

5 Conclusions

The media ecosystem has undergone profound changes in the last decade. Faced with the loss of distribution hegemony in favour of third-party platforms, digital media outlets have had to reinvent themselves. Many have been hampered by dependence on large technology companies that control technology and the channel. This backdrop of changes, led by technological acceleration, has been conducive to a disintermediation that disrupts the value chain of digital content and directly influences business models. Simultaneous to this adjustment is a radical transformation of the scenarios and uses of news consumption. Personalization, ubiquitous access, device interoperability and integration into social dynamics are the key themes. As a result, the media are faced with the challenge of integrating innovation in processes and workflows, associated with flexible and dynamic business models, with innovation in formats and user experience that encourage the creation of value beyond the contribution of the platforms that dominate the distribution channel. Indeed, all of these demands come during an era of great economic uncertainty, and in the context of digital acceleration marked by the pandemic at the beginning of the 2020s.

Implementing modern, multisensory experiences in which all the senses participate, taking advantage of the narrative possibilities of immersive formats and automating aspects of work routines and interaction with the user, as well as integrating voice interfaces, are some of the lines of development that have sought to utilize technology to meet the challenge of an ever-changing media ecosystem. Mobile technologies promote new ubiquitous narratives and open doors to exploit micro-content that generate open information in which the co-creation of content between journalists and the audience is of great value. This process also means the possibility of obtaining and exploiting highly detailed metrics to guide brand and product strategies of the media. But ubiquity is not a lone, independent factor. It is just one of the vertices of a triangle of digital innovation that also includes, and is closely connected to, big data and artificial intelligence.

The ubiquity of mobile devices gives added value to the ability to process data on the located behavior of users. It offers the possibility of adapting content, not only to the identity of the user, but to their physical and socio-cultural situation. At the same time, the potential of utilizing the sensors and expressive capabilities of mobile technology offers a range of opportunities for data visualization formats that have not yet been fully explored [41]. Artificial intelligence also acquires an additional dimension in the context of ubiquitous connectivity. It enables the development of ambient intelligence—the ability to interpret the situations in which the user finds himself and to adapt the news and content services to them. With this, there is great potential for innovation in the forms of interaction with the user and of adaptation to their profile, anticipating ubiquitous journalism articulated on the user's digital identity and on their permanent connectivity [2].

However, with regards to the confluence of mobile technology, big data and AI, the pace with which the media have explored and implemented innovations—both in formats and processes—has been insufficient. The qualitative leap in digitization in the context of the pandemic, said to be three times greater than that expected under normal circumstances, means that these issues need to be resolved urgently.

Funding This article has been developed within the research project *Digital Native Media in Spain: Storytelling Formats and Mobile Strategy* (RTI2018-093346-B-C33), funded by the Ministry of Science, Innovation and Universities (Government of Spain) and co-funded by the European Regional Development Fund (ERDF).

References

1. Abril, G.: Cortar y pegar. La fragmentación visual en los orígenes del texto informativo. Cátedra, Madrid (2003)
2. Aguado, J.M.: Mediaciones ubicuas. Ecosistema móvil, gestión de identidad y nuevo espacio público. Gedisa, Barcelona (2020)
3. Aguado, J.M., Castellet, A.: Contenidos digitales en el entorno móvil: mapa de situación para marcas informativas y usuarios. In: Barbosa, S., Mielniczuk, L. (eds.) Jornalismo e Tecnologias Móveis, pp. 25–50. LabCom, Covilhã (2013)
4. Aguado, J.M., Martínez, I.: La cuarta pantalla: industrias culturales y contenido móvil. In: Aguado, J.M., Martínez, I. (eds.) Sociedad móvil: Tecnología, identidad y cultura, pp. 187–220. Biblioteca Nueva, Madrid (2008)
5. Aguado, J.M., Martínez, I.: El impacto de la tecnología digital en el sector publicitario. Universidad de Murcia, Murcia (2020)
6. Aguado, J.M., Martínez, I., Cañete, L.: Doing things with content: the impact of mobile application interface in uses and characterization of media. In: Serrano-Tellería, A. (ed.) Between the Public and Private in Mobile Communication, pp. 175–198. Routledge, London (2017)
7. Ahonen, T.: Mobile as 7th of the Mass Media: Cellphone, cameraphone, iPhone, smartphone. Futuretext, London (2008)
8. Anderson, M.: Mobile technology and home broadband 2019. Pew Research Center (2019). Available at https://pewrsr.ch/3btCd7C. Accessed 10 Jan 2021
9. Baccin, A., Sousa, M.E., Brenol, M.: A realidade virtual como recurso imersivo no jornalismo digital móvel. In: Canavilhas, J. (ed.) Jornalismo móvel. Linguagem, gêneros e modelos de negócio, pp. 275–287. LabCom, Covilhã (2017)
10. Campbell, S.: Percepciones sobre el uso de teléfonos celulares en espacios públicos: una comparación intercultural. Revista Chilena de Comunicación **1**, 9–31 (2008)
11. Carlan-da-Silveira, S.: Jornalismo ubíquo e smartphones: uma análise das potencialidades nos jornais El País e O Estado de São Paulo. Revista Latina de Ciencias de la Comunicación **28**, 158–166 (2018)
12. Carlson, M.: The robotic reporter: automated journalism and the redefinition of labor, compositional forms, and journalistic authority. Digit. Journal. **3**, 416–431 (2015). https://doi.org/10.1080/21670811.2014.976412
13. Castells, M.: La sociedad red. La era de la información, economía, sociedad y cultura. Alianza Editorial, Madrid (2006)
14. Castells, M.: Comunicación y poder. Alianza Editorial, Madrid (2009)
15. Cebrián, M., Flores, J.: Periodismo en la telefonía móvil. Fragua, Madrid (2011)
16. Cherubini, F., Nielsen, R.K.: Editorial analytics: how news media are developing and using audience data and metrics. Reuters Inst. Study Journal. (2016). https://doi.org/10.2139/ssrn.2739328
17. Christensen, C.: The Innovator's Dilemma: When New Technologies Cause Great Firms to Fail. Harvard Business Press, Boston (1997)

18. Dowling, D., Miller, K.: Inmersive audio storytelling: podcasting and serial documentary in the digital publishing industry. J. Radio Audio Media **26**, 167–184 (2019). https://doi.org/10.1080/19376529.2018.1509218
19. Fidalgo, A.: O celular de Heidegger—Comunicação ubíqua e distância existencial. Revista Matrizes **3**, 81–98 (2009)
20. Gergen, K.J.: The challenge of absent presence. In: Katz, J.E., Aakhus, M. (eds.) Perpetual Contact. Mobile Communication, Private Talk, Public Performance, pp. 227–241. Cambridge University Press, Cambridge, (2002)
21. Geser, H.: Towards a Sociological Theory of the Mobile Phone. Sociological Institute, University of Zurich, Zurich (2004)
22. Gillmor, D.: We the Media: Grassroots Journalism by the People, for the People. O'Reilly Media, Sebastopol, CA (2004). ISBN: 978 0 596007331. http://library.uniteddiversity.coop/Media_and_Free_Culture/We_the_Media.pdf
23. Hermida, A.: Twittering the news: the emergence of ambient journalism. Journal. Pract. **4**, 297–308 (2010)
24. Ishii, K.: Implications of mobility, the uses of psersonal communication media in everyday life. J. Commun. **56**, 346–365 (2006). https://doi.org/10.1111/j.1460-2466.2006.00023.x
25. Islas, O.: La televisión en Internet desde el imaginario de la sociedad de la ubicuidad. Razón y Palabra **60** (2008)
26. Karlsson, M.: The immediacy of online news, the visibility of journalistic processes and a restructuring of journalistic authority. Journalism **12**, 279–295 (2011). https://doi.org/10.1177/1464884910388223
27. Kukkakorpi, M., Pantti, M.: A sense of place: VR journalism and emotional engagement. Journal. Pract. **1**, 1–18 (2020). https://doi.org/10.1080/17512786.2020.1799237
28. Ling, R.: The Mobile Connection. The Cell Phone's Impact on Society. Morgan Kaufmann Publishers, San Francisco (2004)
29. Márquez, I.: El smartphone como metamedio. Observatorio **11**, 61–71 (2017). https://doi.org/10.15847/obsOBS11220171033
30. Martín, A.M.: El móvil como instrumento para la inclusión financiera. Revista Telos **84**, 1–4 (2010)
31. McLuhan, M.: Comprender los medios de comunicación. Las extensiones del ser humano. Paidós, Barcelona (1996)
32. Molyneux, L.: Mobile news consumption. A habit of snacking. Dig. Journal. **6**, 634–650 (2017). https://doi.org/10.1080/21670811.2017.1334567
33. Nel, F., Westlund, O.: The 4C's of mobile news: Channels, conversation, content and commerce. Journal. Pract. **6**, 744–753 (2011). https://doi.org/10.1080/17512786.2012.667278
34. Newman, N., Fletcher, R., Kalogeropoulos, A., Levy, D.A.L., Nielsen, R.K.: Digital News Report 2017. Reuters Inst. Study Journal. (2017). Available at https://bit.ly/3t0IVYG. Accessed 10 Jan 2021
35. Newman, N., Fletcher, R., Kalogeropoulos, A., Levy, D.A.L., Nielsen, R.K.: Digital News Report 2018. Reuters Inst. Study Journal. (2018). Available at https://bit.ly/2N13IMv. Accessed 10 Jan 2021
36. Newman, N., Fletcher, R, Schulz, A., Andi, S., Nielsen, R.K.: Digital News Report 2020. Reuters Inst. Study Journal. (2020). Available at https://bit.ly/38kk8Xt. Accessed 10 Jan 2021
37. Oksman, V., Rautiainen, P.: Toda mi vida en la palma de mi mano. La comunicación móvil en la vida diaria de niños y adolescentes de Finlandia. Revista Estudios de la Juventud **57**, 25–32 (2002)
38. Pink, S., Hjorth, L.: The digital wayfarer: reconceptualising camera phone practices in an age of locative media. In: Goggin, G., Hjorth, L. (eds.) The Routledge Companion to Mobile Media, pp. 488–498. Routledge, New York (2014)
39. Rheingold, R.: Multitudes inteligentes. Gedisa, Barcelona (2003)
40. Rodríguez-Vázquez, A.I., Direito-Rebollal, S., Silva-Rodríguez, A.: Audiencias crossmedia: nuevas métricas y perfiles profesionales en los medios españoles. El Profesional de la Información **27**, 793–800 (2018)

41. Rogers, S.: Mobile is a must for data journalism. IJNET International Journalists' Network (2014). Available at https://bit.ly/3a3IQMZ. Accessed 25 Feb 2021
42. Rosen, J.: What Are Journalists for? Yale University Press, London (1999)
43. Salaverría, R.: Los medios de comunicación que vienen. In: Sádaba, C., García Avilés, J.A., Martínez-Costa, M.P. (eds.) Innovación y desarrollo de los cibermedios en España, pp. 255–263. Ediciones Universidad de Navarra, Pamplona (2016)
44. Salaverría, R., de-Lima-Santos, M.F.: Towards ubiquitous journalism: impacts of IoT on news. In: Vázquez-Herrero, J., et al. (eds.) Journalistic Metamorphosis. Studies in Big Data, vol. 70, pp. 1–15 (2020)
45. Satchell, C., Singh, S.: The mobile phone as the globalizing icon of the early 21st century. In: 11th International Conference on Human Computer Interaction, 22–27 Jul, Springer, Las Vegas, 2005. Available at https://bit.ly/38nVEN0. Accessed 10 Jan 2021
46. Scolari, C.: Hipermediaciones. Elementos para una teoría de la comunicación digital interactiva. Gedisa, Barcelona (2008)
47. Serrano-Tellería, A.: Between the Public and Private in Mobile Communication. Routledge, London (2017)
48. Silva-Rodríguez, A., López-García, X., Toural-Bran, C.: iWatch: the intense flow of microformats of "glance journalism" that feed six of the main online media. Rev. Lat. Comun. Soc. **72**, 186–196 (2017)
49. Steinke, A., Belair-Gagnon, V.: Mobile applications and journalistic work. In: Nussbaum, J. (ed.) Oxford Research Encyclopedia of Communication (2019). https://doi.org/10.1093/acrefore/9780190228613.013.785
50. Túñez-López, J.M., Toural-Bran, C., Cacheiro-Requeijo, S.: Uso de bots y algoritmos para automatizar la redacción de noticias: percepción y actitudes de los periodistas en España. El Profesional de la Información **27**, 750–758 (2018). https://doi.org/10.3145/epi.2018.jul.04
51. Uskali, T.: Towards journalism everywhere: the new opportunities and challenges of real-time news streams in Finland. In: Daubs, M.S., Manzerolle, V.R. (eds.) Mobile and Ubiquitous Media: Critical and International Perspectives, pp. 237–247. Peter Lang, New York (2018)
52. Veglis, A., Maniou, T.A.: Chatbots on the rise: a new narrative in journalism. Stud. Media Commun. **7**, 1–8 (2019). https://doi.org/10.11114/smc.v7i1.3986
53. Vershinskaya, O.: Comunicación móvil como fenómeno social: la experiencia rusa. Estudios de la Juventud **57**, 139–149 (2002)
54. Westlund, O.: The production and consumption of mobile news. In: Goggin, G., Hjorth, L. (eds.) The Mobile Media Companion, pp. 135–145. Routledge, New York (2014)
55. Wheatley, D., Ferrer-Conil, R.: The temporal nature of mobile push notification alerts: a study of European news outlet's dissemination patterns. Dig. Journal. (Online First) (2020). https://doi.org/10.1080/21670811.2020.1799425

Alba Silva-Rodríguez Assistant Professor of Journalism at the Department of Communication Sciences at Universidade de Santiago de Compostela. She is Ph.D. in Journalism and member of Novos Medios research group. She is secretary of the RAEIC journal. As a researcher she focuses on the assessment of digital communication, specially the study of mediated conversation in social media and the evolution of media contents in mobile devices.

Juan-Miguel Aguado-Terrón Professor of Journalism at the School of Communication and Information Studies and Head of the Mobile Media Research Lab at the University of Murcia. Ph.D. in Information Sciences and M.Sc. in Social Research at the Institute of Philosophy and Sociology, Polish Academy of Sciences (Warsaw). Since 2006 his research has focused on the impact of mobile technology in the media ecosystem and the social and cultural background of these transformations.

Information Visualization: Features and Challenges in the Production of Data Stories

Ana Figueiras and Ángel Vizoso

Abstract Information visualization is a communicative discipline that has experienced a continuous growth and development through the centuries. From its origins and first vestiges in the ancient civilizations of Egypt and Mesopotamia to the most up-to-date forms marked by the development of information technology, visual communication has been an effective tool. Furthermore, over the last few years its presence as a journalistic genre has been strengthened, especially with the emergence of new needs arising from the increasing availability of data. This chapter explores some of the challenges that information visualization has to face in our days, like the need to adapt the content to the particularities of different platforms and screens. The authors also address the particularities of interaction in these projects, one of their most salient features due to his usefulness in organizing and supporting storytelling.

1 From Being Engraved on the Rock to the Internet

Information visualization has been used as a tool for human communication in different ways throughout history. Thus, going back 4,000 years, it is possible to see how Mesopotamian, Sumerian and Egyptian civilizations developed visual forms for reporting and preserving information [23]. These first milestones in the visual representation of information, as well as many of the others that followed them, were very linked to emergent disciplines like astronomy or cartography [15]. Hence, history has shown us how humans have employed information visualization in different ways, not only as a method for exchanging knowledge, but also as a helpful tool for understanding the complexity of our world [39:28].

A. Figueiras
iNOVA Media Lab, Universidade Nova de Lisboa, Lisboa, Portugal
e-mail: anafigueiras@fcsh.unl.pt

Á. Vizoso (✉)
Grupo de Investigaci6n Novos Medios, Universidade de Santiago de Compostela, Santiago de Compostela, Spain
e-mail: angel.vizoso@usc.es

J. Vázquez-Herrero et al. (eds.), *Total Journalism*, Studies in Big Data 97,
https://doi.org/10.1007/978-3-030-88028-6_7

Notwithstanding, after centuries of experimentation with the first maps, timelines or navigation charts [20:22, 40:15] it is assumed that modern information visualization started with the contributions of the Scottish engineer and economist William Playfair [5:43]. Some graphic forms published in their main contributions *The Commercial and Political Atlas* (1786) and *Statistical Breviary* (1801) are still very popular and widely used nowadays. It could be the case of pie, line or bar charts [41:60]. Although it existed previous attempts of display statistical information through these—and some other—forms, the final years of the eighteenth century are considered one of the most expansive ages for data visualization. Advances in trading and the development of many modern nations required new forms to register and communicate social, demographic and economic data, fields where visualization proved its usefulness. Thus, in this context, some of the most remarkable contributions to the area were produced, building a solid basis for further works. This could be the case of John Snow's map of the cholera (1854), Florence Nightingale's chart on the death causes during the Crimean War (1858), or Charles Joseph Minard's chart on the Napoleon's Russian military campaign (1869).

All these examples, as well as many others, were turning points of further visualizations due to their ability to deal with a great amount of data, and their communicative effectiveness. These works became suitable not only for highly-specialized audiences, but also for the general public as it was possible to see with the success of visualizations like 1933 Harry Beck's Tube Map [8].

Another way of application for all this accumulated expertise was its use as a journalistic genre, whether as infographics or as information visualization. These are two forms of graphic presentation of the information with differences and similarities. Both infographics and data or information visualizations are visual representations of data. However, one of the main particularities of each of them is that infographics are often made up of multiple visualizations and other elements like pictures or illustrations, among others. On the other hand, data visualizations are the presentation of a single chart, map, graph or diagram [38].

The journey of the journalistic utilization of information visualization started in the British diary *The Times* on 7th April 1806 with an infographic about Isaac Blight's murder [29:110]. Initially, this form of information was very linked to disciplines like design and art rather than journalism. However, during the second half of the twentieth century, a new flowering of visualization took place, represented in journalistic infographics. Finally, one of the last steps of this advance was the exploration of all the features made possible by the development of computers and the arrival of the Internet.

Since the 1990s decade, the media have been exploring the potential of the World Wide Web as a space for the transmission of information. In this pursuit of the innovation information visualization has been playing—and still does—a central role as an element that can attract or catch readers' attention and communicate complex information at the same time [10].

2 Features, Function and Production of Journalistic Information Visualization

News media outlets are not oblivious to the current communicative context, marked by the importance of online communication and the growing presence of social media as spaces to access information. Although journalism has always been living in the middle of a battle for the audience's attention, this fight has become more important nowadays. Today, there are not boundaries for sharing content. Concepts such as trajectory or experience are no longer as relevant, while others, such as visibility or impact have gained presence.

The media are aware of these global phenomena and of the change in the audience's consumption habits. For this reason, they are trying to develop more attractive and impactful forms to attract the public to their content. In this regard, information visualization is being used as one of the formats with a presence in the above mentioned spaces [1] leveraging human preference for attractive and efficient formats [36], as it occurs with some forms of video or photographic contents.

Information visualization is currently experiencing a new golden age. This communicative tool has not changed its nature or conceptualization, which remains to be the visual presentation of complex data and information in a comprehensible manner [42] whose objective is "explore, make sense of, and communicate data" [16]. Nonetheless, it has experienced a metamorphosis process by adapting its narrative to the characteristics of the platform on which it is distributed. From print newspapers to smartphone screens, this way of communication information can play different roles [24], from illustration or accompaniment for the information to share the information by its own [46].

Nowadays, data visualization is "much more than visual representation of data" [17:19]. Its present importance as a journalistic tool, very linked to the growing availability of data [37] has modified some of the main traits of this genre. In this regard, Segel and Heer [34] show how technological development and the emergence of interactive visualization breaks with the supporting or accompaniment function to initiate a new path of works that users can explore, modify and personalize to create a unique experience. Furthermore, these visualizations seek to exploit all technical capabilities of the communication through the Internet by applying characteristics such as animation, interactivity or multimodality, among others [4].

One of the greatest needs for updating in the production of information visualization has been related to the composition of the work teams or units. In the past, the elaboration of infographics was an individual work of professionals more linked to fields like arts or graphic design guided by the needs of journalists and editors [6:8–9]. This idea is almost gone today, because the production of visual journalism—especially the interactive pieces—is built thanks to the contribution of a group of journalists with different areas of expertise and background [11]. Furthermore, this introduces another particularity, since the producers of graphic stories consider themselves to be journalist with full autonomy, fleeing from that label of artists or mere designers of the past [6:175].

Nonetheless, the production of data-based stories has certain needs in terms of time and response capacity to the fast-moving publishing model of our days, where news outlets share information constantly [37]. The production of news graphics—both static and interactive—, is normally more time-consuming than addressing information through text or even through another visual narrative like photography or video. Because of the implementation of the different stages of the production process—data search, analysis and cleaning; design phase and, finally, visualization—, the production of news stories based on information visualization usually follows two speeds. On the one hand, daily works that try to analyze the most up-to-date information and, when possible, breaking news. On the other hand, products that address foreseeable information or events known in advance that can be calmly developed, sometimes including some technical or design features that are not common in the visualizations produced on the same day. Hence, visual journalists try to combine these two strategies to be capable to answer the needs of the topical issues, but without forgetting the quest for experimentation and innovation inscribed in the DNA of information visualization.

Even so, the size of the media outlet can condition how it relates to the information visualization. Larger journalistic brands can deal better with the needs of this form of communication by creating multidisciplinary teams specialized in its production, units that sometimes work as an autonomous department by carrying out their own proposals, and sometimes in collaboration with other units within the newsroom. In contrast, some smaller news outlets have to adapt their routines, workflows and teams to the recent needs and innovations of this format. In a period of crisis for almost the entire media landscape, the optimization of the resources and the reallocation of tasks have become increasingly important. All of that to respond to the new needs and demands of the audience within the framework of the increasingly tight budgets of these organizations. Regarding information visualization, this has led smaller media companies to the redefinition of their journalists' jobs, sometimes after giving them a proper training [30].

Moreover, during the last few years we have witnessed the emergence of tools that facilitate the work of journalists in producing their visualizations. If ten or twenty years ago it was necessary to have extensive programming knowledge in order to create graphics—both static and interactive—, today tools such as Flourish, Datawrapper, Tableau or Carto have changed that reality [27]. These online tools—which usually have also a free version suitable for all audiences—have democratized access to information visualization production in many areas, and news production has been one of them. Thanks to these platforms, even journalists without deep computer or visual knowledge can enrich their news with graphics that help both professionals and readers to communicate and understand large sets of data, respectively. The use of these services makes it possible for the visual teams to carry out more difficult or time-consuming tasks in larger newsrooms. Additionally, brings smaller media outlets the opportunity of using this form of communication for taking advantage of all its effectiveness as communicative and attraction-getter instrument.

3 Interaction as a Mean to Support the Narrative in Information Visualization

Narratives are endemic in human culture, frequently used to "convey information in a memorable form that can engage and establish causal links" [33:361]. Although we cannot pinpoint when Man started this habit of telling stories, we know that their origin is ancient. And, despite having thousands of different forms of storytelling present in our everyday life, what we still instinctively associate with the concept of storytelling the image of an elder narrating an old fairy tale to a child [25].

Nevertheless, storytelling is not restricted to orality or textual depictions. Narrative is transversal to all expressive forms, even to music, which to musical semiotics has also been seen as a form of narrative [26]. Narratives are an intrinsic part of humanity, and each society imprints its identity and culture in its stories. Narratives are also a vital operation of the human brain, bound to the process by which it forms consciousness when interacting with an object [9].

Therefore, it is no surprise that journalism and narrative have walked hand-in-hand for a long, even emerging as a genre of its own. The genre of narrative journalism, also referred to as literary journalism, is often described as the genre that blends techniques from narrative fiction and applies them to non-fiction, maintaining the main characteristics of accurate, well-researched information. The transition to digital rekindled the narrative journalism flame with its potential of deeper audience engagement, coming from the medium's enhanced interaction and immersion. Most often accompanied by a "revolution-esque vocabulary" [21:168], the decline of newspaper circulation and increased public distrust in news media are often highlighted by researchers as problems that can be mitigated by narrative journalism [43]. For instance, while advocating for the benefits of augmented reality as a medium for storytelling in journalism, Pavlik and Bridges stress "the potential [of digital technologies] to transform journalism and the media in several beneficial ways, including new forms of storytelling that might better engage citizens and provide more context, nuance and texture to reported events and issues" [28:4].

However, this long-term relationship has not always been peaceful. Narrative journalism remains the subject of endless debates and tensions between objectivity and subjectivity, partisanship and neutrality and ethics and aesthetics [21, 43]. Due to its potential to provide insights and further exploration of the data that can raise different points of view, visualization can be a tool to mediate the narrative journalism's inclination to subjectivity and aesthetics.

To this day, visualization is more commonly used in journalism to support traditional storytelling forms as extra information or supporting evidence. Nonetheless, its use as an independent storytelling form, with traditional storytelling merely as support, is slowly and steadily growing. This growth can be seen all over the web, often as independent initiatives. However, the most outstanding effort, both in research and practice, unquestionably comes from journalism, an area where there has been a great effort to create multidimensional stories composed of other media besides text. In research, names such as Jeffrey Heer, Edward Segel and Alberto

Cairo stand out, and journalism outlets such as *The New York Times*, *The Washington Post* and *The Guardian* are blazing new paths in exploring the craft of integrating data with a compelling narrative.

Though the concept was introduced more than ten years ago by Edward Segel and Jeffrey Heer in the seminal work *Narrative Visualization: Telling Stories with Data* (2010), Narrative Visualization is still an emerging genre. The concept encapsulates complex data's effective communication to an audience engagingly using visualization and promoting insights, with a structured or semi-structured interpretation path [18].

Segel and Heer [34] identified three narrative structure tactics used in visualization. The first was ordering, which refers to the ways of systematizing the path users take through the visualization, and that can either be prescribed by the author (or linear), have no path suggested at all (random access), or need the user to choose a path among multiple alternatives (user-directed). The second was messaging, which refers to how a visualization communicates observations and commentary to the viewer and might be achieved through short or more substantial text: captions/headlines, annotations, accompanying article, multi-messaging, comment repetition, introductory text, summary/synthesis, among others. With ordering and messaging, interactivity is one of the narrative structure tactics identified, referring to the distinctive means to manipulate the visualization and how the user learns those methods (explicit instruction, tacit tutorial, initial configuration).

Interactivity is still intrinsically part of this process of interlinking narrative in visualization. Although it has been used in information visualization with several purposes, the more common being (1) making the data more engaging or playful and (2) showing the data in manageable portions (by partitioning it, either by browsing or by querying), for narrative visualization, interactivity opened up the possibility of adding new layers of content. These layers of content are vital for visualizations to act as stand-alone narrative news pieces.

Visualizations can, thanks to this added tier of content added most of the time in the form of annotations, present content that adds context, in addition to the data itself. This content's potential as a means to help a user make sense of the data is vast [22]. Nonetheless, interactivity in this scope is not only used to support the narrative. If we define information visualization as "the use of computer-supported, interactive, visual representations of abstract data to amplify cognition" as does Card, Mackinlay and Shneiderman [7: 7], it is practically impossible to dismiss the role of interactivity.

Because visualizations are fundamentally dynamic, the interest in researching interaction techniques in visualization has been ever-growing. In addition to studies on its efficacy, efficiency, or introduction [3, 12, 13], several studies propose information visualization interaction taxonomies. The most well-known is unequivocally the Visual Information-Seeking Mantra by Shneiderman [35]. However, in 2015, we found the necessity to build a broader interaction techniques taxonomy, which can be seen in Table 1, to include more modern interaction techniques, such as participation or gamification [19]. Later in 2020, we derived this taxonomy to the particular case of Spatio-temporal (ST) Visualization since a generic interaction taxonomy cannot

Table 1 InfoVis interaction taxonomy

Interaction technique	Description
Filtering	Only show me the data in which I am interested
Selecting	Mark or track items in which I am interested
Abstract/elaborate	Adjust the level of abstraction of the data
Overview and explore	Overview first, zoom and filter, then details on-demand
Connect/relate	Show me how this data is related
Reconfigure	Give me a different arrangement of the data
Encode	Give me a different representation of the data
History	Allow me to retrace the steps I take in the exploration of the data
Extraction of features	Allow me to extract data in which I am interested
Participation/collaboration	Allow me to contribute to the data
Gamification	Show me the data in a more playful way

Source Figueiras [19]

be applied to analyze ST visualizations without several adjustments [31]. We predict that new derivations are also possible to other particular visualization usage contexts.

Using narratives in the context of visualization, supported by interactivity, is an intricate editorial process characterized by a sequence of rhetorical decisions at various stages. First, to understand how interactivity can support the narrative, we need to understand the commonly used narrative patterns. In 2016, as an outcome of a workshop on Data-Driven Storytelling in Dagstuhl, an interactive browsable collection of narrative design patterns with examples was created. According to Bach and colleagues [2:111], "a narrative pattern is a low-level narrative device that serves a specific intent". In NAPA Cards[1] the 16 patterns (Incorporating the audience, Repetition, Juxtaposition, Breaking the fourth wall, Humans behind the dots, Make a guess, Rhetorical question, Familiar Setting, Call to action, Gradual visual reveal, Defamiliarization, Convention breaking, Speeding up, Concretize, Meaningful use of space and Silent data) are grouped into five higher-level pattern groups: empathy/emotion, engagement, framing, flow and argument/argumentation. This typology derived from an initial open-ended question regarding the purpose of the use of narrative, the existing patterns, and its application. The narrative patterns can be combined or used on their own to create compelling data-driven narratives and are not exclusive for visualization, being easily applied to different data stories.

[1] http://napa-cards.com

To understand how meaningful interaction techniques are commonly used to support these narrative patterns, we searched for a representative visualization example, which can all be seen in Fig. 1, for each narrative pattern and identified which interaction technique is used to support it. The intent is not to bound interaction techniques and narrative patterns but to learn which narrative patterns we could not find any interaction technique supporting it. As can be seen in Fig. 2, we found the use of the narrative techniques Incorporating the audience, Humans behind the dots, Make a guess, Familiar Setting, Call to action, Gradual visual reveal, Defamiliarization, Speeding up and Concretize.

Some of the narrative techniques that we did not find a utilization example that resorts to Interaction, such as Breaking the fourth wall or Rhetorical question, were often found in examples that did not use Interaction to support it. In fact, most of these narrative patterns that we found not to be supported by interaction were catalogued in NAPA Cards in the higher-level pattern group framing, which according to the authors "builds the way facts and events in a story are perceived and understood through narration" [2:116]. These narrative patterns are often achieved with language resources such as Figurative Language or dialogue. This is the case of *How many households are like yours* which uses a rhetorical question in its visualization title that ultimately induces the user to introduce the data for his family and compare it to other American families. This action, however, is not forced or fomented by any interaction technique. Other framing narrative patterns that do not fit the same pattern were Juxtaposition, Convention breaking, Meaninful and Silent data. The only narrative pattern that is not catalogued as framing and that we were not able to find and examples using it resorting to interaction was *repetition*. This narrative pattern was catalogued by Bach and colleagues [2:111–116] both as flow and argument/argumentation. The latter is a linguistic resource that serves "the intent of persuading and convincing audiences" [2:114] and therefore harder to occur supported by interaction techniques.

We also found examples of narrative techniques supported by more superficial interactions such as scroll activated animations. Several examples of Gradual visual reveal would fit this Interaction. In examples such as *How Big is Space* would easily fit this case. In this example by *BBC*, the user pilots a rocket through the Solar System as he/she scrolls the page down. On the way down, the user accesses information such as how close comets have come to our planet, the farthest traveled by a human being and how long the trip on the Starship Enterprise. Here, scroll activated animations are used to provide the narrative sequence.

This type of reflection on how and what interaction techniques are commonly used to support these narrative patterns allows us to have a rule of thumb for the application of both the narrative patterns and the interaction techniques. However, creating compelling visual data stories is not an easy task, and beyond being a team, labor-intensive process, its creation requires expertise, iterations, feedback and essentially a deep understanding of both narrative and visualization. Following the creation of a narrative visualization, there is an implicit decision-making process where several steps have to be taken into account, and their articulation is vital for the success of the visualization. In this process, having clear guidelines can

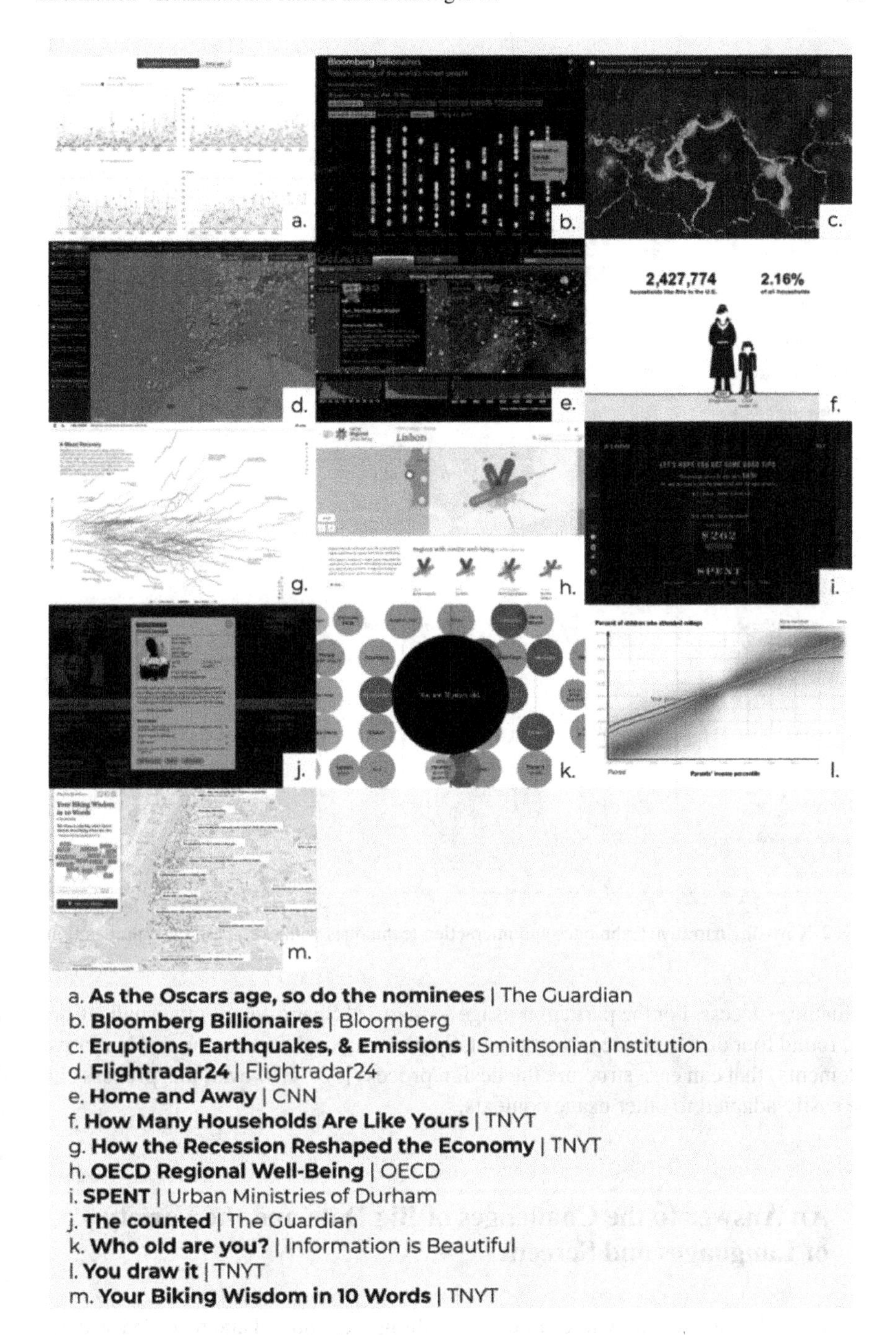

Fig. 1 Gallery of examples used. Own elaboration

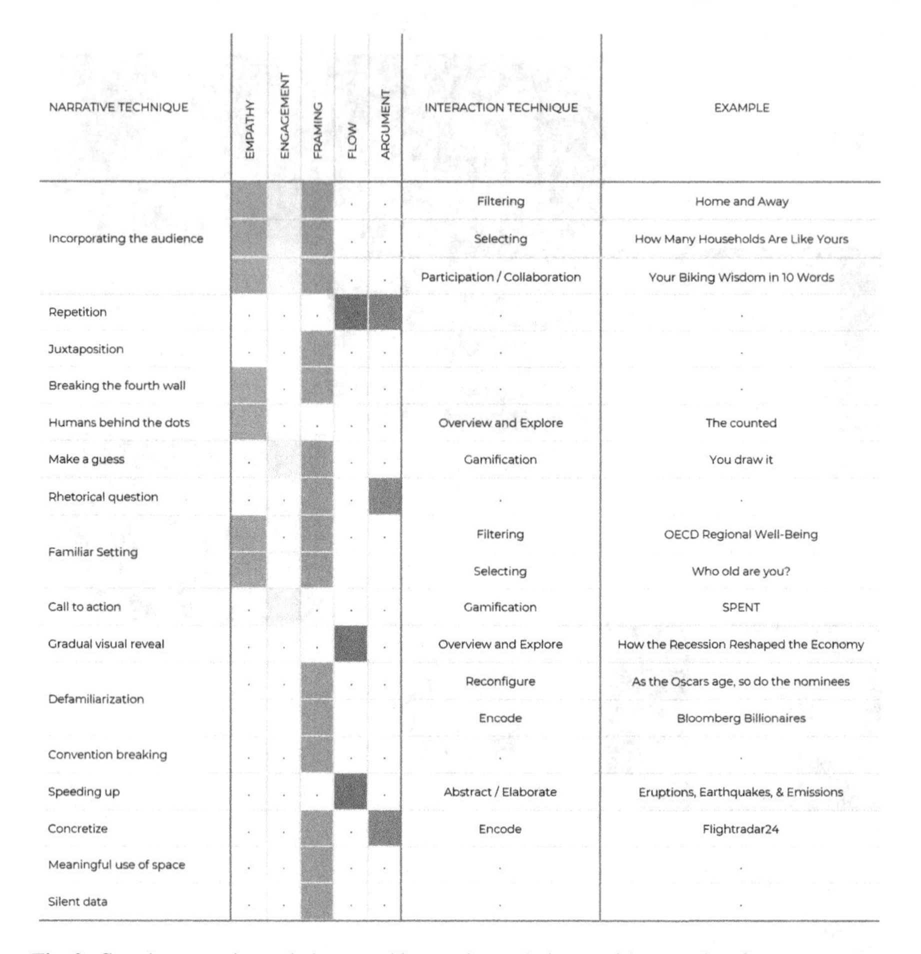

Fig. 2 Crossing narrative techniques and interaction techniques with examples. Own preparation

influence success. For the particular usage scenario of Spatio-temporal visualization, we found four decision categories (Intent, Spatio-temporal, Interaction and Narrative Elements) that can ease structure the design process [32]. However, this process can be easily adapted to other usage contexts.

4 An Answer to the Challenges of Big Data and the Variety of Languages and Screens

Today, we live in a world where almost everything is quantified and recorded as data. Over the years, we have observed how we have stored a large volume of data, but sometimes we find it difficult to relate them and, especially, to communicate them.

In short, our society has been undergoing a continuous process of datafication [14], and tools for the communication of this data are needed.

Journalism has become an indispensable intermediary among large sets of structured data and the audience. Some reports are the consequence of analyzing and studying the reality of these almost infinite data tables with the unique objective of finding the story behind them. Over the last few years we have, therefore, witnessed how information visualization has gained presence in this type of stories. Sometimes visualization is used as a complement that enriches the journalistic pieces. Nonetheless, the importance of information visualization in these cases is growing, and more and more data journalism stories are being produced only by using information visualization.

The emergence of this data-based way of doing journalism has also modified the visuals teams and professionals' routines and characteristics. Until the first years of the twenty-first century, data and stories should be retrieved by journalists themselves. With the emergence of open data portals and other initiatives for sharing large sets of data, news media outlets—and also graphic desks—discovered a new group of sources with a very high informative potential [44:140]. This new reality was both an opportunity and a challenge. Suddenly, visual journalists were able to access not only a new corpus of data, but also a whole host of potential stories to communicate to their audience through information visualization. However, as highlighted before, these changes made necessary the reconfiguration of the newsrooms and the creation of teams with different profiles and backgrounds. All of that in order to give an answer to the high complexity of these data-driven stories.

Another important challenge for information visualization is the adaptation to the changing environments in news consumption. Although this journalistic genre has demonstrated its capacity for adaptation over the years, sometimes it made necessary the adaptation of the production workflow. Once more, there is a difference among larger and smaller news brands. Larger ones, thanks to the size of their visual teams can deal with the production—or the adaptation—of visual content to the different screens. Nonetheless, smaller newsrooms or at least newsrooms with a smaller visualization unit do not have this responsiveness [45:292].

Visual teams and their professionals live in a continuous process of retraining with the unique objective of being able to deal with the most current technologies [44:146]. This quest for updating goes so far that sometimes many of the technologies adopted in the newsrooms enter through the visualization team, and they are the first to test automation or virtual reality in their works [6:12]. In fact, as shown in this chapter, information visualization has experienced a lot of changes and adaptation processes over the last few decades, and it is expected that this need for adaptation will continue to be present in both the renovation of visual languages and the renewal of production processes. All of that with the objective of maintaining its efficiency and ability to communicate complex information without losing all its potential as attention getter.

Funding This article has been developed within the research project *Digital Native Media in Spain: Storytelling Formats and Mobile Strategy* (RTI2018-093346-B-C33), funded by the Ministry of Science, Innovation and Universities (Government of Spain) and co-funded by the European Regional Development Fund (ERDF).

References

1. Amit-Danhi, E.R., Shifman, L.: Digital political infographics: a rhetorical palette of an emergent genre. New Media Soc. **20**, 3540–3559 (2018). https://doi.org/10.1177/1461444817750565
2. Bach, B., Stefaner, M., Boy, J., et al.: Narrative design patterns for data-driven storytelling. In: Riche, N.H., Hurter, C., Diakopoulos, N., Carpendale, S. (eds.) Data-Driven Storytelling, pp. 107–133. CRC Press, Boca Ratón (2018)
3. Boy, J., Detienne, F., Fekete, J.D.: Storytelling in information visualizations: does it engage users to explore data? In: CHI '15: Proceedings of the 33rd Annual ACM Conference on Human Factors in Computing Systems, pp. 1449–1458. ACM, New York (2015)
4. Burmester, M., Mast, M., Tille, R., Weber, W.: How users perceive and use interactive information graphics: an exploratory study. In: 14th International Conference Information Visualisation, pp. 361–368. IEEE, London (2010)
5. Cairo, A.: Infografía 2.0: Visualización interactiva de la información en prensa. Alamut, Madrid (2008)
6. Cairo, A.: Nerd journalism: how data and digital technology transformed news graphics (PhD Thesis). Universitat Oberta de Catalunya, Barcelona (2017)
7. Card, S.K., Mackinlay, J.D., Shneiderman, B.: Introduction. In: Card, S.K., Mackinlay, J.D., Shneiderman, B. (eds.) Readings in Information Visualization: Using Vision to Think, pp. 1–34. Academic Press, San Diego (1999)
8. Cartwright, W.: Rethinking the definition of the word 'map': an evaluation of Beck's representation of the London underground through a qualitative expert survey. Int. J. Dig. Earth **8**, 522–537 (2015). https://doi.org/10.1080/17538947.2014.923942
9. Damásio, A.: The Feeling of What Happens: Body and Emotion in the Making of Consciousness. Houghton Mifflin Harcourt, Boston (1999)
10. de Haan, Y., Kruikemeier, S., Lecheler, S., et al.: When does an infographic say more than a thousand words? Journal. Stud. **19**, 1293–1312 (2017). https://doi.org/10.1080/1461670x.2016.1267592
11. Dick, M.: Interactive infographics and news values. Digit. Journal. **2**, 490–506 (2013). https://doi.org/10.1080/21670811.2013.841368
12. Dix, A., Ellis, G.: Starting simple—adding value to static visualisation through simple interaction. In: Catarci, T., Costabile, M.F. (eds.) Proceedings of the Workshop on Advanced Visual Interfaces AVI, pp. 124–134. ACM Press, L'Aquila (1998)
13. Elmqvist, N., Moere, A.V., Jetter, H.C., et al.: Fluid interaction for information visualization. Inf. Vis. **10**, 327–340 (2011). https://doi.org/10.1177/1473871611413180
14. Engebretsen, M., Kennedy, H., Weber, W.: Visualization practices in Scandinavian newsrooms: a qualitative study. In: 21st International Conference Information Visualisation, pp. 296–300. IEEE, London (2017)
15. Few, S.: Data Visualization Past, Present, and Future (2007). Available at https://bit.ly/3bR4MMq. Accessed 22 Apr 2020
16. Few, S.: Why Do We Visualize Quantitative Data? Visual Business Intelligence (2014). Available at https://bit.ly/2OUsrT9. Accessed 20 May 2018
17. Figueiras, A.: How to tell stories using visualization. In: 18th International Conference on Information Visualisation, pp. 18–26. IEEE, Paris (2014)
18. Figueiras, A.: How to Tell Stories Using Visualization: Strategies Towards Narrative Visualization. Universidade Nova de Lisboa, Lisboa (2016)

19. Figueiras, A.: Towards the understanding of interaction in information visualization. In: 19th International Conference on Information Visualisation, pp. 140–147. IEEE, Barcelona (2015)
20. Friendly, M.: A brief history of data visualization. In: Chen, C., Härdle, W., Unwin, A. (eds.) Handbook of Data Visualization, pp. 15–56. Springer, Berlin (2008)
21. Groot Kormelink, T., Costera Meijer, I.: Truthful or engaging? Surpassing the dilemma of reality versus storytelling in journalism. Digit. Journal. **3**, 158–174 (2015). https://doi.org/10.1080/21670811.2014.1002514
22. Hullman, J., Diakopoulos, N., Adar, E.: Contextifier: automatic generation of annotated stock visualizations. In: Mackay, W.E. (ed.) Proceedings of Conference on Human Factors in Computing Systems, pp. 2707–2716. ACM, New York (2013)
23. Interaction Design Foundation: Information Visualization—A Brief Pre-20th Century History (2017). Available at https://bit.ly/3lnDnVI. Accessed 11 Jul 2019
24. López-del-Ramo, J., Montes-Vozmediano, M.: Construcción comunicativa del reportaje infográfico online de calidad. Elementos constitutivos. El Profesional de la Información **27**(2), 322–330 (2018). https://doi.org/10.3145/epi.2018.mar.10
25. Ma, K.L., Liao, I., Frazier, J., et al.: Scientific storytelling using visualization. IEEE Comput. Graphics Appl. **32**, 12–19 (2012). https://doi.org/10.1109/MCG.2012.24
26. Maus, F.E.: Music as narrative. Indiana Theory Rev. **12**, 1–34 (1991). https://www.jstor.org/stable/24045349?seq=1. Accessed 13 Feb 2021
27. Ojo, A., Heravi, B.: Patterns in award winning data storytelling. Digit. Journal. **6**, 693–718 (2018). https://doi.org/10.1080/21670811.2017.1403291
28. Pavlik, J.V., Bridges, F.: The emergence of augmented reality (AR) as a storytelling medium in journalism. Journal. Commun.Monographs **15**, 4–59 (2013). https://doi.org/10.1177/1522637912470819
29. Peltzer, G.: Periodismo iconográfico. Rialp, Madrid (1991)
30. Reilly, S.: The need to help journalists with data and information visualization. IEEE Comput. Graph. Appl. **37**, 8–10 (2017). https://doi.org/10.1109/mcg.2017.32
31. Rodrigues, S., Figueiras, A.: There and then: interacting with spatio-temporal visualization. In: Khosrow-Shahi, F. (ed.) Proceedings of the 24rd International Conference Information Visualization (IV), pp. 146–152. IEEE, Wien (2020)
32. Rodrigues, S., Figueiras, A., Alexandre, I.: Once upon a time in a land far away: guidelines for spatio-temporal narrative visualization. In: Azzag, H., Lebbah, M., Venturini, G., Banissi, E. (eds.) Proceedings of the International Conference on Information Visualisation, pp. 44–48. IEEE, Paris (2019)
33. Satyanarayan, A., Heer, J.: Authoring narrative visualizations with ellipsis. Comput. Graph. Forum **33**, 361–370 (2014). https://doi.org/10.1111/cgf.12392
34. Segel, E., Heer, J.: Narrative visualization: telling stories with data. IEEE Trans. Visual Comput. Graphics **16**, 1139–1148 (2010). https://doi.org/10.1109/TVCG.2010.179
35. Shneiderman, B.: The eyes have it: a task by data type taxonomy for information visualizations. In: Proceedings IEEE Symposium on Visual Languages, pp. 336–343. IEEE, Boulder (1996)
36. Sjafiie, S.S.L., Hastjarjo, S., Muktiyo, W., Pawito.: Graphic visualization in printed media: How does the use of technology influence journalism culture. J. Komunikasi Malaysian J. Commun. **34**, 373–385 (2018). https://doi.org/10.17576/JKMJC-2018-3404-22
37. Smit, G., de Haan, Y., Buijs, L.: Visualizing news. Digital Journal. **2**, 344–354 (2014). https://doi.org/10.1080/21670811.2014.897847
38. Stat Silk (2019) The Real Difference between Infographics and Data Visualizations. Statsilk.com. Available at https://bit.ly/2QbvsPP. Accessed 6 Jun 2019
39. Tufte, E.R.: The Visual Display of Quantitative Information. Graphics Press, Cheshire (2001)
40. Tufte, E.R.: Visual Explanations. Graphics Press, Cheshire (1997)
41. Unwin, A.: Good graphics? In: Chen, C., Härdle, W., Unwin, A. (eds.) Handbook of Data Visualization, pp. 58–78. Springer, Berlin (2008)
42. Uyan Dur, B.I.: Data visualization and infographics in visual communication design education at the age of information. J. Arts Humanities **3**, 39–50 (2014). https://doi.org/10.18533/journal.v3i5.460

43. van Krieken, K., Sanders, J.: What is narrative journalism? A systematic review and an empirical agenda. Journalism (Online First) (2019). https://doi.org/10.1177/1464884919862056
44. Veira-González, X., Cairo, A.: From artisans to engineers. how technology transformed formats, workflows, teams and the craft of infographics and data visualization in the news. In: Toural-Bran, C., et al. (eds.) Information Visualization in the Era of Innovative Journalism, pp. 134–153. Routledge, London (2020)
45. Vizoso, Á., López-García, X.: Diferencias y similitudes en la presentación de infografía en las app y las versiones en línea de El País y The New York Times. In: Canavilhas, J., Rodrigues, C., Giacomelli, F. (eds.) Narrativas jornalísticas para dispositivos móveis, pp. 279–301. Universidade da Beira Interior, Covilhã (2019)
46. Zwinger, S., Zeiller, M.: Interactive infographics in German online newspapers. In: 9th Forum Media Technology FMT 2016 and 2nd All Around Audio Symposium 2016, pp. 54–64. St Pölten, Austria (2016)

Ana Figueiras Research scientist at iNOVA Media Lab, where she coordinates the research line in Information Visualization and Data Analysis. Before joining iNOVA, she received her Ph.D. in Digital Media from Universidade Nova de Lisboa (UT Austin Program | Portugal). Her main research area is information visualization, focusing on visual forms of storytelling and on identifying the best techniques for incorporating narrative elements into visualization. She is also interested in new methodologies for performance and experience evaluation in visualization.

Ángel Vizoso Degree in Journalism, Master in Political Communication and Ph.D. candidate at Universidade de Santiago de Compostela (USC). Researcher at Novos Medios research group (USC), and visiting scholar at Universidade Nova de Lisboa (Portugal). His areas of study are mainly information visualization, fact-checking and journalistic production for online media.

Emerging Journalisms: From Intuition to Prediction and the Constructive Approach

Xosé López-García, Carlos Toural-Bran, and Jorge Vázquez-Herrero

Abstract Journalism lives in permanent change. The advent of the Internet and communication and information technologies has driven a metamorphosis that is fueling reinvention in a network society scenario awaiting the transformations that the Internet of Things and intelligent automation will bring. Journalistic intuition has given way to a renewal of techniques that allow prediction and greater social utility through the construction of a better-informed society. This chapter analyzes the current scenario of change with a historical perspective and with special attention to the impact of technology during recent decades. The text describes the passage from a more intuitive stage to the current stage, which is more predictive and has more assets, through emerging journalisms, to provide constructive journalistic models that bring added value to citizens and serve the public interest through proposals and solutions to the issues that journalists are reporting.

1 Introduction

Despite changes in the boundaries of the profession and in the media ecosystem, with radical transformations in the organization and dissemination of news, journalism remains the most established and widespread part of society responsible for generating and pooling knowledge in all areas of life [18], even in the digital era. Journalism past and present aims at observing, analyzing and communicating facts that involve social relevance and public interest, consolidating it as a professional ideology that undergoes constant construction and change [9]. Journalistic values are still very important because they are the essence of the profession, at the same time, in each historical moment it is also relevant who selects the news, for whom, in what medium, by what means and with what resources [20]. The elements of journalism [23] do not hinder renewed journalistic manifestations in the network society, on the contrary, data indicates that new forms of journalism are flourishing in different parts

X. López-García (✉) · C. Toural-Bran · J. Vázquez-Herrero
Universidade de Santiago de Compostela, Santiago de Compostela, Spain
e-mail: xose.lopez.garcia@usc.es

J. Vázquez-Herrero et al. (eds.), *Total Journalism*, Studies in Big Data 97,
https://doi.org/10.1007/978-3-030-88028-6_8

of the world. The future of digital journalism is, thus, unstable and to some extent fragile, but hopeful [10] and evolving.

The meanings and visions of journalism, as a form of cultural production in charge of connecting real stories with the public, emerge through the metajournalistic discourse, which must connect three components—actors, audiences and topics—with the processes of definition, limits of the work, and legitimization [2]. These reflections and debates—necessary at each historical stage, within the framework of a given social, political and economic context—must now be framed in the new technological scenario, which has reshaped the journalistic field by importing new values with the incorporation of people outside the professional field [34]. In the new communicative ecosystem, the logics of the oldest and the newest media coexist and compete, which defines a hybrid ecosystem [5], which brings us to the third decade of the third millennium, with the Internet of Things and intelligent automation as protagonists.

Technology-driven change has introduced profound transformations in journalism in the twenty-first century [31], including the emergence of multimedia production [8, 17], the importance of user-generated content and the participation of active audiences [3, 21, 37], data journalism [1, 39], immersive journalism [22, 36], and transmedia journalism [14, 15], among others. The reinvention of journalism moves hand in hand—technologically and socially—with techniques and work systems that have allowed it to improve verification through fact-checking strategies [29]. Despite the difficulties and experiences that show the inexistence of rapid radical changes [27], transformation has facilitated a move from simple intuition to prediction and the elaboration of a useful and constructive nonfiction discourse for the public interest and society.

2 From Consolidation to New Challenges

Scientific research in the field of digital journalism has demonstrated not only the consolidation of this specialty, but new challenges on the agenda of researchers and updated proposals that are useful for the construction of a better-informed society. At the beginning of the third decade of the third millennium, research on digital journalism is an established and developing discipline, although it faces several methodological and thematic challenges [35]. In fact, different voices, from the empirical study of the complexity of journalism in today's society, have warned of the need for a critical reflection that feeds a greater openness to what is happening in the field of communication and society in general. The aim of these reflections is to stimulate the search for models that escape from rigid frameworks and poorly grounded exclusions in order to better understand the multiple and diverse experiences and practices present in the field [43], which will undoubtedly lead to renewed, more interdisciplinary, transdisciplinary and multidisciplinary investigations of journalism in flux and about the process of reinvention.

The evolution of the digital communication ecosystem in the last two decades of the twenty-first century has shown a great diversity of communicative experiences in the network society, where media promoted by the traditional industry (today's digital or matrix migrants) coexist and compete with digital native media. The so-called 'high technology'—based on cyber-physical systems that primarily combine physical infrastructure with software, sensors, nanotechnology and digital communications technology—has positioned the different methods of communicative mediation in a new scenario [26]. And, if threats and slow progresses were not enough, the crisis caused by the COVID-19 pandemic arrived in 2020, which has driven the definitive priority for digital among all communicative actors.

The breeding ground created in the new communication ecosystem has encouraged a clear trend since the beginning of the 2010s, in light of the social web and new platforms, as well as in the shadow of changes in the technological framework, which point to major renovations in the organizations involved, an increase in the plurality and diversity of models and the opening of new territories for communication actors. This rich variety has demanded a response from research towards methods and epistemologies that would allow not only a deeper understanding of organizations [6] but also of the journalistic field, where studies of the multiple and diverse experiences of journalistic initiatives show the difficulties of media entrepreneurship and its renewed formulas, with insights beyond the journalistic discipline itself [11].

Certainly, journalism today is much more than digital technology, but it is difficult to understand journalism today without the digital dimension because the media and journalism have both fully entered onto the path traced by this technological dimension. Perhaps this is why the field of digital journalism has been defined and horizons and challenges have been set which, as we have pointed out, with good judgment, advocate the incorporation of renewed approaches and contemporary ideas that investigate the practice of digital journalism [13], its characteristics and casuistries. The diversity of journalistic cultures that coexist in today's world, through the diverse perceptions held by journalists about their role and responsibility in society [19] and the complexity of the current digital communication ecosystem, demand not only works from our field with renewed perspectives and methodologies, but other views from other disciplines, with their own methodologies and approaches.

3 The Basis for Reinvention

The metamorphosis that journalism has undergone has led to new practices, in many cases through renewed journalistic movements that seek their own ways of practicing the profession, both in the existing media—with more or less traditional models—and through the creation of new media platforms. The reconfiguration of the communication ecosystem has encouraged initiatives that feed the reinvention of journalism, which now face renewed challenges through audacious formulas to fulfill its tasks in the new social, political and economic context of the network society [44].

Studies on the evolution of journalism point out that a dispersion and hybridization of journalistic activities is taking place, some very evident in the professional media and others less visible but important for the understanding of how information is constructed, distributed and used today [12]. The hybrid media system that characterizes the communication ecosystem—where old and new logics collide and interact [5]—has favored experimentation through innovative journalistic proposals that, in addition to updating old techniques, have jumped on the options brought about by current technologies such as virtual reality, augmented reality, 5G and blockchain, among others.

After the consolidation of mobile and ubiquitous communication in the 2010s, predictions point to a horizon for journalism in which it will exist hand-in-hand with artificial intelligence, big data, new visual interfaces and current technologies [30], which will have an impact on new journalistic directions. Social networking sites, such as Facebook and Twitter, as well as other social networks that have strongly entered the communication ecosystem, will maintain their key role in the diet of online news and informative pieces [24].

Journalism in the time of social networks not only seeks to establish its space and maintain its central role in society, but also aspires to do so with current tools and with renewed narratives that add value to messages and offer guarantees of truthfulness and quality. To this end, journalism has incorporated automated verification systems and fact-checking strategies that seek to reinforce the credibility of its messages in order to regain the trust of citizens who, for various reasons, have lost faith in journalistic brands.

4 Emerging Journalisms

Just as muckrakers, New Journalism, precision journalism or civic journalism previously revitalized the field and offered new ways of understanding and practicing journalism, today data, immersive, predictive, constructive or solutions journalism provide ways to make dreams of other possible journalisms come true. The transformation and revitalization of journalism in the digital era has staged a variety of proposals that, with a wide range of technological tools, have placed information in different formats and channels within this ubiquitous scenario.

Since the beginning of the digital era, the technological and social environment has shown new attributes for information professionals to elaborate journalistic content [4]. At the same time, the lack of a well-defined professional identity has weakened journalists' capacity for independence and confused journalism with various infotainment and public relations practices [7], which has fed a scenario favorable to the loss of credibility for journalism.

Despite the aforementioned threats to quality journalism in the context of the changes and transformations of the first digital decade of the twenty-first century, the lessons learned from the experiences harvested in the past by different journalistic movements have suggested new paths through current techniques and served as a guide for the consolidation of data journalism, new narrative journalism, consortiums for global investigative journalism, interactive journalism and immersive journalism, among others, and for the search for renewed paths for a journalism that is more committed to the communities and citizens it seeks to inform. This avenue has been opened by data journalism, showing the possibilities for telling stories from large amounts of information [38] and, with some limitations, for modeling predictions [32].

Data journalism, database journalism, predictive journalism and information visualization have established models to bring added value to news stories, which, with other approaches and technologies, were then followed by immersive journalism, transmedia narratives and other interactive modalities. It was the beginning of a 'high tech' path, but which required the solid quality of journalistic technique and precepts for the contribution of models that lead to renewed journalistic movements and media models, such as hyperlocal networks that have appeared in recent years in different scenarios [25]. Robotics, drones and blockchains show evidence of tools and options that feed experimentation and innovation in the journalistic field today.

From that experimentation came renewed trends in nonfiction storytelling [40], experimentation with virtual reality brought immersive journalism to the media landscape, with renewed narrative formulas and challenges for journalistic ethics [33], and from such experimentation emerged strategies to combat misinformation [41] and platforms that show paths to recover the credibility of journalism [42]. All of this occurred in the shadow of 'high technology' and a new social, political and economic scenario in the 2020s.

Another movement has also emerged, ranging from those who come under the umbrella of positive journalism to those who opt for constructive or solutions journalism. This is the search for ways, without abandoning the essence of journalism, to offer messages that are useful for citizens when facing the challenges of today's society based on examples, cases and experiences reported in journalistic pieces. Its strategy consists of applying technologies that allow the best possible documentation of the issues, the best sources and, with the support of technology and from commitment, to offer messages that, when possible, show solutions or options for relevant improvements for society as a whole.

The chosen path includes the application of techniques from positive psychology to the construction of news [28]. It is the option of what is configured as a journalistic movement with three variants, involving some nuances, but with a clear objective of differentiating itself from the current dominant journalism model. It is one of those movements that has aroused interest in sectors of the profession concerned with the negative bias of many news reports and which has found an echo—in some countries—in a significant number of new media companies committed to change.

The renewed conceptions of journalism, its role in society and high technology have given rise not only to new journalistic practices, but also to renewed perspectives for journalism and, therefore, to the demand for renewed approaches in journalistic research. The practice of journalism in European digital native media is showing moves in that direction, as the way news is produced and consumed in the digital age is breaking down the boundaries that once divided professionals, citizens and activists, while establishing new frontiers and renewed alliances. It is more than technological determinism [16]. The research indicates that it is a new journalistic scenario that has new actors and new journalistic forms in the digital communication ecosystem of today's network society, with migrant media and digital natives as central axes of its operation.

5 Conclusion

Today's journalism, undergoing constant change and adaptation to new social and technological scenarios, maintains its activity while the voices and initiatives for its reinvention multiply. There are also voices advocating new approaches to research on the complexity of digital journalism. The process of metamorphosis that journalism has undergone in the last two decades has kept many debates alive, ranging from the need to reinforce its professional identity to the different visions of the professionals themselves about their role in society, or the old and new ethical debates. Despite the loss of credibility in different societies and the existing limitations to create scenarios that guarantee the autonomy of professionals and their independence, the role of journalism is vital for well-informed plural and democratic societies.

The transition to the digital scenario—convulsed and complicated by severe economic crises such as the financial crisis of 2008 and the structural crisis of 2020 resulting from the COVID-19 pandemic, and by the breakdown of the old business models that characterized the media industry for more than a century—has led to the search for new ways of journalistic practice, guided by the essential characteristics of digital journalism and by the emergence of digital native media, which coexist with legacy media or digital migrants and with renewed initiatives promoted by society to ensure the practice of journalism, both in an amateur way and with strategies that only sometimes favor journalistic quality.

The experimentation of recent years and the arrival of 'high technology' to the journalistic scenario have both posed renewed challenges to a journalism that needs to regain credibility and show its strengths to provide citizens with better information. Many current initiatives, with more or less success, try to overcome a past and recent stage—marked by some improvisation and with too many competences delegated to the journalistic sense of smell—to another in which, by means of current techniques derived from the hand of human talent and current technological tools, allow the elaboration of journalistic pieces of higher quality. It is not a question of burying the journalistic sense of smell, which is still necessary, but of wrapping it in methods, techniques and strategies that allow the construction of journalistic works that provide added value in a scenario of hyperabundance of messages and tangles of misinformation.

Research suggests that the rigorous and quality information that democratic societies need will only come from professionalized journalism that is adapted to this new scenario, committed to society and developed through strategies that ensure social involvement and collaboration. High technology is an opportunity for the journalism of the present and the future. Journalistic movements, new or established, are vitamins for the revitalization of the journalistic field. Emerging journalism, with its search for renewed models and practices, with more or fewer proposed solutions, constitutes an incentive for such constant adaptation and reinvention.

There is no doubt that there are several ways and options to guarantee the future of journalism. There is no single path. And that is the richness that emerging journalisms brings, with techniques that allow the construction of more rigorous pieces and the elaboration of more predictive messages and useful proposals to meet the demands and needs of citizens.

Funding This research has been developed within the research project *Digital Native Media in Spain: Storytelling Formats and Mobile Strategy* (RTI2018-093346-B-C33), funded by the Ministry of Science, Innovation and Universities (Government of Spain) and the ERDF structural fund.

References

1. Borges-Rey, E.: Unravelling data journalism. A study of data journalism practice in British Newsrooms. Journal. Pract. **10**(7), 833–843 (2016). https://doi.org/10.1080/17512786.2016.1159921
2. Carlson, M.: Metajournalistic discourse and the meanings of journalism: definitional control, boundary work, and legitimation. Commun. Theory **26**(4), 349–368 (2016). https://doi.org/10.1111/comt.12088
3. Carpentier, N.: Media and Participation. A Site of Ideological-Democratic Struggle. Intellect, Bristol (2011)
4. Casero, A.: Contenidos periodísticos y nuevos modelos de negocio. Evaluación de servicios digitales. El Profesional de la Información **21**(4), 341–346 (2012). https://doi.org/10.3145/epi.2012.jul.02
5. Chadwick, A.: The Hybrid Media System. Politics and Power. Oxford University Press, Oxford (2013)
6. Corbett, A., Cornelissen, J., Delios, A., Harley, B.: Variety, novelty, and perceptions of scholarship in research on management and organizations: an appeal for ambidextrous scholarship. J. Manage. Stud. **51**(1), 3–18 (2014). https://doi.org/10.1111/joms.12032
7. Dader, J.L.: Periodismo en la hipermodernidad: Consecuencias cívicas de una identidad débil (y algunas vías de reconstrucción). Textual Vis. Media **1**(2), 147–170 (2009)
8. Deuze, M.: What is multimedia journalism? Journal. Stud. **5**(2), 139–152 (2004). https://doi.org/10.1080/1461670042000211131
9. Deuze, M.: What is journalism? Professional identity and ideology of journalists reconsidered. Journalism **6**(4), 442–464 (2005). https://doi.org/10.1177/1464884905056815
10. Deuze, M.: Considering a possible future for digital journalism. Revista Mediterránea de Comunicación **8**(1), 9–18 (2017). https://doi.org/10.14198/MEDCOM2017.8.1.1
11. Deuze, M., Witschge, T.: Beyond Journalism. Polity Press, Cambridge (2020)
12. Domingo, D.: News practices in the digital era. In: Witschge, T., et al. (eds.) The SAGE Handbook of Digital Journalism. Sage, London (2016)
13. Eldridge, S.A., II., Hess, K., Tandoc, E.C., Jr., Westlund, O.: Navigating the scholarly terrain: introducing the digital journalism studies compass. Digit. Journal. **7**(3), 386–403 (2019). https://doi.org/10.1080/21670811.2019.1599724
14. Gambarato, R.R., Alzamora, G.C.: Exploring Transmedia Journalism in the Digital Age. IGI Global, Hershey (2018)

15. Gambarato, R.R., Tárcia, L.P.T.: Transmedia strategies in journalism. An analytical model for the news coverage of planned events. Journal. Stud. **18**(11), 1381–1399 (2017). https://doi.org/10.1080/1461670X.2015.1127769
16. García-Orosa, B., López-García, X., Vázquez-Herrero, J.: Journalism in digital native media: Beyond technological determinism. Media Commun. **8**(2), 5–15 (2020). https://doi.org/10.17645/mac.v8i2.2702
17. George-Palilonis, J.: The Multimedia Journalist. Storytelling for Today's Media Landscape. Oxford University Press, Oxford (2012)
18. Godler, Y., Reich, Z., Miller, B.: Social epistemology as a new paradigm for journalism and media studies. New Media Soc. **22**(2), 213–229 (2020). https://doi.org/10.1177/1461444819856922
19. Hanitzsch, T., Hanusch, F., Ramaprasad, J., De Beer, A.S.: Worlds of Journalism: Journalistic Cultures Around the Globe. Columbia University Press, New York (2019)
20. Harcup, T., O'Neill, D.: That is news? News values revisited (again). Journal. Stud. **18**(12), 1470–1488 (2016). https://doi.org/10.1080/1461670X.2016.1150193
21. Holton, A., Lewis, S.C., Coddington, M.: Interacting with audiences. Journal. Stud. **17**(7), 849–859 (2016). https://doi.org/10.1080/1461670X.2016.1165139
22. Jones, S.: Disrupting the narrative: immersive journalism in virtual reality. J. Media Pract. **18**(2–3), 171–185 (2017). https://doi.org/10.1080/14682753.2017.1374677
23. Kovach, B., Rosenstiel, T.: The Elements of Journalism. Three Rivers Press, New York (2007)
24. Kümpel, A.S.: The Matthew effect in social media news use: assessing inequalities in news exposure and news engagement on social network sites (SNS). Journalism **21**(8), 1083–1098 (2020). https://doi.org/10.1177/1464884920915374
25. López-García, X., Negreira-Rey, M.-C., Rodríguez-Vázquez, A.I.: Cibermedios hiperlocales ibéricos: el nacimiento de una nueva red de proximidad. Cuadernos.info **39**, 225–240 (2016). https://doi.org/10.7764/cdi.39.966
26. López-García, X.: Panorama y desafíos de la mediación comunicativa en el escenario de la denominada automatización inteligente. El Profesional de la Información **27**(4), 725–731 (2018). https://doi.org/10.3145/epi.2018.jul.01
27. Lough, K., McIntyre, K.: Transitioning to solutions journalism: one newsroom's shift to solutions-focused reporting. Journal. Stud. **22**(2), 193–208 (2021). https://doi.org/10.1080/1461670X.2020.1843065
28. McIntyre, K., Gyldensted, C.: Constructive journalism: an introduction and practical guide for applying positive psychology techniques to news production. J. Media Innovations **4**(2), 20–34 (2018). https://doi.org/10.5617/jomi.v4i2.2403
29. Mena, P.: Principles and boundaries of fact-checking: journalists' perceptions. Journal. Pract. **13**(6), 657–672 (2018). https://doi.org/10.1080/17512786.2018.1547655
30. Newman, N.: Journalism, media and technology. Trends and predictions 2020. Reuters Institute for the Study of Journalism, Oxford (2020)
31. Pavlik, J.: The impact of technology on journalism. Journal. Stud. **1**(2), 229–237 (2000). https://doi.org/10.1080/14616700050028226
32. Pentzold, C., Fechner, D.: Data journalism's many futures: diagrammatic displays and prospective probabilities in data-driven news predictions. Convergence **26**(4), 732–750 (2019). https://doi.org/10.1177/1354856519880790
33. Pérez-Seijo, S., López-García, X.: La ética del Periodismo Inmersivo a debate. Hipertext.net **18**, 1–13 (2019). https://doi.org/10.31009/hipertext.net.2019.i18.01
34. Reese, S.D.: Theories of journalism. In: Nussbaum, J. (ed.) Oxford Encyclopedia of Communication. Oxford University Press, Oxford (2016). https://doi.org/10.1093/acrefore/9780190228613.013.83
35. Salaverría, R.: Periodismo digital: 25 años de investigación. Artículo de revisión. El Profesional de la Información **28**(1) (2019). https://doi.org/10.3145/epi.2019.ene.01
36. Shin, D., Biocca, F.: Exploring immersive experience in journalism. New Media Soc. **20**(20), 2800–2823 (2017). https://doi.org/10.1177/1461444817733133

37. Singer, J.B.: The political j-blogger: "Normalizing" a new media form to fit old norms and practices. Journalism **6**(2), 173–198 (2005). https://doi.org/10.1177/2F1464884905051009
38. Stalph, F.: Classifying data journalism. Journal. Pract. **12**(10), 1332–1350 (2018). https://doi.org/10.1080/17512786.2017.1386583
39. Uskali, T., Kuutti, K.: Models and streams of data journalism. J. Media Innovations **2**(1), 77–88 (2015). https://doi.org/10.5617/jmi.v2i1.882
40. Vázquez-Herrero, J.: Tendencias en no ficción interactiva: perspectiva de los productores a través de un estudio Delphi. adComunica **19**, 61–82 (2020). https://doi.org/10.6035/2174-0992.2020.19.5
41. Vázquez-Herrero, J., Vizoso, Á., López-García, X.: Innovación tecnológica y comunicativa para combatir la desinformación: 135 experiencias para un cambio de rumbo. El profesional de la información **28**(3) (2019). https://doi.org/10.3145/epi.2019.may.01
42. Vizoso, Á., Vázquez-Herrero, J.: Fact-checking platforms in Spanish. Features, organisation and method. Commun. Soc. **32**(1), 127–144 (2019). https://doi.org/10.15581/003.32.1.127-144
43. Witschge, T., Deuze, M.: From suspicion to wonder in journalism and communication research. Journal. Mass Commun. Q. **97**(2), 360–375 (2020). https://doi.org/10.1177/2F1077699020912385
44. Zelizer, B.: What Journalism Could Be. Polity Press, London (2017)

Xosé López-García Professor of Journalism at Universidade de Santiago de Compostela (USC), Ph.D. in History and Journalism (USC). He coordinates the Novos Medios research group. Among his research lines there is the study of digital and printed media, analysis of the impact of technology in mediated communication, analysis of the performance of cultural industries, and the combined strategy of printed and online products in the society of knowledge.

Carlos Toural-Bran Assistant Professor at the Department of Communication Sciences, Universidade de Santiago de Compostela (USC). Ph.D. in Communication Sciences (USC). Chief of Staff of the Rector (USC) and also President of AGACOM (Galician Association of Communication Researchers) and secretary of Novos Medios research group. He has also been Vice-dean of the Faculty of Communication Sciences (2014–2019).

Jorge Vázquez-Herrero Assistant Professor at the Department of Communication Sciences, Universidade de Santiago de Compostela (USC). Ph.D. in Communication and Contemporary Information (USC). He is a member of Novos Medios research group and the Latin American Chair of Transmedia Narratives (ICLA–UNR). He was visiting scholar at Universidad Nacional de Rosario, Universidade do Minho, University of Leeds and Tampere University. His research focuses on the impact of technology and platforms in digital journalism and narratives.

Big Data and Information Disorders

Misinformation Beyond the Media: 'Fake News' in the Big Data Ecosystem

Ramón Salaverría and Bienvenido León

Abstract In recent years, the term 'fake news' has become popular as a paradigm of mis- and disinformation. The term tends to put this phenomenon within the framework of media organizations. However, the problem is more complex, since it also involves other entities dedicated to deliberately producing and spreading falsehoods, as well as social networks and large internet platforms that work as global carriers of such misleading content. This chapter analyzes the evolution of disinformation in this expanded framework, contextualizing the production, dissemination and consumption of deceptive content beyond the media. Drawing upon a historical review of the mis- and disinformation phenomena over the last few centuries, it examines the more recent transformation experienced by purposefully false content on the internet and big data ecosystem.

1 The Complex Ecosystem of Mis- and Disinformation

In 2016, Oxford Dictionaries declared 'post-truth' as the international word of the year. With this, it recognized the pervasiveness reached by that term in public discourse, where objective facts had less and less influence than emotions when shaping citizens' opinions. Months later, on January 22, 2017, during a Meet the Press interview, the US Counselor to the President Kellyanne Conway defended White House Press Secretary Sean Spicer's false statement about the attendance numbers of Donald Trump's inauguration as President of the United States, calling them 'alternative facts'.

'Post-truth' and 'alternative facts' are not the only terms used in recent times to designate realities halfway between truth and lies. In fact, the most widespread term is another one, which brings the problem closer to journalism: 'fake news'. With its

R. Salaverría (✉) · B. León
Universidad de Navarra, Pamplona, Spain
e-mail: rsalaver@unav.es

B. León
e-mail: bleon@unav.es

J. Vázquez-Herrero et al. (eds.), *Total Journalism*, Studies in Big Data 97,
https://doi.org/10.1007/978-3-030-88028-6_9

reference to the news, this term attributes to the media a large part of responsibility in spreading the falsehoods that circulate in the public sphere. But is 'fake news' just a problem of the media?

The recent popularization of this term has a main protagonist. On December 10, 2016, a month after being elected 45th president of the United States, Donald J. Trump published a tweet where, for the first time in that network, he used the expression 'fake news'. He did it to accuse a medium, CNN, of manipulation by publishing falsehoods. It would not be the last time. In his four-year tenure, Trump used this disdainful term hundreds of times to disavow countless media and individuals both on Twitter and in public appearances. With his constant repetition, Trump turned those two words into a common expression around the world. The result is that 'fake news' has become a label used indiscriminately to designate all kinds of deception, fraud or falsehood, even when the lie does not come from the media [21].

As we can see, the complexity of the phenomenon of disinformation begins with its own terminology. In fact, none of the three terms mentioned—post-truth, alternative facts, fake news—seem to accurately designate the set of phenomena and realities linked to the public dissemination of falsehoods. Faced with these controversial terms, consensus is growing in using two other alternative concepts: 'disinformation' and 'misinformation'. Used for at least half a century [9], the first concept designates the intentional communication of lies and the second, the involuntary dissemination of falsehoods. The effects of both coincide to a large extent, since they translate into the communication of false information to the public, but it is clear that there is a key difference between lying and making a mistake.

In recent years, the study of the deliberate or involuntary dissemination of false content has received a notable boost, within the framework of the emerging discipline of so-called "information disorders" [36]. These studies are revealing the complex ecosystem of elements—technological, social, psychological, political and economic, to name just a few—that lie behind the phenomena of disinformation. They are showing that the public dissemination of falsehoods is not limited to the realm of the news media. There are many other actors interested in poisoning society with certain agendas and frames of mind, through hidden strategies of manipulation. In a context of growing disintermediation, the shaping of public opinion depends more and more on social networks, ruled by dynamics of influence and gregariousness. In this new ecosystem, where journalistic media have transferred part of their influence to new digital platforms, the public dissemination of falsehoods responds to new patterns that force us to analyze disinformation beyond the media.

1.1 Typology of Fake Content

Since the time of Aristotle and Plato, the nature of truth and its limits with respect to falsehood have been the subject of an endless debate. In today's society, characterized by the prominence of social networks, determining where the truth ends and falsehood begins continues to be a controversial issue, which has transcended the sphere of Philosophy and reaches the daily lives of all citizens.

The "post-truth era" [18] has made the concept of truth a contested issue in public debate. Rather than an objective category, truth has become relative for many. Those who relativize the truth tend to consider as right what coincides with their opinion, preferences or ideology, while they dismiss as false everything that challenges, contradicts or opposes them. Taking advantage of this trend, many organizations and political leaders, especially those aligned in populist positions of any sign, have normalized the use of falsehood as a political weapon, which is many times used without limit [34]. According to a report by *The Washington Post* [32], during his four-year presidency, Donald J. Trump accumulated more than 30,000 false claims, exaggerations and lies.

The diversity of fake content is indeed very wide. It ranges from the simple joke or meme, which can be based on a more or less innocent distortion of reality, to the fabrication of deceptive content with the support of very sophisticated technologies, such as so-called deep fakes.

All this wide range of falsehoods is divided according to a key criterion: the presence or absence of intent to deceive. Sometimes, indeed, certain messages are false and confuse the public, but they do not respond to an intention to deceive; they are simply due to carelessness or inadvertent confusion on the part of their authors. It is the kind of unintentional misrepresentation that we know as misinformation. Other times, the falsification of the messages is deliberate and their authors manipulate the elements with the intention of deceiving. This second category is what is properly known as disinformation.

Based on these two main concepts, recent research identifies different types of falsehoods. In one of the most cited classifications, Wardle [35] classified falsehoods into four categories—satire or parody, misleading content, imposter content and fabricated content—, to which she adds three falsification procedures—false connection, false content and manipulated content. The false content categorized by Wardle can appear, in principle, on any platform: from journalistic media to social networks. Tandoc Jr and colleagues [31], for their part, focus their typology on false content that is specifically detected in news media. They distinguish between six types of 'fake news': news satire, news parody, news fabrication, photo manipulation, advertising and public relations, and propaganda. Brennen and colleagues [8] propose a more conceptual typology, based on three falsification procedures. Specifically, they distinguish between satire or parody, reconfiguration and fabrication. The first case refers to messages whose falseness responds to a satirical or humorous intention; frequently, the falsehood of these messages is not intended to mislead the recipients, but rather to ridicule or laugh down the subject or person they are about. The contents falsified by reconfiguration are those that start from a real content to place it in a manipulated context. A frequent example of this type of falsehood is real photos, videos or audios that are shown in deceptive contexts. Finally, the third type corresponds to the absolute fabrication of falsehoods. In this case, all the elements that make up the message are designed with the purpose of deceiving.

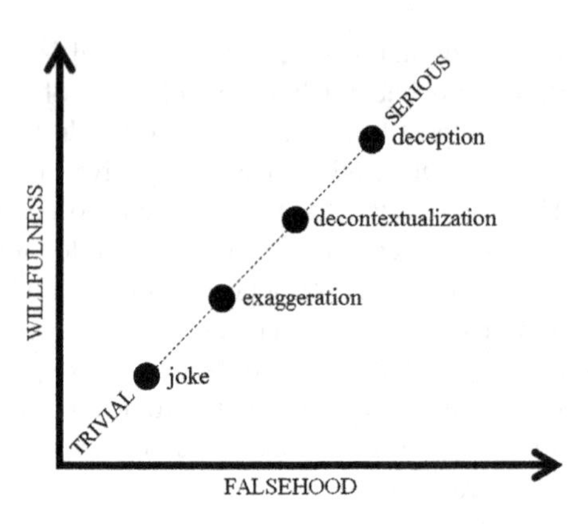

Fig. 1 Hoax severity diagram. *Source* Salaverría et al. [24]

As can be seen, recent theory coincides in understanding the types of disinformation not so much as a more or less long list of concrete forms of falsehoods, but rather as a *continuum* that begins with satirical or parodic falsehoods, to a certain point innocent, and it ends in the area of informative fabrications, the most serious deceptions. Under this same prism, Salaverría and colleagues [24] propose a "hoax severity diagram" (see Fig. 1), which shows this longitudinal character of misinformation.

The form of the fake content also varies according to the platforms. A misleading message posted on a video portal is not the same as a disinformation thread distributed through a messaging application. Sometimes fake content is dissimulated by taking the appearance of trustworthy messages. This occurs, for example, in certain digital media that constantly publish false content, but simulate reliable news publications. Misleading content, in its intention to go unnoticed, is adapted to the specific format and language of each platform.

1.2 Factors Contributing to the Dissemination of Falsehoods

Beyond the fact that some people or organizations try to deceive society, disinformation is also favored by various technological, psychosocial, political and media factors. Altogether, they set up an ideal environment for misinformation to spread.

Among the factors that contribute to the spread of misinformation, the first is of a technological nature. The popularization of messaging applications and social networks has contributed to accelerating the dissemination of false or unverified content. The algorithmic systems on which social networks are based often contribute to this process of disseminating suspicious content. Since these algorithms tend to reward the most controversial and shared content, by giving them greater visibility, social media technology ends up being a propitiatory factor for misinformation.

It is true that these same networks, worried about their growing reputational problem, have begun to implement automated systems for detecting and mitigating false content. To do this, they have hired verification organizations in charge of verifying the potentially false content. Likewise, they are implementing artificial intelligence systems aimed at the early detection of disinformation campaigns. However, the business model of social networks continues to be based on the constant participation or engagement of their users. Therefore, the temptation remains for these companies to promote the circulation of any type of content, including unquestionably false, since they are all financially profitable.

Second, psycho-sociological factors must be mentioned. Contemporary psychological research has proven that people are by nature reluctant to change their minds, so we tend to look only at the evidence and arguments that support our point of view and ignore any information that contradicts it. It is, in short, the phenomenon baptized by the British psychologist Peter C. Wason as confirmation bias [37].

It is not only that we tend to pay attention to what we believe from the outset. It also happens that the evidence to the contrary, powerful and irrefutable as it may be, is incapable of fully convincing us. Studies of confirmation bias have indeed shown that certain beliefs remain even when the initial evidence that supposedly supported them is overturned. In those cases, at least a part of the initial belief remains after the rebuttal is complete.

This psychological bias is amplified in environments with high social interaction, such as social networks. When individuals are surrounded by a mass that is aligned with their initial postulates, they adopt a gregarious behavior and are even more reluctant to change their mind, no matter how unquestioned the evidence that contradicts it may be [2]. This type of psycho-sociological factors helps to understand the growing political polarization of the networks [10], while they also explain the difficulty of neutralizing some deceptive messages.

Over the last decade, many studies have focused on these phenomena. Those that have received the most attention are the so-called echo chamber effect [16] and the filter bubbles effect [20]. Several studies have shown that, contrary to the supposed global and open deliberation that networks would make possible, they have produced a selective exposure to certain types of content and opinions, those that coincide with the frames of mind and ideology of subjects [29]. Nevertheless, several studies have shown that these effects might have been magnified, as their incidence is more limited [12].

Political motivations must also be mentioned among the factors conducive to the rise of disinformation. Certain leaders and political organizations from different countries have not hesitated to use false statements and arguments in favor of their positions, in political debates and electoral campaigns. In an increasingly ideologically polarized public sphere, they have realized the power of manipulation to mobilize their supporters. Undoubtedly, turning falsehood into a political weapon contributes to expanding disinformation.

Finally, we must also mention different phenomena related to the news media and media literacy. To begin with, due to the economic crisis and the harsh transformation towards digital models, many media outlets today have more limited teams of journalists and less experienced staff. This makes it easier for some false messages to get past the professional filter of journalists and become, now quite properly, 'fake news'.

To this internal problem of journalism, the phenomenon of news saturation is added. Citizens show a tipping point from which they stop paying attention to the news, in an effect known as news avoidance [27]. This rejection comes especially when bad news accumulates, as has happened during the COVID-19 pandemic. Getting the truthful messages to reach the public when they turn their backs on all kinds of news is a very complicated task.

Furthermore, an insufficient media literacy of a good part of the citizenship limits their ability to distinguish truth from opinion, as well as to identify credible sources in relevant topics. The paucity of educational programs geared towards media literacy makes misinformation more likely to take hold.

2 Publishing Lies, from Pasquinades to Social Media

The dissemination of deliberately false information has been a common practice throughout history. Successive technological developments have been decisive factors in creating different forms and modes of dissemination, ranging from sheets printed on paper in the Middle Ages, to messages on social networks in the twenty-first century. We review, below, the main forms of disinformation in recent history and how they have used the different technological tools at all times.

The printing press made possible the proliferation of forms dedicated to disseminating news about public figures, mostly false, which were pasted on the walls of city streets, in public view. Early examples include sonnets that tried to undermine the reputation of some candidates for the pontifical election in 1522, and which were pasted in the center of Rome on a figure known as Pasquino. This experience seems to be the beginning of the genre of "pasquinades" [11]. They generally filed a complaint against the political authority, were usually satirical in character, were written in verse and designed with different fonts and colors, in an attempt to increase their impact among the public [22].

This type of publication took various forms and names. One of the best known were the so-called libels, defamatory writings against people or institutions that had a wide echo in many countries. Although the first examples appear on the occasion of the Lutheran reform and the religious wars of the sixteenth century, libels multiplied in subsequent centuries and were prohibited by law in several countries [14].

Since the dawn of the periodical press in the seventeenth century, there were publications of a "mundane" nature, which mixed "in a chaotic way courtesan chronicles, military victories and news, with sensational stories", often false [17:13]. Printed and illustrated gazettes, called canards, generally based on false content, were very popular in Paris. During the French Revolution, the canards became instruments of political propaganda. Among the deliberately false information they included, it is worth highlighting that referring to Queen Marie Antoinette, whose dissemination apparently contributed to her becoming the hated figure who would finally be executed in 1793 [11].

Along the nineteenth century, the so-called popular press was developed, characterized by including abundant doses of sensationalism, which enjoyed great success in countries such as the United States, England and France. At the end of the same century, this type of press reached its peak in the newspapers of the American businessman William Randolph Hearst, which have gone down in the history of journalism as a paradigm of the 'yellow press'. It has been defined as "journalism without a soul, without ethics, where the principle of provocation of the news was carried to the extreme of inventing it, if it suited the interests of the newspaper" [25:112].

Hearst's newspapers took an active part in campaigns whose objective was to manipulate public opinion. The most relevant was the Spanish-American War, which broke out in 1898, and which has been described as the "war of Mr. Hearst", for the decisive role he played in triggering the armed conflict [13:151]. The United States declared war on Spain after the explosion of the battleship USS Maine in Havana Bay. For months before, the Hearst press had been reporting alleged suffering caused by the Spanish colonizers to the Cuban people. When the Maine battleship exploded, Hearst sent journalist Fredric Remington to Cuba, who is said to have claimed "I will provide the war" [19:407]. Although it is probable that the sinking of the Maine was due to an accidental explosion in its interior, most of the Americans believed that the Spanish were responsible. From that point on, the tabloid press continued its campaign of agitation, which contributed to the United States declaration of war on Spain on April 25, 1898 [19:383].

According to Rodríguez Andrés [23], the term disinformation began to be used at the beginning of the twentieth century. Apparently, the Bolshevik police used the expression *disinformatzia* to designate actions aimed at preventing the consolidation of the communist regime in the Soviet Union. The term began to be used more and more frequently in that country, as indicated by the fact that it was included in the *Dictionary of the Russian Language*, in 1949.

For its part, the Soviet Union used disinformation as a means to establish proletarian goals throughout the world. Both in the USSR and in satellite countries, such as German Democratic Republic, Czechoslovakia, Hungary, Poland and Bulgaria, departments dedicated to generating disinformation were created [3]. The Soviet secret services (KGB) used disinformation to discredit capitalist countries and help establish communist regimes. To achieve this, deception and manipulation were used, with the ultimate goal of muddying international relations and discrediting regimes and people in the capitalist world. According to Volkoff [33:170], "within that total and incessant conflict that the Soviet Union maintained against the capitalist and bourgeois world, disinformation was a particularly effective weapon, a capital instrument to condition individuals". The same KGB training manuals clearly explained the intended use of disinformation:

> The strategic disinformation helps the execution of the tasks of the State and is directed to mislead the enemy with regard to the basic questions of State policy, the economic situation, the military and the scientific-technical achievements of the USSR; It also tends to mislead the politics of certain imperialist countries regarding their relations with each other and also mislead the special operations of the State security organs [3:368].

During World War I (1914–1918), a powerful new technology entered the scene: cinema. Although the cinema had already been used as an instrument of propaganda during the Spanish-American War in Cuba (1898) and the Second Boer War (1899), it is from the Great War that it acquires greater importance. During this conflict, several countries used the cinema as a means of propaganda, both through fiction films and cinematic newscasts. Thanks to these films, many citizens received information that was often biased, if not false, according to the interests of their respective governments and in response to the needs imposed by the development of the war. For example, when the entry of the United States into the war seemed inevitable, the movie *To Hell with the Kaiser* (1918) was released in theaters, which portrayed German soldiers as "brutish thugs" [19:461].

In the Soviet Union, the cinema also played a prominent role as an instrument of propaganda. In 1919, Lenin nationalized the film industry and created a training school for filmmakers. The Soviet state viewed cinema as the most important of all the arts and gave it an essential role in building communist society.

In Germany, the cinema was used in Nazi propaganda, even before the rise to power of the National Socialist Party in 1933 [15]. His early propaganda efforts focused on spreading his vision of World War I, emphasizing the responsibility of leftists, Jews and politicians in the defeat of Germany [19:463]. Subsequently, the propaganda, led by the minister of the branch, Joseph Goebbels, became a decisive instrument to the Nazi regime, which objectives were achieved by spreading abundant false information, such as blaming the Jews for the problems suffered by Germany.

Radio also was key for Nazi propaganda efforts. Goebbels considered that this medium was an essential instrument to achieve his objectives, to the point that he used government resources to promote the production of low-cost receivers and thus expand the number of listeners. At the same time, German broadcasters were granted high-power equipment, whose waves even crossed national borders [1].

Since the beginning of World War II (1939), all the conflicting countries launched their propaganda messages, using all available technologies. Among the many cinematographic documentaries made, some American films stand out, such as the *Why We Fight* series (1942–1945), as well as those made by British authors such as Humprey Jennings [4]. As for radio, the German example was quickly followed by the *BBC* and later by other countries, filling the airwaves with often false information.

During the Cold War, disinformation became of key importance, for both communist and western countries. The United States created various disinformation services with similar goals to those of the communist bloc, which ended up being unified in the USIA (United States Information Agency). According to Charles Wick, director of this agency in the 1980s, information became the most important element of American foreign policy [28:81]. Apparently, the actions of USIA had a notable influence on the politics of rival countries.

Television was also a medium used by propaganda agencies, not only in totalitarian regimes, but also in democratic countries. In the United States, during the Cold War there was what has been called "a partnership between government information officers and network news producers to report and sell the Cold War to the American public" [7:2], which caused many news stories to be scripted, when not directly produced, by the government.

Over the last quarter of the twentieth century, digitization boosted a paradigm shift in all areas of communication, which meant an increase in disinformation in different fields, especially in international politics. In this new technological and social context, the dissemination of falsehoods acquired dimensions unknown until then, thanks to the enormous capacity of digital media to produce content and disseminate it quickly. Although the efficacy of disinformation remains a controversial issue [6], it is widely accepted that it makes today's society more vulnerable and poses a threat to democracy [26].

Social networks are a key factor in the growth of misinformation in different aspects of political and social activity, with potentially serious consequences. The biggest example is found in the global COVID-19 pandemic, which began in 2020, about which social networks contributed to spread abundant disinformation [24].

All in all, the historical evolution of propaganda and the manipulation of public opinion reveals that the news media have always been a key part of the disinformation ecosystem. But they are not the only actors. In fact, in the current scenario of social networks and big data, other actors assume an essential role.

3 Disinformation Beyond the Media

The current ecosystem of disinformation is very complex. In it, they interact from large intelligence agencies of countries that seek to destabilize their rivals, to simple citizens who, due to evil or clumsiness, contribute to the visibility of hoaxes and conspiracy theories on their social networks. As in the past, the news media continues to play a key role in this ecosystem, as any falsehood that gets published on television, radio, newspapers or in digital media has a great reach. In return, the professional media and, especially, fact-checking organizations play a key role in debunking misinformation.

Who is producing fake content today? There are two great profiles. In the first place, that of those people and organizations that spread lies voluntarily and even professionally. Second, those who unwittingly contribute to the spread of these falsehoods.

The first profile is the most dangerous and corresponds to deliberate agents of disinformation. This category includes government-dependent intelligence organizations that design campaigns to destabilize countries and regions by disseminating falsehoods. Sometimes they rely on the creation of a large number of fake profiles generated by bots, which contribute to an artificial multiplication of false or destabilizing messages. On the occasion of the COVID-19 pandemic, for example, the US State Department accused Russia of creating a network of thousands of social media accounts dedicated to posting negative messages and conspiracy theories against the United States [5].

Another profile of organizations that produce manipulated or, directly, false messages are political parties. Especially in electoral periods, some of these organizations activate campaigns to stimulate public controversy, discredit rivals and strengthen the cohesion of supporters. The use of this type of techniques has been widely detected, for example, in the electoral campaigns of Donald J. Trump [30]. There is also a network of social movements, think-tanks and more or less suspicious organizations, dedicated to the public dissemination of conspiracy theories and dubious content.

Finally, some companies professionally devoted to creating false content are also key players in deliberate disinformation. These are known as 'content farms', devoted to the manufacture of sensational or totally false content whose purpose is to generate large volumes of traffic, using clickbait techniques [38]. These companies seek to make a profit through the dissemination of any type of content, without worrying about its authenticity or other ethical considerations.

The deliberate action of all these agents of disinformation is reinforced by the collaboration, often unintentional and inadvertent, of many other social actors. First of all, this category includes the billions of users of social networks and instant messaging applications, who, unwittingly, contribute to the dissemination of false content originally spread by others. In the belief that these contents are true or because, being aware of their possible falsehood, they align with their mindset or ideology, many users of social networks act as public speakers of disinformation. Special mention should be made of celebrities and public personalities who, perhaps in an effort to increase their notoriety, echo information without rigor and theories without scientific basis.

Among the involuntary contributors to the phenomenon of disinformation, we should also mention the large Internet platforms. Among them, especially social networks, whose business model is largely based on engagement, that is, on people consuming and sharing content non-stop. Often, fictional content and conspiracy theories spread faster than true information and denials. Faced with this dynamic, the corrective actions taken by internet platforms have been, in general, not very proactive. In 2018, large Internet platforms such as Facebook, Google or Twitter pledged to make political advertising more transparent and to introduce mechanisms to fight disinformation in the face of the May 2019 European elections. This commitment did not come voluntarily from the platforms, but following a requirement from the European Union.

Big data technologies and, in particular, the algorithms that rank the contents of large social networks are a black box. Its criteria for highlighting or hiding content from the eyes of its billions of users are completely opaque. Therefore, it is unknown to what extent effective firewalls are applied to mitigate the spread of disinformation campaigns and messages. Algorithmic systems tend to promote highly shared content, so they can function as unwanted contributors to the dissemination of false content. However, not everything in these technologies is negative. On the other side of the coin, these same algorithmic systems are enabling the development of early detection tools, which identify and mitigate false content in the initial stages of its dissemination.

Mis- and disinformation continues to be, in short, a problem in which the news media play a key role: both in its dissemination and, fortunately, increasingly in its neutralization too. However, media organizations are not the only actors. In a context of fast transformation of journalism and public communication, digital networks and other social actors are key to understanding the increasingly complex ecosystem of disinformation.

Funding This chapter is a result of two research projects: (1) *RRSSalud*, funded by the BBVA Foundation, within the Grants for Scientific Research Teams—Economy and Digital Society, 2019; and (2) IBERIFIER—*Iberian Digital Media Research and Fact-Checking Hub*, funded by the European Commission under the call CEF-TC-2020–2 (European Digital Media Observatory), grant number 2020-EU-IA-0252.

References

1. Acevedo, M.: La utilización de la radio en la Segunda Guerra Mundial. Creación Producción Diseño Comunicación **25**, 89–91 (2009)
2. Allcott, H., Gentzkow, M.: Social media and fake news in the 2016 election. J. Econ. Perspect. **31**(2), 211–236 (2017). https://doi.org/10.1257/jep.31.2.211
3. Álvarez, J.T., Secanella, P.M.: Desinformación. Diccionario de Ciencias y Técnicas de la Comunicación. Ediciones Paulinas, Madrid (1991)
4. Barnouw, E.: El Documental: Historia y Estilo. Gedisa, Barcelona (1996)
5. BBC: Coronavirus: Russian media hint at US conspiracy. Available at https://bbc.in/3lvAXo0 (2020). Accessed 12 Jan 2021
6. Benkler, Y., Faris, R., Roberts, H.: Network Propaganda: Manipulation, Disinformation, and Radicalization in American Politics. Oxford University Press, New York (2018)
7. Bernhard, N.: US Television News and Cold War Propaganda, 1947–1960. Cambridge University Press, Cambridge (2003)
8. Brennen, J.S., Simon, F.M., Howard, P.N., Nielsen, R.K.: Types, sources, and claims of Covid-19 misinformation. Reuters Institute for the Study of Journalism Factsheet, April 2020. Available at https://bit.ly/3vT1Abf (2020)
9. Burnam, T.: The Dictionary of Misinformation. Thomas Y. Crowell, New York (1975)
10. Conover, M., Ratkiewicz, J., Francisco, M., Gonçalves, B., Menczer, F., Flammini, A.: Political polarization on Twitter. Proc. Int. AAAI Conf. Web Social Media **5**(1), 89–96 (2011)
11. Darnton, R.: The True History of Fake News. The New York Review of Books, New York (2017)
12. Dubois, E., Blank, G.: The echo chamber is overstated: the moderating effect of political interest and diverse media. Inf. Commun. Soc. **21**(5), 729–745 (2018). https://doi.org/10.1080/1369118X.2018.1428656

13. González, A.J.: La guerra de los corresponsales. Revista Biblioteca Nacional José Martí **1**, 151–162 (2018)
14. Guillamet, J.: De las gacetas del siglo XVII a la libertad de imprenta del XIX. In: Barrera, C. (ed.) Historia del Periodismo Universal, pp. 43–76. Ariel, Barcelona (2004)
15. Hull, D.: Film in the Third Reich. University of California Press, Berkeley (1960)
16. Jamieson, K.H., Cappella, J.N.: Echo Chamber: Rush Limbaugh and the Conservative Media Establishment. Oxford University Press, Oxford (2008)
17. Langa-Nuño, C.: Claves de la historia del periodismo. In: Reig, R. (ed.) La dinámica periodística perspectiva, contexto, métodos y técnicas, pp. 10–40. Asociación Universitaria Comunicación y Cultura, Sevilla (2010)
18. Levitin, D.J.: Weaponized Lies: How to Think Critically in the Post-Truth Era. Dutton, New York (2017)
19. Manning, M.J., Wyatt, C.R.: Encyclopedia of media and propaganda in Wartime America. ABC-CLIO, Santa Barbara (2011)
20. Pariser, E.: The Filter Bubble: What the Internet is Hiding From You. Penguin, New York (2011)
21. Quandt, T., Frischlich, L., Boberg, S., et al.: Fake news. In: Vos, T.P., Hanusch, F. (eds.) The International Encyclopedia of Journalism Studies. Wiley-Blackwell, Hoboken (2019)
22. Revilla Orías, P.A.: Pasquines reformistas, pasquines sediciosos: aquellas hojas volanderas en Charcas (siglos XVIII-XIX). Revista Ciencia Cultura **22–23**, 33–43 (2009)
23. Rodríguez Andrés, R.: Fundamentos del concepto de desinformación como práctica manipuladora en la comunicación política y las relaciones internacionales. Historia Comunicación Social **23**(1), 231–244 (2018). https://doi.org/10.5209/HICS.59843
24. Salaverría, R., Buslón, N., López-Pan, F., León, B., López-Goñi, I., Erviti, M.-C.: Desinformación en tiempos de pandemia: tipología de los bulos sobre la Covid-19. El Profesional Información **29**(3), (2020). https://doi.org/10.3145/epi.2020.may.15
25. Sánchez Aranda, J.J.: Evolución de la prensa en los principales países occidentales. In: Barrera, C. (ed.) Historia del Periodismo Universal, pp. 77–117. Ariel, Barcelona (2004)
26. Schia, N., Gjesvik, L.: Hacking democracy: managing influence campaigns and disinformation in the digital age. J. Cyber Policy **5**(3), 413–428 (2020). https://doi.org/10.1080/23738871.2020.1820060
27. Skovsgaard, M., Andersen, K.: Conceptualizing news avoidance: towards a shared understanding of different causes and potential solutions. Journalism Stud. **21**(4), 459–476 (2020). https://doi.org/10.1080/1461670X.2019.1686410
28. Solbès, J.: Media business: Argent, Idéologie, Désinformation. Messidor, Paris (1988)
29. Spohr, D.: Fake news and ideological polarization: filter bubbles and selective exposure on social media. Bus. Inf. Rev. **34**(3), 150–160 (2017). https://doi.org/10.1177/0266382117722446
30. Swire, B., Berinsky, A.J., Lewandowsky, S., Ecker, U.K.H.: Processing political misinformation: comprehending the Trump phenomenon. Roy. Soc. Open Sci. **4**(3), (2017). https://doi.org/10.1098/rsos.160802
31. Tandoc, E.C., Jr., Lim, Z.W., Ling, R.: Defining 'fake news.' Digit. Journal. **6**(2), 137–153 (2018). https://doi.org/10.1080/21670811.2017.1360143
32. The Washington Post: Trump's false or misleading claims total 30,573 over 4 years. Available at https://wapo.st/3bUuqzX (2021). Accessed 25 Jan 2021
33. Volkoff, V.: La Désinformation, Arme de Guerre. Julliard, Paris (1986)
34. Waisbord, S.: Why populism is troubling for democratic communication. Commun. Cult. Critique **11**(1), 21–34 (2018). https://doi.org/10.1093/ccc/tcx005
35. Wardle, C.: Fake news: it's complicated. First Draft. Available at https://bit.ly/3bR7sJR (2017). Accessed 12 Jan 2021
36. Wardle, C., Derakhshan, H.: Information disorder: toward an interdisciplinary framework for research and policy making. Council of Europe. Available at https://bit.ly/3cxipzm (2017). Accessed 12 Jan 2021

37. Wason, P.C., Johnson-Laird, P.N.: Psychology of Reasoning: Structure and Content. Harvard University Press, Cambridge (1972)
38. Zannettou, S., Sirivianos, M., Blackburn, J., Kourtellis, N.: The web of false information: rumors, fake news, hoaxes, clickbait, and various other shenanigans. J. Data Inf. Qual. **11**(3), (2019). https://doi.org/10.1145/3309699

Ramón Salaverría Full Professor of Journalism at the School of Communication, University of Navarra (Pamplona), where he heads Digital Unav - Center for Internet Studies and Digital Life. Author of over 250 scholarly publications, his research focuses on digital journalism and disinformation, with many international comparative studies. His most recent book is *Journalism, Data and Technology in Latin America* (published by Palgrave Macmillan in 2021), a comprehensive analysis of the innovation trends of digital media in Latin America.

Bienvenido León Associate Professor of Science Journalism and Television Production, is Head of the Department of Journalism and coordinator of the Research Group on Science Communication at the University of Navarra (Pamplona). He has published over 90 peer-reviewed articles and 23 books as author or editor, including *Communicating Science and Technology Though Online Video* (published by Routledge in 2018). He is also the director of LabMeCrazy Science Film Festival.

Big Data and Disinformation: Algorithm Mapping for Fact Checking and Artificial Intelligence

David García-Marín, Carlos Elías, and Xosé Soengas-Pérez

Abstract This chapter looks at the intricate relationship between journalism and mathematics (big data, algorithms, data mining) as a tool to verify information and fight disinformation. The first section focuses on the relationship current students have with techniques such as big data or artificial intelligence and their ideas on applying them to their profession. The second section maps which universities and researchers in the world are looking into that relationship, how they approach it and where they publish their results. A relevant result is the presence of engineers in those studies, as well as Asian-origin researchers. Finally, we present results that show the increasingly close relationship between different disciplines such as computational linguistics, artificial intelligence and big data to solve the challenge of fake news and disinformation.

1 Introduction

Big data, that is the gathering of information from millions of data sources, and disinformation are closely related because they feed each other: it is the basis of information campaigns in social media, both with real or fake data. It helps find narratives and makes them more appealing to hyper-segmented audiences. It is used both in election campaigns—since 2012 with Obama, till the Trump election in 2016 and the Cambridge Analytica scandal—and in the dissemination of fake news (for instance, through Russian bots). But there is also a positive side to it: companies such as Google and official institutions such as the European Union are using artificial

D. García-Marín
Universidad Rey Juan Carlos, Madrid, Spain
e-mail: david.garciam@urjc.es

C. Elías (✉)
Carlos III University of Madrid, Madrid, Spain
e-mail: carlos.elias@uc3m.es

X. Soengas-Pérez
Universidade de Santiago de Compostela, Santiago de Compostela, Spain
e-mail: jose.soengas@usc.es

J. Vázquez-Herrero et al. (eds.), *Total Journalism*, Studies in Big Data 97,
https://doi.org/10.1007/978-3-030-88028-6_10

intelligence to score websites and news content and establish the veracity of the data. That is, using big data, data mining and natural language processing as fact-checking techniques.

If there is something that defines communication in the twenty-first century compared to the 20th is the influence of disruptive technologies such as big data, artificial intelligence and the huge influence of social media. This has to be quantified. Another relevant phenomenon is that of disinformation, defined as a deliberate strategy to disseminate untruth in order to modify behaviors and opinions. This chapter will focus on how those elements are interconnected and aims to demonstrate that in order to fight disinformation and improve journalism you don't need just journalists, you need their good intentions and professionalism. They must have an increasing knowledge of engineering and, at the same time, engineers must know more about communication. It is the symbiosis of both professions that can better fight disinformation.

Big data is defined as the possibility of obtaining information or behavior patterns based on huge amounts of data generated by digital technologies. Journalistic knowledge is obtained from this, as those patterns and trends can have relevant news that the power may want to conceal [10]. Advances in computation not only allow for the accumulation of trillions of data, but also its management and the questioning of that data in order to have some answers. Artificial intelligence is also related to this. Thus, those technologies are used to obtain serious journalistic information, but also to verify the growing number of fake news.

In order to see how those technologies can improve journalism, we will start initially with an exploratory section about how young students of communication value those technologies. This is very relevant because those tools belong to an academic discipline very different to typical journalism students, who normally come from the humanities. Those technologies are deeply related with mathematics and engineering. After looking at this issue and despite the differences on those cultural fields—humanities and sciences—we will see that there is a desire to unify them in the interest of truthful information and journalism. Then we will look into mapping what has been done in the world in this regard. Thus, we have carried out a systematic review of the best experiences and research on those synergies. An important topic we will see later on in the chapter is the studies coming from Asia, where school curricula do not separate sciences from humanities as western schools do [11]. This helps those professionals learn to manage tools to create stories through big data, algorithms and artificial intelligence, but also to do the opposite process: once a piece of information is received, they apply the technology described to check its veracity.

2 Humanities versus Sciences: Journalists, Big Data and Artificial Intelligence

There are numerous studies allowing us to see the views information professionals have on the implementation of artificial intelligence instruments in editorial rooms and in the journalist workflow. In general, those studies highlight that professionals do not believe robots will fully replace editors, but that they will collaborate with them, especially doing mechanical tasks. This is good news as they happily accept working with those technologies. Similarly, they highlight that the quality of automated information has important gaps, mainly lack of contrast, of interpretation, the non-existence of humanity and sensitivity and bad drafting [5]. They therefore adopt a narrative that rejects machines replacing humans, but welcome human–machine cooperation in the field of information production.

Nevertheless, research on opinions regarding the specific use of those tools for fact-checking is nowadays still scarce. In order to solve this deficit and with the aim of making a first approach to this subject of study, we interviewed young working journalists recently graduated from 21 Spanish universities. Using a questionnaire, we asked them to share their views on the application of artificial intelligence and big data in the fight against fake news. Why was the research focused on young journalists? This group is strategic when analyzing social representation on disinformation as they are not only consumers of traditional media and information in digital environments, but mainly because they are the media professionals who will have to live with this problem professionally in the short, mid and long term. Moreover, most of their university studies took place during the dawn of this phenomenon, from 2016, and so their perspective is somehow halfway between the academic world they were part of in the last few years and the professional world. All in all, they are the first generation of journalists that access editorial departments in a media ecosystem highly contaminated by disinformation and fake news.

The questionnaire used the technique of semantic differential that allows us to extract the meaning certain concepts have for the respondents by breaking them down into bipolar attributes that participants can value on a scale [8] similar to a Likert scale. More specifically, respondents had to give a value from 1 to 5 to artificial intelligence according to several variables: (1) efficacy, (2) simplicity, (3) economic cost, (4) suitability to all types of situations, contexts and topics, (5) innovative nature, (6) response speed, (7) neutrality and (8) experience required to manage it. Participants did the same test for data journalism and fact-checking in order to compare the results obtained for these three instruments against disinformation. Before doing so, a reliability test was done on the questionnaire (Cronbach's test) and the result value was 0.703. Although that value is not extremely high, we can consider it sufficient to test the internal coherence of the tool [14], especially so when executing exploration studies such as ours.

The results of this research show that for young journalists the main values of artificial intelligence in the fight against disinformation are (1) its innovative nature and (2) its efficacy (in fact, the three instruments—data journalism, fact-checking

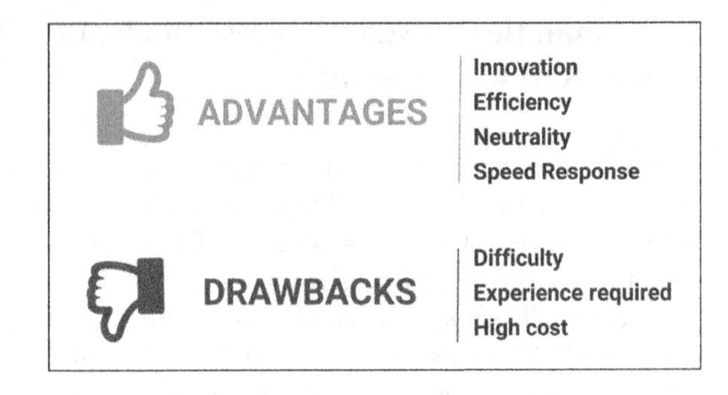

Fig. 1 Advantages and drawbacks of artificial intelligence as a tool against disinformation, according to Spanish young journalists. Own elaboration

and artificial intelligence—were considered really effective, with data journalism being the best valued instrument in this variable). The third best valued feature was its neutrality (Fig. 1). The latter is especially relevant because, when comparing these results with those obtained for fact-checking and data journalism, artificial intelligence is considered to be the least biased approach. Contrary to this, the worst results for artificial intelligence are down to the difficulties to learn the new necessary skills to master it and the high experience required to apply it. The latter is important because it shows that communication studies must add technology subjects and also that engineering studies must include media communication topics.

The field work for this research was carried out in 2020 and included an inferential statistics study to establish significant differences in the perception of artificial intelligence, data journalism and fact-checking as effective instruments against disinformation. The multiple comparisons between variables and tools showed significant differences in four out of eight variants analyzed: economic cost, experience required, response speed and innovation. Artificial intelligence is considered significantly more costly in financial terms than fact-checking and data journalism. Participants in this study stated that more expertise is needed when designing artificial intelligence tools than when applying fact-checking strategies, whilst data journalism is considered significantly slower in its response to disinformation than fact-checking and artificial intelligence. Finally, data journalism is perceived as a less innovative practice in the fight against these problems than the other two tools.

These results allow us to conclude that, at least amongst the new generations of journalists, there is a high level of confidence in algorithm technology as a weapon against untruth, especially due to its immediacy and neutrality. In this sense, it is interesting the viewpoint that artificial intelligence (the most technological tool of the three) is a tool with less bias than data journalism and fact-checking. On the other hand, this does not conceal a slightly techno-utopic tendency to forget that the design of any technological solution always has the legacy of its creators' values. That is why the algorithm's transparency is a key aspect that will foster confidence on those solutions. It is crucial that those machine-learning techniques must be easy to interpret and explain. Algorithms must be able to justify their 'decisions', making public the complete set of data, regulations and mathematical formulas used to reach the final result. In this sense, to understand the foundations of artificial intelligence as a solution to the fake news problem—and therefore, to give it credibility—, we need to analyze in depth what is the logic that supports the creation of those tools and how the most advanced research centers of the world are investigating the effectiveness of those solutions.

3 Big Data Against Disinformation: A Journalism and Engineering Curriculum

What is the logic that explains the development of algorithmic solutions to fight disinformation? What has been researched exactly in this field? What are the most relevant centers, studies and approaches? To have more accurate knowledge on the research interests on artificial intelligence applied to this problem and to develop an initial mapping of the empirical knowledge obtained, a detailed review of scientific literature published between the years of 2016 and 2020 in the Web of Science, the main repository of technical-scientific publications, was performed. A total of 605 articles were selected after carrying out several searches for the terms 'misinformation', 'disinformation', 'fake news' and 'post-truth' using different Boolean operators.

Although this research comes from a total of 44 countries, most are concentrated in the Anglo-Saxon world (Fig. 2). The United States is the country with the greatest number of studies (45.0%), followed by far by the United Kingdom (10.6%) and Canada (5.6%). Some of the most common topics are those related to research on cognitive and ideological biases that partly explain the dissemination of disinformation (19.7%), and the consequences those fake news have in society (9.4%) (Table 1).

Nevertheless, the studies on solutions and/or detection of disinformation content are the most common category (23.8%). In this category, we include specific research on artificial intelligence and automated fact-checking (Fig. 3). They are recent studies, given that most of the papers, at least those that have greater impact, date from 2019 and 2020. A very striking fact is that the studies on solutions to improve

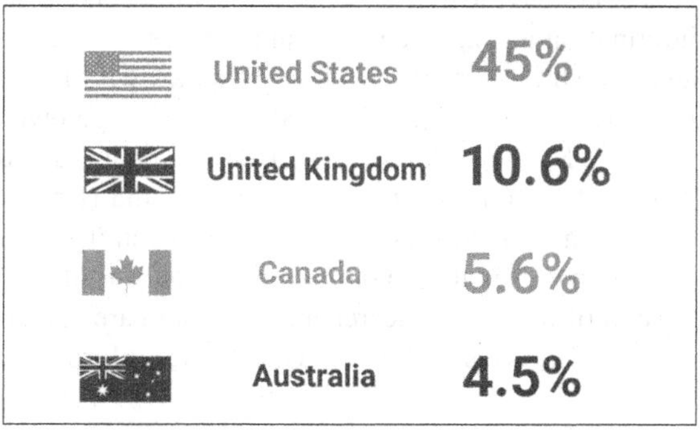

Fig. 2 Countries where most research on disinformation is carried out (2016–2020). Own elaboration

Table 1 Most investigated topics related to disinformation (2016–2020). Own elaboration

Topic	%
Solutions	23.8
Cognitive biases	19.7
Production of disinformation	13.9
Consequences/impact	9.4
Spread	7.8
Causes of disinformation	7.6

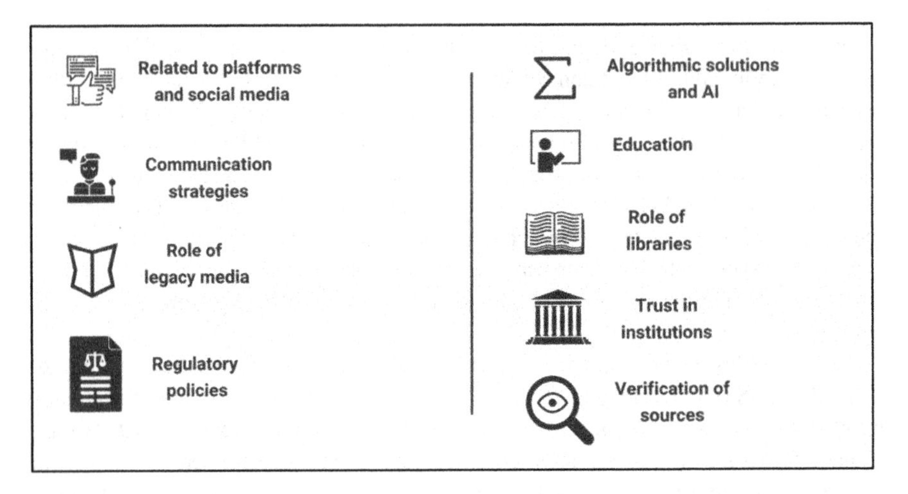

Fig. 3 Most investigated solutions against disinformation (2016–2020). Own elaboration

the identification of fake news are not published in typical communication journals but in journals on IT engineering, engineering in general, applied mathematics and IT systems management. In the field of communication, only journals on the technology of media have published studies of this type. The authors of those papers are not linked to the field of journalism or communication in general, but rather to departments related to IT engineering, computational sciences, machine learning, mathematics, research in complex networks and technology research institutes and information management. The universities those experts come from are mainly US and British universities. Of the 15 studies with greater impact, there is work done in the United States (Arizona, Stanford, Georgetown, Simon Fraser University and Carnegie Mellon), Canada (Brandon University) and the United Kingdom (Oxford). Two studies come from universities in India (Bennet University) and from South Korea (Chosun University). Despite the centralization of this research line in North American universities, a more detailed analysis of the authors shows relevant details. Several of the main researchers in the field are Asian talents that research from US centers. This is the case for Qian Li (Stanford), Kai Shu (Arizona), Qian Chen

(Brandon University, Canada) and Fatemeh Torabi Asr (Iranian researcher from the University Simon Fraser, alumni of Shiraz University, the old University of Pahlaví).

This data is interesting for several reasons. On the one hand, it shows that a multidimensional phenomenon such as disinformation needs to have approaches coming from different areas of knowledge, beyond the borders of communication. This argument is strengthened by the fact that the best-indexed scientific journals in the field of communication have hardly published results of this type until present, even though disinformation is a problem, in first instance, related to communication and information. Moreover, it shows the relevance of STEM knowledge (Science, Technology, Engineering and Mathematics) in eastern countries compared to the stagnation of European research centers, lagging behind in terms of production of this type of knowledge, with the exception of the main British universities. This highlights the excellent results of the educational policy of China and other Asian countries such as South Korea and Singapore of not separating sciences from humanities in higher secondary education [11], thus fostering having university students with subjects such as IT engineering, data mining and mathematics as well as history of literature, journalistic genres or corporate communication.

Lastly, it is clear how, when applying artificial intelligence to fight disinformation, US universities have been able to attract talent to develop innovative projects with advanced technology in order to respond to emerging social problems. A clear example is that of the University of Columbia, the best in the world for journalism and the one that awards the Pulitzer Awards and that has designed a double degree in Journalism and IT Engineering. This would be unthinkable in countries such as Spain, not only because of how students are separated in higher secondary education, but also due to the lack of relation between humanities and science schools in Spain.

4 Mapping of Algorithms Against Disinformation

What artificial intelligence applications have been empirically tested against disinformation? Which lines of development have been researched to date and what are the future possibilities? First of all, there are numerous studies focused on the creation of large databases and guidelines to prepare tools in order to implement future algorithmic developments against fake news (Fig. 4). One of the first challenges that had to be overcome initially on automatic fact-checking was the lack of data with which to train the models. Online repositories with large volumes of information started appearing in 2017 due to the growth in the number of fact-checking websites [20]. Right at the onset of those developments we find a database made of 106 *Politifact* checks, one of the main US fact-checkers [26]. Other repositories, such as EMERGENT, provided 300 false statements using Twitter and the Snopes verifier as sources [13].

Those initial data corpora, although useful, showed a serious problem: they were too limited in information to train machine-learning models. The quantitative leap came about in 2017 with the collection of a corpus of 12,800 checks coming from

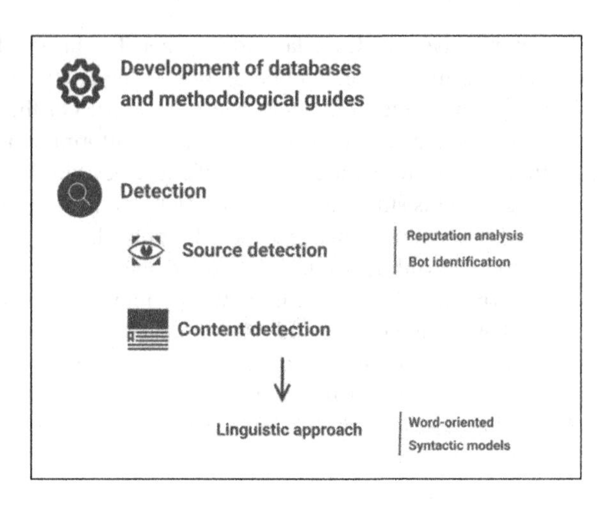

Fig. 4 Approaches to the study of artificial intelligence as a tool against disinformation. Own elaboration

Politifact [27]. Since then, the volume of databases has grown to reach figures close to 200,000 statements collected by Thorne and colleagues [24].

The creation of large-volume information repositories has not been the only great challenge. The quality of data provided is also key to design algorithmic solutions against falseness. Asr and Taboada [2] champion that the prior creation of databases must be made up of samples of false and real information on a balanced distribution of potential topics. In research done in 2019, they reviewed the databases available till date and presented a new repository, called MisInfoText. This tool showed a set of complete text available to artificial intelligence creators against fake news, as well as a set of tags of previously allocated veracity based on the manual assessment of the credibility of those articles. Moreover, they carried out a modelling experiment on topics to detect subjects that databases elude or mention briefly, as well as imbalance sources in the sets of data available till date, in order to guide future efforts.

We consider the study of Papadopoulou and others [21] in that same line. It includes a database of 380 videos generated by users and over 5000 shared and edited versions of those videos, as well as a set of checking experiments to be used for future comparisons. From a methodological approach, Lara-Navarra and colleagues [18] drafted a useful guide for the automation of autonomous algorithm-building processes to detect false information online, specifically in the field of health.

A second set of studies is related to the detection of misinformative sources and content in digital environments. Those methods have shown empirically their effectiveness in the field of fake news detection, especially when combining several of them from a hybrid perspective as well as presenting a promising potential to classify misinformative content in the social media [16]. Within this type of research, we can find two subtypes: (1) the analysis of the origin of disinformation, with special attention to the bots that operate in social media, and (2) the detection of misinformative content.

4.1 Ways to Verify Sources of Disinformation

The fight against disinformation based on the origin has been mainly done in two ways. The first one is based on a reputation analysis of the source of disinformation, where the algorithms represent numerically the level of credibility of the content producer, taking as reference several groups of data: keywords, sentences, topics or content created by the users/media to be analyzed, opinions or comments received and, even, the trust of the IP or the domain from which the information comes.

The second one is the detection of bots that have their field of action in the social media. Bots are robotic accounts "programmed to artificially amplify messages or trends" [22:148]. They are more active in Twitter, where it is estimated that between 9 and 15% of the accounts are automated [25]. Although they can be used for laudable goals, as it allows for the programming of conversations (chatbots) or the issuing of informative content, they are also being used to massively disseminate political propaganda, false information and hate messages. They have a permanent and systematic nature, they imitate human traits and expressions, they produce in mass parasocial messages and actions (likes, retweets), they have a special tendency to produce content in controversial discussions or sensitive issues that polarize population and "they align with fake news, urban legends and conspiration theories" [22:149]. They allow for a single person to manage a high volume of automated accounts. They tend to be used in astroturfing pseudo-campaigns (actions in social media that simulate to be created by a large volume of users but that in fact are activated by multiple bots) or to hijack hashtags (an action that basically saturates a legitimate conversation with the production of malicious content to pervert the original sense of the conversation) [1].

The bots behind these operations are detectable with algorithmic tools easy to use at a user level. For instance, Botometer is a simple tool available online[1] that, just by introducing the name of the suspicious Twitter account, establishes the possibilities (in a range from 0 to 5) of the account being a bot. Similar solution such as Botcheck.me have a binary logic: the results establish if the account analysis shows or not automated content production patterns. On the other hand, the solutions tested by Beskow and Carley [3], who propose a sort of 'multimodel toolbox' to detect bots with several levels of data granularity, are of greater complexity.

4.2 Computational Linguistics Against Disinformation

Misinformative content detection algorithms use, above all, linguistic models to classify information. Although most fake news are expressed in order to be realistic, that is, to seem truthful, in many cases malicious content creators leave traces of the untruthfulness of their statements, given that there are some aspects of language that are difficult to control, such as the frequency of words or the excessive use of terms

[1] Available at https://botometer.osome.iu.edu.

and expressions that provoke negative emotions [12]. The goal of this linguistic approach is to find predictive deception cues that can be found in the content of a message [7]. The simplest approach is to add and analyze the frequency of the words used to show signs of deceit. Given that this is the most affordable method to implement, it has a serious problem: it considers each word as an isolated unit, without taking into consideration the contextual references around them. Sometimes, those word-oriented approaches are not enough to identify false information. The relationship between words in a text needs to be established. The key is to stop looking into the words and start analyzing the sentences. Syntax models use that approach; sentences are transformed into a set of rules, a kind of tree that draws the syntactic structure of a text. Based on that information, the algorithm establishes the confidence of the text according to its internal logic. Syntactic patterns can be used to distinguish arguments based on emotions from those based on facts through the association of argumentation patterns learnt from real samples of truthful and untruthful information.

Those algorithms have evolved incredibly over the last few years and have reached some effectiveness in detecting untruth. For example, Li and colleagues [19] have developed a detection system based on a convolutional neural network with several levels able to calculate the weight of specific words considered 'sensitive' for each piece of news. Those researchers carried out several fake news detection experiments about cultural issues to compare this system with several other previously existing models. The results showed their model's reliability. The proposal by Kwon and others [17] is also of great interest. Their study was divided into two parts. First, they demonstrated through statistical processes that the linguistic features of disinformation in Twitter act as a good indicator of fake news in an early stage. Based on those results, the second part of their study proposed a new fake content classification algorithm that manages to be highly accurate both on early stages of dissemination but also in the long term. Their findings not only provide new ideas to explain the theories of online fake news dissemination, but also to identify them early on.

The detection of fake images is also considered using methods based on artificial intelligence. Bondi and colleagues [4] created an algorithm able to recognize if an image has been altered and where did the manipulation take place, using the traces left by some camera models on images. Their logic is very simple: all image pixels that have not been touched must appear as if they had been recorded by the same device. If the content had images coming from other cameras, those camera's traces could be found. That same approach was used by other artificial intelligence experts such as Zhang and Ni [29], whose experiments replicate the model previously described.

5 Conclusion

This chapter has looked into the convergence needed between journalism and computational sciences in order to fight disinformation, through the development of algorithm models and artificial intelligence. After exploring the views of young communication professionals on the complex intertwining of algorithmic solutions available to date, we moved on to the drafting of a map of the most important research being done in this field.

Different approaches to automated fact-checking, coming from the largest technological hubs of the planet (Asian/North-American axis) in the field of IT engineering, take on a great variety of processes. First of all, studies for the creation of large databases are essential and key for the design of those models. From a detection perspective, artificial intelligence allows us to establish the credibility of information sources based on their reputation analysis, at the same time as it offers a strong solution to identify false profiles (bots), able to develop campaigns against the stability of certain countries and to create false trends that support specific ideologies and political figures. The other large battle front is the detection of disinformation content through the use of computational linguistics (based on semantic and syntax models) and of non-linguistic methods to detect potential manipulations present in images (videos or photographs).

The developments described in this chapter have been implemented in the last few years by large technological platforms in the fight against misinformative content. Facebook, as well as hiring an army made up of 5000 employees in charge of identifying offensive messages, has also invested in artificial intelligence and machine learning tools to detect fake, untruthful content, especially related to computational propaganda [15, 28], in order to respond to the crisis generated by Cambridge Analytica. One of those algorithm engines included in the platform's operation is Deeptext, an invention based on computational linguistics that reads and understands text content coming from several thousands of posts per second in more than 20 different languages [6].

Although to a lesser extent and not as widely adopted, blockchain technology is also becoming a powerful player in the fight against digital disinformation. Considered a disruptive technology [23] in multiple fields (education, finances, politics) including journalism, this technology opens up promising paths in the fight against disinformation. The total traceability made possible by this technology and the impossibility to modify content created via block chains incredibly helps the implementation of checks [9] and offers the security that there has been no manipulation in the process of information dissemination.

There is no doubt that, over the next few years, many analyses on the potential of these developments to fight fake news will be made. Just like the latest technologies are used by disinformation producers to deceive, manipulate and attack democratic systems, they can be adopted as defense mechanisms to identify and create counter-narratives that can foster the inherent journalistic and scientific ideals: the quest for truth and making it public.

Funding This research is also part of the Jean Monnet Chair *EU, Disinformation and Fake News*, supported by the Erasmus + Program of the European Commission.

References

1. Aparici, R., García-Marín, D.: La posverdad. Una Cartografía de los Medios, las redes y la Política. Gedisa, Barcelona (2019)
2. Asr, F.T., Taboada, M.: Big data and quality data for fake news and misinformation detection. Big Data Soc. **6**(1), (2019). https://doi.org/10.1177/2053951719843310
3. Beskow, D.M., Carley, K.M.: It's all in a name: Detecting labeling bots by their name. Comput. Math. Organ. Theory **25**, 24–35 (2019). https://doi.org/10.1007/s10588-018-09290-1
4. Bondi, L., Lameri, S., Güera, D., Bestagini, P., Delp, E.J., Tubaro, S.: Tampering detection and localization through clustering of camera-based CNN features. In: IEEE Conference on Computer Vision and Pattern Recognition Workshops, 21–26 July, Honolulu (2017). https://doi.org/10.1109/CVPRW.2017.232
5. Calvo-Rubio, L.M., Ufarte-Ruiz, M.J.: Percepción de docentes universitarios, estudiantes, responsables de innovación y periodistas sobre el uso de inteligencia artificial en periodismo. El Profesional Información **29**(1), (2020). https://doi.org/10.3145/epi.2020.ene.09
6. Choy, M., Chong, M.: Seeing through misinformation: a framework for identifying fake online news. Available at https://arxiv.org/abs/1804.03508 (2018). Accessed 14 Jan 2021
7. Conroy, N.K., Rubin, V.L., Chen, Y.: Automatic deception detection: methods for finding fake news. Proc. Assoc. Inf. Sci. Technol. **52**(1), 1–4 (2016)
8. Corbetta, P.: Metodologías y técnicas de investigación social. McGraw Hill/Interamericana de España, Madrid (2007)
9. Dickson, B.: How blockchain helps fight fake news and filter bubbles. The Next Web. Available at https://bit.ly/2OLsNvq (2017). Accessed 20 Jan 2021
10. Elías, C.: El selfie de Galileo. Software social, político e intelectual del siglo XXI. Península, Barcelona (2015)
11. Elías, C.: Science on the Ropes. Decline of Scientific Culture in the Era of Fake News. Springer, Cham (2019)
12. Feng, V.W., Hirst, G.: Detecting deceptive opinions with profile compatibility. In: Mitkov, R., Park, J.C. (eds.) Proceedings of the Sixth International Joint Conference on Natural Language Processing, pp. 338–346. AFNLP, Nagoya (2013)
13. Ferreira, W., Vlachos, A.: Emergent: a novel data-set for stance classification. In: Knight, K., Nenkova, A., Rambow, O. (eds.) Proceedings of the 2016 Conference of the North American Chapter of the Association for Computational Linguistics: Human Language Technologies, pp. 1163–1168. ACL, San Diego (2016)
14. Frías-Navarro, D.: Apuntes de consistencia interna de las puntuaciones de un instrumento de medida. Universidad de Valencia. Available at https://bit.ly/3l7yVKs (2019). Accessed 20 Jan 2021
15. Iosifidis, P., Nicoli, N.: Digital Democracy, Social Media and Disinformation. Routledge, New York (2020)
16. Kaliyar, R., Goswami, A., Narang, P., Sinha, S.: FNDNet—A deep convolutional neural network for fake news detection. Cogn. Syst. Res. **61**, 32–44 (2020). https://doi.org/10.1016/j.cogsys.2019.12.005
17. Kwon, S., Cha, M., Jung, K.: Rumor detection over varying time windows. PLoS ONE **12**(1), (2017). https://doi.org/10.1371/journal.pone.0168344
18. Lara-Navarra, P., Falciani, H., Sánchez-Pérez, E., Ferrer-Sapena, A.: Information management in healthcare and environment: towards an automatic system for fake news detection. Int. J. Environ. Res. Public Health **17**(3), 1066 (2020). https://doi.org/10.3390/ijerph17031066

19. Li, Q., Hu, Q., Lu, Y., Yang, Y., Cheng, J.: Multi-level word features based on CNN for fake news detection in cultural communication. Pers. Ubiquit. Comput. **24**, 259–272 (2020). https://doi.org/10.1007/s00779-019-01289-y
20. Lowrey, W.: The emergence and development of news fact checking sites: institutional logics and population ecology. Journalism Stud. **18**(3), 376–394 (2019). https://doi.org/10.1080/1461670X.2015.1052537
21. Papadopoulou, O., Zampoglou, M., Papadopoulos, S., Kompatsiaris, I.: A corpus of debunked and verified user-generated videos. Online Inf. Rev. **43**(1), 72–88 (2019). https://doi.org/10.1108/OIR-03-2018-0101
22. Redondo, M.: Verificaci6n digital para periodistas. Manual contra bulos y desinformaci6n internacional. Editorial UOC, Barcelona (2018)
23. Tapscott, D., Tapscott, A.: Blockchain Revolution: How the Technology Behind Bitcoin is Changing Money, Business and the World. Random House, New York (2016)
24. Thorne, J., Vlachos, A., Christodoulopoulos, C., Mittal, A.: FEVER: A large-scale dataset for fact extraction and verification. In: Walker, M., Ji, H., Stent, A. (eds.) Proceedings of the 2018 Conference of the North American Chapter of the Association for Computational Linguistics: Human Language Technologies, vol. 1, pp. 809–819. ACL, New Orleans (2018)
25. Varol, O., Ferrara, E., Clayton, A.D., Menczer, F., Flammini, A.: Online human-bot interactions: Detection, estimation, and characterization. Available at https://arxiv.org/abs/1703.03107 (2017). Accessed 21 Jan 2021
26. Vlachos, A., Riedel, S.: Fact checking: task definition and dataset construction. In: Danescu-Niculescu-Mizil, C., Eisenstein, J., McKeown, K., Smith, N.A. (eds) Proceedings of the ACL 2014 Workshop on Language Technologies and Computational Social Science, pp. 18–22. ACL, Baltimore (2014)
27. Wang, W.Y.: "Liar, liar pants on fire": a new benchmark dataset for fake news detection. In: Barzilay, R., Kan, M.-Y. (eds.) Proceedings of the 55th Annual Meeting of the Association for Computational Linguistics, pp. 422–426. ACL, Vancouver (2017)
28. Wooley, S.: The Reality Game: How the Next Wave of Technology Will Break the Truth and What We Can Do About It. Endevour, London (2020)
29. Zhang, R., Ni, J.: A dense U-net with cross-layer intersection for detection and localization of image forgery. In: IEEE International Conference on Acoustics, Speech and Signal Processing, pp. 2982–2986. IEEE, Barcelona (2020)

David García-Marín Ph.D. in Sociology, professor and researcher at Rey Juan Carlos University (Madrid), where he teaches new technologies and information society in the Bachelor's Degree in Journalism. He is a guest lecturer in various Master's Degree programmes at the National University of Distance Education (UNED, Spain), where he has also chaired several courses on disinformation. His latest book is *La posverdad. Una cartografía de los medios, las redes y la política* (published by Gedisa in 2019).

Carlos Elías Ph.D. in Social Sciences. He is a former chemical scientist and journalist (in EFE News Agency) and now is full professor of Journalism at the Carlos III University of Madrid, where he won The Excellence Prize for Young Researchers in 2012. In 2019 he obtained one of the Jean Monnet Chairs (focused on "Disinformation and Fake News"). He has been visiting scholar at the London School of Economics and at Harvard University. His latest book is *Science on the Ropes* (published by Springer, 2019).

Xosé Soengas-Pérez Full Professor of Audiovisual Communication and Head of the Department of Communication Sciences at Universidade de Santiago de Compostela (USC). He is also a member of Novos Medios research group (USC). His research is focused on the analysis of information, specially in radio and television news content.

From Misinformation to Trust: Information Habits and Perceptions About COVID-19 Vaccines

Carmen Costa-Sánchez and Carmen Peñafiel-Saiz

Abstract The arrival in Spain of the vaccine against SARS-CoV-2 has presented numerous challenges in terms of citizen communication and journalism. Uncertainty about the virus and illness predominate. There is an infodemic that is adding to the spread of unverified information which can also pose a danger to health. Vaccines are bursting onto this complex stage accompanied by competitive maneuvers from pharmaceutical companies that do not help to increase trust. Spanish public opinion in this respect is evolving and maintains an attitude dependent on the generated media context. Some recommendations of interest are compiled in order to foster trust in the vaccines and cultivate at all times a link of understanding with public opinion.

1 Introduction: Emergency Communication in the Face of SARS-CoV-2

The year 2020 will go down in history as the year of a world health crisis with extremely adverse effects on the population caused by COVID-19, an infectious coronavirus disease discovered at the end of 2019, representing one of the most important world crises in recent history, causing a great impact not only from a health care viewpoint, but also social and economic. Since then, the population has experienced frustration, anguish, home confinement or geographical perimeter lockdowns, social distancing, together with serious effects on economic prospects.

This pandemic has put health systems all around the world under great pressure. The rapid rise in demand facing health centers and professionals is posing a constant threat in the two waves we have already had and has overloaded some health systems, impeding their efficiency. "The best defense against any outbreak is a solid

C. Costa-Sánchez (✉)
University of A Coruña, A Coruña, Spain
e-mail: carmen.costa@udc.es

C. Peñafiel-Saiz
University of Basque Country, Bilbao, Spain
e-mail: carmen.penafiel@ehu.eus

J. Vázquez-Herrero et al. (eds.), *Total Journalism*, Studies in Big Data 97,
https://doi.org/10.1007/978-3-030-88028-6_11

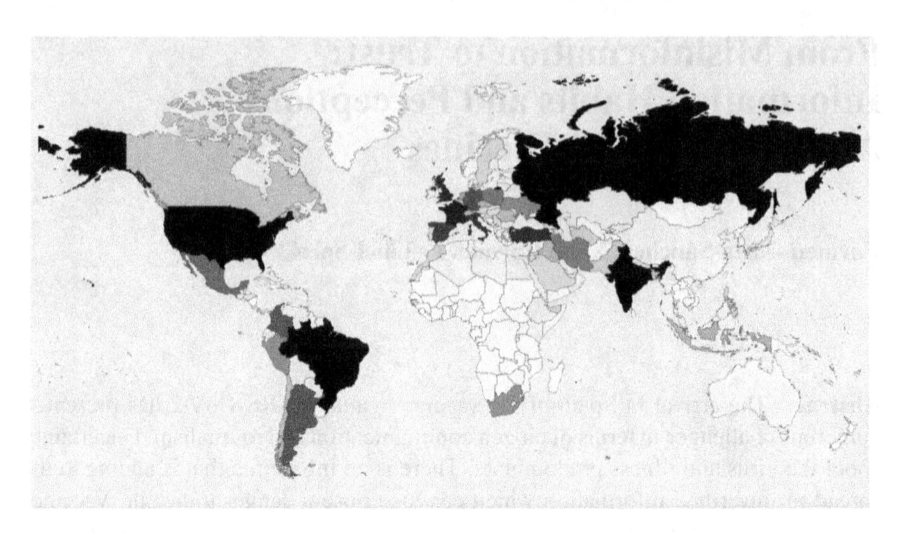

Fig. 1 Number of infected people throughout the world (the darker the color, the larger the number). *Source* RTVE.ES with data updated at 5:50 h in the Spanish peninsula on 27th December 2020

health system—highlights the Director General of the World Health Organization (WHO), Tedros Adhanom Ghebreyesus. COVID-19 is highlighting the fragility of many health systems and services all over the world, and is forcing countries to take difficult decisions about the best way to meet the needs of their citizens".[1]

In December 2020, the pandemic registered around 73 million infected people, over 41 million cured and around 1,700,000 dead throughout the world (Fig. 1).

The SARS-CoV-2 coronavirus has already produced 1,854,951 cases in Spain and 49,824 dead, according to the Ministry of Health update on 24th December 2020. The population has increasingly demanded information about the pandemic in order to try to control the situation. "It is a tool that can reduce uncertainty and anxiety, or, on the contrary, increase panic and chaos" [3]. Information about SARS-CoV-2 has developed within what we call 'Crisis Communication' or 'Emergency Communication'.

Communication in a health crisis of such magnitude forms part of crisis management and its aim is to minimize the impact of negative consequences with a communication plan aimed at the general public. In a global world where communication is almost instantaneous, communicative strategies and tactics that should be applied must be present in all an organization's activities and processes, with a view to maximizing the effectiveness of both the institutional and social response.

For their part, communication professionals must possess transparent and accurate information about the pandemic in order to use it appropriately, thus fulfilling their educational and informative function. Every message should be as realistic as possible, with updated information, which demands frequent, rapid and transparent communication with society, thus generating trust.

[1] WHO [27]. Available at https://bit.ly/3ljV0W8.

In this sense, at the beginning of the pandemic the WHO set up a website containing guidance for the public and updated information. It has published daily reports about the situation since 21st January 2020, it also updates information on Twitter,[2] issues press releases and holds press conferences periodically, and offers information about the WHO response in different countries [19].

If the first half of 2020 was characterized by actions in the face of the disease and the health crisis, the second half has been marked by the vaccine war. The WHO has had to update its guidelines in order to balance the demands of the direct response to COVID-19 with the need to continue providing essential health services and reduce the risk of the system collapsing. Therefore, different countries have to determine which essential services they prioritize in their effort to maintain the continuity of service provision and introduce strategic changes to guarantee that increasingly limited resources provide the maximum benefit to the population. Among the essential services that countries must offer is the coronavirus vaccine and the rollout of different treatments to accelerate the recovery of the sick (steroids—dexamethasone, Remdesivir, convalescent plasma, ACE and BRAS inhibitors...), contact and asymptomatic case tracking, distancing and protection measures [20:684]. Thus, governments, scientists and pharmaceutical companies have become involved in the race for the search for a specific vaccine that can put an end to an unsustainable situation. In relation to this, it is important to consult the WHO's *'Solidarity' clinical trial about COVID-19 treatments* [26].

There are currently as many as 261 vaccine projects at different points in the research or trial phase. The most advanced of these vaccine production projects are those of Pfizer, Moderna, Sputnik V and Oxford-AstraZeneca, which are being used to vaccinate the population. This fact has generated great hopes in the move towards controlling the coronavirus pandemic (although in December 2020 there was already talk of a new strain emerging in the UK and other countries).

In December 2020 the United Kingdom and Russia signed a co-operation agreement between scientists from both countries in the joint fight against the coronavirus and they will test a combination of the Oxford-AstraZeneca vaccine and the Russian Sputnik V with a view to improving the immune response.

At the same time, both Pfizer and Moderna are already testing their coronavirus vaccines against the new strain. Moderna has announced additional tests to confirm the expectation that the vaccine already in existence is also effective against the new strain. In this sense, Pfizer has given assurance that they are "generating data" about blood samples from people immunized with their vaccine "to see whether they are able to neutralize the new strain".[3]

The year 2021 has begun almost everywhere in the world with a majority vaccination of the population in addition to continuing the advancement of the creation and dissemination of scientific knowledge. The Spanish government expects to vaccinate 2.5 million people in the initial stage of the COVID-19 vaccination plan that runs

---

[2] See the Twitter account of Dr. Tedros Adhanom Ghebreyesus, director of the WHO (https://twitter.com/DrTedros) and the account of the WHO itself (https://twitter.com/who).

[3] Available at https://bit.ly/3bS1Q2k.

from January to March, with the second stage being from April to June and the third from that date forward. This final phase will involve the start of the vaccination of the general population.

The priority designated to the different groups in Spain is based on four types of danger: risk of exposure to the disease, risk of death, risk of socio-economic impact and the transmission of the coronavirus. Taking into consideration these four types, the Government has come up with a vaccination campaign with 18 priority groups on a scale from high to low vulnerability. The aim is to immunize the elderly and care home personnel, health workers and the disabled during the first phase.

1.1 Pandemic and Infodemic

The *infodemic* or the *information pandemic* involves generating an excess of inexact information, manipulative or falsified that refers to a specific case. On this occasion, the WHO has applied this concept to the informative treatment given to the coronavirus SARS-CoV-2. This 'information pandemic' can generate confusion and force decisions taken more to calm the population rather than for their effectiveness.

We live in a society in which misinformation and an over-abundance of information prevail, situations that complicate the codification of information [2]. One of the greatest obstacles is fake news. For Waisbord [25] this is malicious information created to influence and destabilize society and institutions, and to generate uncertainty and anxiety among the general public. Coinciding with Casero [3] and Benett and Livingston [2]:

> Misinformation has its origins in the loss of credibility in the traditional media, something which aids its extension and boosts its effects on credulous citizens. Furthermore, this phenomenon is associated with the growth in alternative information sources linked to populism and the far right, that seek geopolitical objectives and interests via the generation of chaos and confusion through information [3:3].

According to Divina Frau-Meigs, misinformation attempts to deliberately deceive people and is designed to be difficult to detect. This poses two challenges to democratic societies: control of the impact of misinformation that leads to a polarization of opinions and a reduction in public debate; the promotion of informational resilience in the population beyond the source verification practiced by a limited number of journalists.[4]

The ethical media attempts to guarantee the verification of information by contrasting sources and by the traceability of the news elaboration process. Whilst not a majority practice, there are more and more media entities developing innovative verification tools and which are active members of organizations such as: Cross-Check, First Draft, Comprova and International Fact-Checking Network (IFCN). Fact checking methods are effective in limiting erroneous perceptions and have become

[4] In an intervention in the framework of the VII International AE-IC Congress that took place in Valencia between November 28 and 30, 2020.

one of the most relevant activities in the journalistic profession over the last few years due to the ease of spreading the news via social media [20:699]. Verified information lies at the heart of the work of any professional that chooses to put honesty first.

Social media also has a great interest in misinformation, since it has become part of its business model to a degree, and it deliberately plays with rumor and 'clickbait' in order to create tendencies like 'trending now'.

Fake news, erroneous or malicious, can spread through information highways at supersonic speed nowadays. A research team from MIT (Massachusetts Institute of Technology), led by David [11], has scientifically confirmed in *Science* magazine that fake news spreads much more quickly than authentic news, with people not bots being mainly responsible for the dissemination of misleading information. The MIT experts, Sinan Aral, Soroush Vosoughi and Deb Roy, analyzed 126,000 stories broadcast on Twitter between 2006 and 2017, with over 4.5 million tweets among some 3 million people. The team used assessments from six independent data verification organizations to classify the authenticity or falseness of these stories, estimating that there are some 48 million bots on Twitter and 60 million on Facebook (bot: programs that automatically replicate tweets).

Furthermore, it has been shown that, whilst the truth is rarely spread to more than 1000 people, just 1% of the most viral fake news is routinely shared among 1000 and 100,000 people:

> There is worldwide concern over false news and the possibility that it can influence political, economic, and social well-being. To understand how false news spreads, Vosoughi et al. [24] used a data set of rumor cascades on Twitter from 2006 to 2017. About 126,000 rumors were spread by 3 million people. False news reached more people than the truth; the top 1% of false news cascades diffused to between 1000 and 100,000 people, whereas the truth rarely diffused to more than 1000 people. Falsehood also diffused faster than the truth. The degree of novelty and the emotional reactions of recipients may be responsible for the differences observed [24:1146].

For their part, David [11] point out that falsehood is spread further, quicker, more deeply and widely than truth in all information categories, and that the effects have been more notorious for false political news than for fake news about terrorism, natural disasters, science, urban myths or financial information.[5]

The rise of fake news highlights the erosion of long-standing institutional bulwarks against misinformation in the internet age. Concern over the problem is global. However, much remains unknown regarding the vulnerabilities of individuals, institutions and society to manipulations by malicious actors. A new system of safeguards is needed [11:1094].

Regarding the information about treatments and vaccines to fight against the pandemic, there have been two factors that have hindered the spread of scientific knowledge among society: fake news and anti-vaccine movements. Together they have formed a joint attack on the trustworthiness of the institutions and the vaccines themselves.

[5] Available at https://bit.ly/3rVX3Cg.

Some of the rules we can use to detect fake news are: (a) to go beyond the headline—no matter how striking or sensationalist a message may be, it might not be real; (b) we must check the content of the news item, verify whether the narrated facts contain precise or ambiguous data, whether it is partial news or an opinion article, even if it is written in a sarcastic or humorous tone and, most importantly, whether the news is from a recognized and trustworthy source; (c) carry out an Internet search to see whether the official media or other sources are repeating the same message; (d) not automatically share without verifying or comparing with reliable sources, and even search for information about the authorship of the article, etc. [15].

The anti-vaccination or denier movements are made up of a sea of people who deny the existence of the coronavirus and the health crisis or attribute the pandemic to an attempt to manipulate the population. They promote conspiracy theories and smear campaigns and attack governments, scientists and health institutions. Above all they are movements that appear on social media and their organizers are followed on Twitter. Judy Mikovits is one of the protagonists of viral videos filled with untruths about the coronavirus, face masks, vaccines, etc. [20]. The same is true of the denier's association Medics for Truth in Spain: both their videos and those in which one of their spokespersons Natalia Prego appears, have had hundreds of thousands of views on YouTube and shares on Facebook. They claim that the coronavirus is nothing more than a flu, they are against lockdowns, and, what's more, they stand against the use of face coverings. In these videos the same false conspiracy theories are repeated, with hundreds of thousands of views.

What is happening in Spain is a reflection of what is taking place in many other countries (Brazil, USA, Germany, etc.). Both those in favor and those against the vaccine provoke great debate in the public health sector. Therefore, it is important to include the media debates and governmental decisions presented by experts in public health and medicine. Scientific journalists often quote different experts and consider that scientists are particularly credible and more reliable than other sources [16]. Professors Catalán-Matamoros and Peñafiel-Saiz [5] state that scientific journalists tend to give priority to institutional sources within the Government and business spheres, and that little attention is paid to citizens in the construction of the news.

2 Information Habits in Times of Pandemic: The Spanish Case

Spain is one of the countries in Europe hardest hit by COVID-19 and since the beginning of the state of alarm and during the situation of emergency, the population has radically changed its habits and daily routines due to home confinement (decreed by the Government), social distancing and when faced with the informative uncertainty about the pandemic.

Before we come to 2020, let's look at some of the data from 2019 to see the changes in Spanish consumers, who opt for a complementarity between the traditional and digital media: they increasingly choose the Internet over television; the popularity of mobile devices and social media has led to an increasingly more fragmented and multi-screened consumption of the media; OTT platforms have increased over 40% in one year [1].

According to data collated by Statista [23], the average time spent on consuming different media in Spain in 2019 indicated that television was the most consumed medium with 212 min, followed by the Internet, with approximately 161 min per day. Radio came in third with 97 min. Some 7.8 min was spent on daily papers and the remaining media received less than a minute a day: weekly magazines (0.8), cinema (0.8), monthly magazines (0.6) and supplements (0.4).[6]

In times of pandemic, according to a study by Montaña and colleagues [17:155] and the company Havas Media that surveyed the Spanish population (N = 1.500 participants) between March 13 and March 30, 2020, television stands out as the most used medium when it comes to getting information about the virus, ahead of digital media. The authors also point out that "infoshow" television programs have higher viewing figures than traditional news programs. The least used mediums for getting information about the evolution of COVID-19 are the internet, social media, radio, information from relatives or friends and the printed press. Whilst the radio does not stand out as a medium used for staying informed, it is considered the most credible, together with television.

Television therefore has recovered its younger viewing public that it seemed to have lost to digital media, thus consolidating a media panorama which is more and more complex and competitive.

The consumption of television and the digital media rocketed during lockdown. The websites of digital daily newspapers received 45% more visits and increased their traffic 100%; the digital audience of online radio has grown 112% [22] and has become one of the mediums most trusted by its audience, and 93% of single users chose live online television. Television has also become the main source of entertainment and information [17:159].

The most relevant increases can be seen among young people between the ages of 13 and 24, by time slots—mornings, early afternoon and evenings are the times which have most increased, around 25% of the total population. On the other hand, news programs have also increased in their daily consumption by 65% [17:159].

In this context, one of the conclusions that Casero [3] arrives at is that COVID-19 has reconnected the sector of the population that was least interested and most distanced from information. Anther of this author's conclusions is that the general public, when faced with a complex and risky informative context, opts for established sources of information with a long trajectory. Television especially is the medium that has obtained the highest levels of consumption and credibility.

[6] Available at https://bit.ly/3vusc21.

3 In the Mirror: Public Opinion and Perceptions Surrounding COVID-19 and Vaccines in Spain

By the middle of 2020, the vaccine against the virus offered a degree of hope. Uncertainty about whether some kind of cure or effective treatment could be achieved, and in what timescale, opened up an uneasy, hopeful and worried social panorama without a clear return date to the 'old normality'.

Diverse public and private initiatives were revealed worldwide, which embarked on a veritable race in the pharmaceutical development of an antiviral cure and reflected the fight for geopolitical leadership. China and Russia marked positions in their own development and application of their vaccines before the end of phase 3 clinical trials (Nius 2020).[7] Every so often, however, information would transcend about obstacles—standstills in the development of the main research—that fueled citizen concern while the disease continued advancing. Some of this information was related to unexpected side effects in trial participants: *Covid-19 Vaccine: myelitis paralyzes the AstraZeneca trial* (Medical Writing 2020)[8]; *Johnson & Johnson paralyzes the trial of its vaccine due to an "inexplicable" reaction in a participant* (Economía Digital 2020[9]).

On November 9, 2020, extremely important world news was released. BioNTech and Pfizer issued a press release publishing intermediate data from the phase 3 trial they had been carrying out with their candidate for the COVID-19 vaccine, formally called BNT162b2. Analysis of this preliminary data revealed that the vaccine was 90% effective. A new informative phase of the pandemic was opened up, in which the vaccines and vaccination health policies now became the protagonists of the social, media and citizen agenda. Immediately afterwards, the rise in company share prices and the fact that Pfizer's CEO received a considerable sum of money from a programmed share sale, generated certain suspicion about the vaccine (*Pfizer CEO sold 62% of his shares for 4.7 million following the vaccine*, (El Confidencial 2020).[10] A few days later, the US company Moderna announced that its experimental COVID-19 vaccine was 94.5% effective. Thus began a war of figures between Pfizer and Moderna to get the most effective vaccine against the virus (*Pharmaceutical War: Pfizer raises the effectiveness of its vaccine to 95%*, Economía Digital 2020b[11]). All this via press releases that bypassed the habitual conduits of scientific communication and the publication of the relevant studies.

Spanish public opinion watched this world battle for the vaccine with concern and mistrust. A study carried out by the FECYT in July 2020 revealed that some 32% of the population would be reluctant to receive a COVID-19 vaccine for reasons related with the speed of the research and the importance of not being among the

[7] Avaliable at https://bit.ly/3tU7tn1.

[8] Avaliable at https://bit.ly/3vgCamx.

[9] Avaliable at https://bit.ly/3sJYIuo.

[10] Avaliable at https://bit.ly/3gEbWqa.

[11] Avaliable at https://bit.ly/32Mlmba.

first to be vaccinated, whilst 12.7% feared the possibility of negative side effects. However, trust in the vaccines in general was relatively high, in line with studies carried out prior to the pandemic: the vast majority of Spanish citizens considered that vaccines are effective at preventing diseases (a mere 4.4% disagreed) and are safe (6% disagreed).

More recently, in November 2020, is the third round of results from the Cosmo-Spain study, coordinated by the Instituto de Salud Carlos III from the National Center for Epidemiology (CNE), with the collaboration of the National Center of Tropical Medicine (CNMT) and promoted by the World Health Organization (WHO). The study is being carried out in 31 other countries and seeks to discover what knowledge and perception of risk the population has in relation to the COVID-19 pandemic [14].

According to the most recent data, the percentage of people in Spain willing to receive the COVID-19 vaccine if it were available tomorrow is 39%, which shows a decrease with respect to round 2 (43%). There has also been a drop of 7% in the members of the population that say they would not have doubts about the vaccine, if it was recommended. The main reasons given for not wanting to receive the vaccine are: "I would have a second or a third, but not the first" (52%) and "It could pose a risk to my health" (48%).

Furthermore, the study survey revealed that daily television news broadcasts, Internet and the national press are still the sources of information most consulted by the Spanish population. The most trusted information continues to be that provided by health professionals, the WHO and the Ministry of Health. Once again, social media is quite often consulted, although trust in them is low, as is also true of the Internet in general and TV debate programs. TV news broadcasts have the highest relevance index at 0.47 (frequency x trust), followed by the Internet with 0.38, although these values are a little lower than in round 2.

The Center for Sociological Investigations (CIS), an institutional reference source about the state of Spanish public opinion, shows an evolution in social perception about the vaccine (Table 1). Furthermore, this perception is affected by the gender/age variable. In the Barometer for the month of December 2020 (the most recent), some 40.5% of the Spanish population would be prepared to receive the vaccine the moment it is available [6]. Some 23.6% would do so if there were sufficient guarantees (it has been tested, it is reliable, its origin, there is sufficient information, or on the advice of the scientific-health authorities). Meanwhile, 28% would not get vaccinated immediately. This picture reveals a considerable lack of trust in the vaccine. By gender, women are more reluctant than men. By age range, those between the ages

Table 1 People prepared to be vaccinated immediately by age

	18–24	25–34	35–44	45–54	55–64	≥ 65
Yes immediately	35.6	32.3	39.1	41.3	41.1	45.3
Not immediately	31.2	40.8	32.2	31.2	24.5	18.2

Source CIS [6], December Barometer. Own elaboration

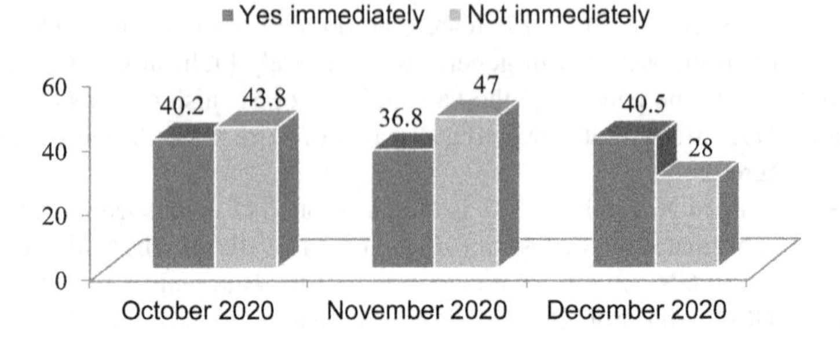

Fig. 2 Evolution of the two main blocks of Spanish public opinion regarding the vaccine. *Source* CIS [6–9]. Own elaboration

of 25 and 34 who would accept the vaccine are outnumbered by those who wouldn't—this being the only age range in which this occurs. The general tonic appears to be that the older the person, the greater their trust and the lower their reluctance.

As was anticipated, the data shows that the general situation in this respect has evolved.

In the Barometer for the month of November [7], 36.8% of the population were prepared to receive the vaccine immediately, whilst 47% were not. In the month of October, 40.5% of the population would accept it but 43.8% wouldn't. What can be observed in this evolution is that the main variations in Spanish public opinion are produced in the group that would not agree to be vaccinated immediately rather than in the group that would (40% or a little less). From November to December there is almost a 20-percentage point drop in those who would not accept the vaccine, fundamentally shifting from a clear negative to a variety of doubts (Fig. 2).

When the question is qualified and three response options are given (immediately, under no circumstances or wait to see the effects of the vaccine), a clear indicator of the mistrust within Spanish society emerges. Whilst 32.5% would agree to be vaccinated immediately, 55.2% would prefer to wait to see the effects of the vaccine [6]. Given that the interviews for this latest monographic study were carried out between November 23 and 26, 2020, news of the market trading of Pfizer's CEO, together with the battle for the effectiveness of the vaccines had emerged and may be behind this rise in mistrust. In December, the beginning of the vaccination campaign in other countries appears to offer a more favorable panorama in this respect within Spanish public opinion.

4 Communication for Trust: Recommendations

The pandemic brought about by COVID-19 has generated a landscape of multiple changes and uncertainty on an international scale. Politicians and governments, medical professionals, scientists and health workers, together with the general public, have been challenged by a highly contagious threat which is still largely unknown.

So much so that even today the side effects of the SARS-CoV-2 on our organism and the after effects of having the disease remain a mystery. If we add to this the high-speed race to obtain a vaccine, with abnormal developmental procedures, it is easy to understand the doubts concerning the COVID-19 vaccine [13]. Even more so in the context of the *infodemic* [3, 10].

Given the landscape and perception of public opinion, communication and adapted communication strategies can fulfill an important role in the transmission of appropriate messages to the general public [18]. In order for the vaccination campaign to work, some 70–80% of the population must be vaccinated. Clear communication about the benefits, risks and importance of the COVID-19 vaccines in order to generate public trust is one of the elements defined as key by the European Commission in the preparation for vaccination strategies against COVID-19 and the rollout of the vaccines [12].

In this sense, the media, scientists, health workers and the authorities are the most prominent actors as emitters of information and opinion leaders in this field. It involves planning and implementing persuasive and awareness-raising communication strategies with one singular characteristic—the time factor. As opposed to other public health campaigns in which the aim is to achieve behavioral changes over time, in this case plans must be made within a time scale covering the present and very near future (the upcoming months). Below are some recommendations about managing communication at the service of trust in the vaccine.

Firstly, transparency must apply at all times. Information about vaccination in other countries which are ahead might help to reassure the population group that would prefer to wait due to possible side effects of the vaccine. Without being alarmist, it should be recognized that side effects are a possibility, but that they are limited, and the positive effects outweigh the negatives. Scientists and health professionals have a very important informative role in this sense. A message of confidence in the European Medicines Agency and its work could counterbalance the lack of trust provoked by the pharmaceutical companies themselves. Furthermore, transparency will be the clear tool to public trust in a context that continues to be highly uncertain. Important information about the vaccines is still lacking: their real degree of effectiveness, how long immunity lasts or their possible role in preventing infection and the spread of the disease. Recognizing what is *not* known could, paradoxically, help to encourage trust in the professionals and in what *is* known.

Secondly, it is important that political leaders and governments on a national and international level show responsibility in terms of their statements and actions. Their words cross borders, their actions provide an example. Public trust will be greater if well-known professionals and prominent leaders are seen to be adopting responsible patterns of behavior and accepting the vaccine in a natural and normal way.

Thirdly, the general public's information habits should be taken into account. Television and digital media have become the reference information sources. Messages should be channeled through those media that guarantee their arrival to an objective public. Among the younger members of the public, who are the most skeptical regarding the vaccine, digital channels should be promoted. A positive communication tone or the use of humor could be ingredients which have been under utilized up to now and yet are of great use to institutional campaigns that have largely used fear, given that communication campaigns about risk often recur to methods of persuasion based on fear and possible harm to the population [21]. Social media's commitment to verified information is especially interesting (*Twitter has announced that it will remove false information about the COVID-19 vaccine*, Nius 2020).

More specifically regarding messages and emotion management, recognizing fears, anger and other negative emotions whilst stressing strict safety standards and the effectiveness of the development of the vaccine, whilst encouraging people's self-efficacy via vaccination may help to increase trust [4].

Likewise, messages that raise awareness about the manipulation of negative emotions via misinformation campaigns may help to inoculate against the effects of misinformation concerning the vaccines. Conversely, positive emotions need to be provoked in terms of helping the community to restore health and well-being, and consequently take the decision be vaccinated against the most important disease of our time. Adaptation to the audience's existing emotional state is recommended, instead of a single approach for all [4]. In this sense, agile communication strategies in real time will be the most appropriate.

References

1. AIMC: Marco General de los Medios en España. AIMC. Available at https://bit.ly/3qNmRPD (2020). Accesed 27 Dec 2020
2. Bennett, W., Livingston, S.: The disinformation order: Disruptive communication and the decline of democratic institutions. Eur. J. Commun. **33**, 122–139 (2018). https://doi.org/10.1177/0267323118760317
3. Casero-Ripollés, A.: Impact of Covid-19 on the media system. Communicative and democratic consequences of news consumption during the outbreak. El Profesional Información **29**(2), (2020). https://doi.org/10.3145/epi.2020.mar.23
4. Chou, W., Budenz, A.: Considering emotion in COVID-19 vaccine communication: Addressing vaccine hesitancy and fostering vaccine confidence. Health Commun. **5**(14), 1718–1722 (2020). https://doi.org/10.1080/10410236.2020.1838096
5. Catalán-Matamoros, D., Peñafiel-Saiz, C.: Media and mistrust of vaccines: a content analysis of press headlines. Rev. Lat. Comun. Soc. **74**, 786–802 (2019). https://doi.org/10.4185/rlcs-2019-1357
6. CIS: Barómetro de diciembre 2020. Avaliable at https://bit.ly/3qQLHON (2020a). Accesed 02 Jan 2021
7. CIS: Barómetro de noviembre 2020. Avaliable at https://bit.ly/3lp8pfU (2020b). Accesed 02 Jan 2021
8. CIS: Barómetro de octubre 2020. Avaliable at https://bit.ly/3eKRdQK (2020c). Accesed 02 Jan 2021
9. CIS: Efectos y consecuencias del coronavirus (II). Avaliable at https://bit.ly/3s4VfXC (2020d). Accesed 02 Jan 2021
10. Costa-Sánchez, C., López-García, X.: Comunicación y crisis del coronavirus en España. Primeras lecciones. El Profesional Información **29**(3), (2020). https://doi.org/10.3145/epi.2020.may.04

11. Lazer, D., Baum, M., Benkler, Y., Berinsky, A.J., Greenhill, K.M., Menczer, F., et al.: The science of fake news. Science **359**(6380), 1094–1096 (2018). https://doi.org/10.1126/science.aao2998
12. European Commission: Communication from the commission to the European Parliament and the Council. Preparedness for COVID-19 vaccination strategies and vaccine deployment. Avaliable at https://bit.ly/3qWhOwh (2020). Accesed 20 Dec 2020
13. Funk, C., Tyson, A.: Intent to get a COVID-19 Vaccine Rises to 60% as Confidence in Research and Development Process Increases. Pew Research Center. Available at https://pewrsr.ch/3rOit4i (2020). Accesed 20 Dec 2020
14. Instituto de Salud Carlos III: Monitorización del comportamiento y las actitudes de la población relacionadas con la COVID-19 en España (COSMO-SPAIN): Estudio OMS. Avaliable at https://bit.ly/3eKRZ06 (2020). Accesed 20 Dec 2020
15. Martínez, L.A.: Cómo saber si una noticia es una fake news. Proconsi. Available at https://bit.ly/3cFuUsX (2019). Accessed 20 Dec 2020
16. McIntosh, J.: Source choice in agricultural news coverage: impacts of reporter specialization and newspaper location, ownership, and circulation (PhD Thesis). Texas A&M University. Avaliable at https://bit.ly/3rOizJc (2006). Accesed 19 Dec 2020
17. Montaña, M., Ollé, C., Lavilla, M.: Impact of the Covid-19 pandemic on media consumption in Spain. Revista Latina Comunicación Social **78**, 155–167 (2020). https://doi.org/10.4185/RLCS-2020-1472
18. Moreno-Montoya, J.: El desafío de comunicar y controlar la epidemia por coronavirus. Biomedica **40**, 11–13 (2020)
19. Peñafiel-Saiz, C., Ronco-López, M., Castañeda-Zumeta, A.: Ecología comunicativa en tiempos del coronavirus SARS-CoV-2. Del momentum catastrophicum al virtus veritas. Revista Española de Comunicación en Salud Extra 1 (2020a).
20. Peñafiel-Saiz, C., Castañeda-Zumeta, A., Ronco-López, M.: El Tsunami COVID-19 y la información sobre vacunas contra el coronavirus. In: de Vicente, A., Abuin, N. (coord.) La comunicación especializada del siglo XXI, pp. 683–705. Mc Graw Hill, Madrid (2020b)
21. Prior, H.: Comunicación pública de riesgo en tiempos de pandemia: Las respuestas de Portugal a la COVID-19. Más Poder Local **41**, 6–11 (2020)
22. Rodero, E.: La radio: el medio que mejor se comporta en las crisis. Hábitos de escucha, consumo y percepción de los oyentes de radio durante el confinamiento por el Covid-19. Profesional Información **29**(3), (2020). https://doi.org/10.3145/epi.2020.may.06
23. Statista: Promedio de tiempo diario destinado al consumo de medios de comunicación en España en 2020, por tipo. Available at https://bit.ly/3aCoOcB (2020). Accesed 19 Dec 2020
24. Vosoughi, S., Roy, D., Aral, S.: The spread of true and false news online. Science **359**(6380), 1146–1151 (2018). https://doi.org/10.1126/science.aap9559
25. Waisbord, S.: Truth is what happens to news: on journalism, fake news, and post-truth. Journalism Stud. **19**(13), 1866–1878 (2018). https://doi.org/10.1080/1461670X.2018.1492881
26. WHO: Ensayo clínico 'Solidaridad' sobre tratamientos contra la COVID-19. Avaliable at https://bit.ly/3bOFiiG (2020a). Accesed 19 Dec 2020
27. WHO: La OMS publica directrices para ayudar a los países a mantener los servicios sanitarios esenciales durante la pandemia de COVID-19. Avaliable at https://bit.ly/3bR5V6N (2020b). Accesed 19 Dec 2020

Carmen Costa-Sánchez Associate Professor of Communication Strategies in the Department of Sociology and Communication Sciences at the School of Communication Sciences, University of A Coruña. Her research interests include health communication and public relations. She has authored the book *Hospital Communication. Communication strategies on health area* (published by Comunicacion Social) and numerous papers related to this field.

Carmen Peñafiel-Saiz Professor at the Department of Journalism, University of the Basque Country (Bilbao). She is involved in a wide variety of research projects within the field of health communication. Her latest research works focused on vaccine coverage in media and health information in times of COVID-19. Her research has been published in journals such as *Human Vaccines & InmunotherapeuticsPerspectives in Public Health* and *Health Communication.*

Models, Professionals, and Audiences

Transformation of Local Journalism: Media Landscapes and Proximity to the Public in Spain, France and Portugal

María-Cruz Negreira-Rey, Laura Amigo, and Pedro Jerónimo

Abstract This chapter deals with the transformation of local media in the European context, focusing the study on the digital scenario in Spain, France and Portugal. The respective media maps are used to determine the current weight of regional, local and hyperlocal digital media. Based on the characteristics of the media ecosystem in each country, we seek to identify common trends and specific characteristics of the global mutations in journalism in these countries. Special attention is paid to the way in which the media seek to (re)create proximity with their audiences through their editorial models, the forms and channels they establish for interaction with the public or the creation of collaborative relationships.

1 Introduction

The ecosystem of the local media has undergone a profound transformation in recent years. In different countries and media contexts, the media scene has been transformed from the local and hyperlocal level, giving rise to common development trends, new media models, new ways of interacting with the audience, and shared challenges.

The changes—concomitant with the Internet boom in the last 20 years—have led to a "crisis in journalism" [22] marked by a transformation in public information practices as well as in the methods and means of producing and disseminating information [8, 41, 52, 55]. At the local level, this situation is reflected in the loss of the hegemonic position of the traditional media as a source of information in their territory of diffusion, cuts in investment and in personnel; the absorption of local media

M.-C. Negreira-Rey (✉)
Universidade de Santiago de Compostela, Santiago de Compostela, Spain
e-mail: cruz.negreira@usc.es

L. Amigo
Université de Neuchâtel, Neuchâtel, Switzerland

P. Jerónimo
Universidade da Beira Interior, Covilhã, Portugal

J. Vázquez-Herrero et al. (eds.), *Total Journalism*, Studies in Big Data 97,
https://doi.org/10.1007/978-3-030-88028-6_12

by (inter)national groups; and the appearance of new forms of digital journalism. The crisis of traditional local media [21] has raised alarm about the emergence of media deserts [13] and the deficit of the democratic role that local media play in local communities [1, 6].

Local media remain a reliable source of information and are in demand by citizens [40]. In crisis situations, such as the one experienced as a result of the COVID-19 pandemic, local media reinforce their mission as truthful informers at the service of the community's neighbors. Thus, local media have had to transition to the digital model, which is increasingly mobile and multiplatform, in the face of a small market and increasingly limited business structures [30]. This has led them to seek a balance between a printed model that is still profitable in the local scenario and the diversification of revenue streams in the digital arena [31].

In the digital scenario, local and hyperlocal digital media have experienced a remarkable growth. Observed at an international level, this growth is linked to the low cost for the production and distribution of content on the net, the high demand for local information, or the change in audience behavior towards more active roles [46]. These new information projects, born on the Internet, are generally characterized by maintaining an information objective linked to the community, constituting an alternative information source to the traditional media, covering the gaps forgotten by the large communication groups and maintaining a participatory relationship with the audience [26, 36, 53]. The rise of the emerging model of hyperlocal media has been studied by researchers from several countries. Thus, these digital media have already been mapped in the United States [28], the United Kingdom [25], the Netherlands [34], Sweden [42], Finland [29] and Norway [24].

In this chapter we approach the realities of the local digital media in Spain, France and Portugal. Focusing on digital natives, we observe the current state of the local media in the three countries and reflect some common trends, models and strategies of development of these media. The naturally close link that these media maintain with their community leads them to experiment with various forms of interaction and participation with the audience, which are also addressed in this work.

2 Spain

Local journalism in Spain is going through a phase of growth and experimentation, in which the development of local and hyperlocal media and the search for new models in the digital scenario have been decisive. Local media, which have a great tradition and weight in the Spanish media scene, have renewed themselves on the Internet and have developed new journalistic approaches, new professional and business models, as well as new forms and channels to interact with audiences.

To understand the current situation of local digital journalism in Spain, it is convenient to contextualize its development in recent decades. In the 1980s, the regional and local press experienced a renewed boom due to the political-administrative and social transformation that the country experienced following its democratic transition [23]. The privatization of the press and the expansion of television and radio channels gave rise to a new media scenario that, in the 1990s, was dominated by large regional and local press groups, such as Prensa Ibérica, Grupo Moll and Grupo Correo [23]. In Spain in 1990 it was possible to find 95 regional or local newspapers, 427 local radio stations and 113 local television stations [35].

With the first journalistic initiatives on the Web from 1995 onwards, the first digital media of proximity also appeared. These included *El Correo Gallego*, *Diario Vasco, Barcelona Televisió*, and the digital natives *Vilaweb* and *Proyección TV*. Both the local digital media and the pure players were adapting in these years to the new editorial possibilities offered by the Web, a process that also required new business strategies.

The economic and the media industry crisis suffered from 2008 onwards truncated this adaptation process and accelerated media renewal in the areas of proximity. Job destruction in the journalistic sector has motivated many professionals to undertake their own projects, many betting on the creation of local digital native media [3, 4]. In 2008 the first hyperlocal digital media appeared in Spain, such as *A Voces de Carabanchel*, *Hortaleza en Red*, *OMC Radio Villaverde* and *Zona Retiro* [20]. In general, hyperlocal media are usually legally incorporated as limited companies, partnerships or cooperatives; they establish collaborative relationships with other media and local associations; they are active in social networks and they hardly reach economic profitability [4].

The scenario of local journalism in Spain in 2018 comprised a total of 1148 local and 62 hyperlocal digital media news outlets [39]. The importance of digital media of proximity becomes evident if one considers that a total of 3065 active digital media were mapped in the country at that time [48].

In the map of local and hyperlocal Spanish digital media, it was observed that their distribution was not homogeneous throughout the country. These media tend to be concentrated in the regions with the highest media density, with most hyperlocals being found in the large urban capitals of Madrid and Catalonia. Despite being digital media, radio and paper are platforms with an important weight among these media—which sometimes causes their digital editions to preserve the timing of the publication of the printed edition. These media are, for the most part, privately owned and make relevant use of the co-official languages of autonomous communities to strengthen identity links with members of the local community [39].

In relation to digital native media—those that are only published and distributed on the Internet [37]—only 31.2% of the locals and 40.6% of the hyperlocals can be described as such. In this chapter we focus on these media, reflecting the different models they develop and how they build their relationships with their audiences based on a series of case studies [38].

Among the digital natives with hyperlocal reach, *Somos Malasaña* stands out as one of the first initiatives of this kind in Spain. It was founded with the aim of creating a network of hyperlocal media covering different neighborhoods in the city of Madrid. In 2019, Grupo Somos edited *Somos Malasaña, Somos Chueca* and *Somos Chamberí*. In the same year the national newspaper *elDiario.es*, also a digital native, integrated these outlets into its editorial project. After this integration, the group expanded its coverage to a fourth district with *Somos Tetuán*. In its development, the project went from depending economically on advertising from local advertisers, to diversifying its sources of income by providing the service of organizing events, to finally benefiting from the *elDiario.es* subscription model.

El Español is another example of the association of local and hyperlocal media with national newspapers. This national pure player integrated other local digital natives, such as *Quincemil* and *Treintayseis*—which cover different areas of Galicia—and *Navarra.com*, and other matrix media such as the Canarian *Diario de Avisos* and the Catalan *Crónica Global*. This type of associations provides local information projects a greater scope, helping them to diversify their sources of income and provide them with greater stability through direct funding from subscribed users.

Like Grupo Somos in terms of its digital native condition and its informative purpose is the Seville hyperlocal project that publishes *Triana al Día* and *Nervión al Día*. These media are promoted by four partners who, due to the high instability of the sector, are forced to remain as self-employed workers and combine their occupation in the media outlets with other professional activities. Although becoming a stable project over time (it was founded in 2008), it faces obstacles that are common among local and hyperlocal digital media: high dependence on advertising from small advertisers; insufficient support from public administration; lack of economic resources to be able to expand the human team and produce more informative content; and the difficulty of establishing alliances with other media and proximity initiatives.

Local and hyperlocal digital media are generally characterized by limited human resources. Some examples are the digital natives *Sahagún Digital* (where only its founder works), *Vigo É* (where two journalists work) and *Fuencarral-El Pardo* (maintained in their free time by its founder and a collaborator). Lack of human resources limits their capacity to produce information and update their websites, improve their editions, and innovate with new formats. Despite the growth of local and hyperlocal digital media, they need formulas and strategies that allow them to achieve economic sustainability and professional stability.

If many of these media manage to maintain their activity over time is because of their mission to inform and serve the hyperlocal community [36]. The promoters of this type of media usually prioritize the civic and social objectives of their projects over the economic ones [26]. Their community orientation is also reflected in their participatory relationship with the audience, through which users can take active roles in the information production process [19].

The study of the Spanish local and hyperlocal digital media [38] has shown that this interaction and participation occurs through the media websites and their profiles in social networks. In their web editions, the most frequent forms of interaction are the possibility of establishing direct contact with the media outlets, accessing their social profiles, sharing content through different channels, extending information with related content, commenting on news, publishing user generated content (UGC), and maintaining exclusive sections for this UGC. Social networks have become essential platforms not only for the dissemination of informative content, but also as a channel of communication and interaction with the audience. It is here where users comment on current events, alert media journalists about what is happening in the neighborhood and about what they demand in the form of news coverage, and express their opinions about the contents of the media.

The citizens of the community are integrated into the information production process, but more as information sources than as producers of the information. Although most local and hyperlocal digital media are open to publishing user-generated content, this is usually limited to opinion, reporting or creative content. Some examples of this participation of citizens can be found in *Somos Malasaña*, which maintains a series of thematic blogs updated by collaboration with users, and which has also produced the documentary film *Sol de Malasaña*, recorded by citizens. *Nervión al Día* and *Triana al Día* have produced some interactive pieces in collaboration with citizens, such as a map of the most dangerous points on the cycle paths in the city or a gastronomic guide. Alternatively, *Fuencarral-El Pardo* maintains a section where it publishes literary stories by users, who participate in a contest promoted by a neighborhood association. Although it is not a digital native, the case of *Goiena* is noteworthy, as it maintains a space on its website called Community. In this space, any registered and authorized user can freely publish their content and edit their personal space, which can be followed by other readers.

3 France

The advent of Web 2.0, known as the 'collaborative' Web [47], and the introduction of digital and mobile technologies have been profoundly and constantly transforming the French media landscape. These changes concern both information consumption practices—which have become plural, fragmented, mobile, etc.—and ways of producing and distributing information—which are now reticular, participative, immediate, etc.—, shaking up the economic equilibrium of the field. This context of strong change also extends to the emergence of new actors and new journalistic approaches towards the audience. Online-only news sites—generally called 'pure players' in France—made their appearance alongside websites developed by other forms of press. National pioneers included *Rue89*, a generalist news website created in 2007, and *Mediapart*, an investigative news site launched in 2008. In order to ensure their economic health, *Rue89* bet on free access and counted on revenue from advertising, while the latter based its income on subscription fees. Both founded by former journalists of the 'legacy' news media, these news websites have been a reference point for their local counterparts, even if they focus on different territories. On a local scale, local news sites launched at that time included *Tout Metz* in Lorraine (in 2006), *94 Citoyens* in Essonne (2007), and the local franchise of *Rue89* in Marseille (2008) among others. A few years later, a second generation of pure players was born, including *Le Téléscope d'Amiens* in 2012, *Corse Net Infos* in Corsica in 2016, and *Mediacités* in Lille in 2016, followed by editions in Lyon, Toulouse and Nantes in 2017. Several of these digital media were launched by young journalists thanks to crowdfunding campaigns, basing their business models on subscriptions rather than advertising. With light and flexible organization, low apparent hierarchy, adherence to the project, and continuous experimentation of their product the management of these online publishers is similar to that of start-ups [49].

It is in a media scene strongly dominated by regional or national daily newspapers and local national public television stations that these Internet-native news media tried to position themselves. Some managed to make their mark but often on the fringes (*Marsactu* in Bouches du Rhône, *Rue89* editions in Strasbourg, Lyon, Marseille and Bordeaux, *94 Citoyens* in Val-de-Marne), while others disappeared after a few years of struggle (*Essonne info, Le Télescope d'Amiens, Carré d'info*). Despite their dynamism, pure players are aware of their vulnerability that results from the fragility of their business model. While the trend is the adoption of the subscription business strategy, the constant changes in the media landscape force online news sites to (re)think, diversify and combine ways of monetizing content [15].

According to a collaborative inventory of local news media coordinated by Ouest MédiaLab,[1] by February 2020, 1512 local news outlets were available to inform the French. Among them 16.7% were digital news media, while the vast majority (63.0%) were print newspapers. Ouest Média Lab's inventory[2] does not give insight into the scope of the geographical territory covered by these news media that would allow identification of possible news deserts in France. Considering that they are potentially available from potentially anywhere, for online media the choice of a territory is part of an editorial strategy that takes into account the target audience, the type of news covered, and possible partnerships, among other considerations. However, most French pure players choose to cover large cities, except those conducting investigative journalism that covers a wider territory. As French researchers Bousquet and Smyrnaios [12] note, "all news sites covering news at an infra-national level have adopted an administrative geographic unit. Thus, the frameworks set by local authorities in France seem very difficult to overcome from the point of view of professional journalistic information".

For Bousquet and Smyrnaios [11] there are three main editorial models of local digital news media. The first and rarest, based on *Mediapart*'s model, is local investigative journalism that aims to create debate in the local public space. News websites that adopt this approach (like *Marsactu* recently or *Dijonscope* and *Le Télescope d'Amiens* in the past) reach an audience interested in politics and public life in general, and they usually have regional expansion. The second model presented by the authors can be associated with the *Rue89* style. Its goal is to cover current events

[1] The collaborative inventory was coordinated by Ouest MédiaLab for the Festival de l'Info Locale. As explained on their webpage, the authors understood a local news outlet as "any news outlet that deals with the current events of an area ranging from the neighborhood up to the region. [The authors have] done so without setting limits in terms of nature of publication (online, radio, television, print) nor of economic models (free of charge or with a fee, nonprofit, private or public), nor of the quality of the published contents. The basic unit of this inventory is the editorial staff. That is to say, a team of journalists broadcasting information collected by them (for the most part) on the territory surrounding their place of work".

[2] The inventory is accessible at https://www.festival-infolocale.fr/medias-locaux-aidez-nous-a-vous-compter/.

in an original way—by choosing a specific news angle or by reporting on subjects neglected by other news outlets. It also encourages user participation in news making while maintaining a high standard of journalism. This editorial model targets young and educated inhabitants of large cities, whose profile is close to that of journalists [10]. Examples are *Rue89* in Lyon and Strasbourg or former *Grand-Rouen* in Rouen and *Carré d'infos* in Toulouse. The third editorial model described by the authors is close to that of the regional daily press. News and life about local organizations are covered in an editorial magazine-style. Online media such as *Aqui.fr* in Aquitaine and former *Ariegenews* in Ariège have adopted this model.

Beyond the specificities of journalistic models, the common feature of online news sites is that they place audience participation at the heart of their editorial line. Embodied by pure players, participatory journalism—understood as the overall process of an audience engaging with journalists in the construction and dissemination of information [43, 50, 54]—experienced a major rise in the mid-2000s. Journalistic productions were fueled by contributions of an audience that plays the role of commentator, source, witness, news producer, or expert. The integration of the contributions depended on how much journalists actually opened up the editorial process to the audience [44]. Also, the inclusion of user contributions in the editorial process corresponds to digital media's ambition to differentiate themselves from legacy news outlets as well as to target new audiences by inviting citizens to play an active role. The profiles of these non-professional contributors provides insight into their engagement. In addition to taking a critical stance towards mainstream media [2], "the possession of a strong cultural capital [...] as well as a marked interest in politics in the broad sense are discriminating factors that explain the propensity of Internet users to participate in the development and dissemination of news content" [51]. However, some years later, the movement waned due to the difficult integration of participatory approaches into rather rigid professional organizations and practices [18, 27].

Nonetheless, there is still a demand for a rather horizontal relationship between the audience and journalists, which echoes people's information practices enabled through social networks. Moreover, the context of mistrust towards news media and journalists that has grown year-on-year to the point of becoming a "big misunderstanding" [14] that pushes online media to question their relationship to the audience. Recent research [45] on current local news media strategies towards their audiences shows that news outlets seek to connect with audiences mostly by inviting them to contribute with content that may fuel the editorial production, such as comments on news articles, messages or questions related to a specific subject. Another widespread strategy consists of promoting dialogue between journalists and the audience through thematic online groups, editorial conferences open to all, and meetings focusing on the production of information. After narrowing down the database used in this research to online-only French local media and studying these media's websites, we see that the main strategies to engage with audiences are pretty consistent: developing initiatives that enhance dialogue with journalists and including users in the news making process through content contribution. For instance, *Rue89 Strasbourg* organizes thematic debates related to current news, to a film screening or to address journalistic aspects of the news outlet, such as the redesign of its visual identity. *Médiacités*, for its part, hosts online debates to discuss issues such as the promises made by local elected officials. Both media also call for citizens' contributions (questions, messages, witnesses) regarding matters concerning their territory. Furthermore, pure players such as *Streetpress* or *Sans A*, create online groups using social media platforms in order to boost exchanges with their audience. If the first strategy is a legacy or a revival of participatory journalism, particularly for news outlets that have built their editorial project on users' content contributions, the latter seems to be a response to the growing disconnection of citizens towards news media and journalists. Dialogue, whether online or face-to-face, appears to be a step towards re-humanizing journalistic work and building a more accessible, reciprocal relationship between journalists and the French audience.

4 Portugal

Local journalism[3] in Portugal is still in a period of transition to digital and adapting to a new reality. The public global health problem of COVID-19 is a challenge to all sectors of society and in particular to the media that operate in small territories. While, on the one hand, information is more necessary than ever, on the other hand, purchasing power decreases, as does the main source of revenue for these media (advertising). Other contemporary challenges also exist, such as disinformation, which is particularly critical in contexts of proximity, since regional, local or hyperlocal media traditionally have fewer resources, strongly influencing fact-checking abilities [16].

In Portugal, the press occupies a relevant place in the media ecosystem, especially in districts, cities and smaller places. Regarding the concepts applied to the media of proximity, it is common to find various denominations, both at the legislative level and in academic studies. For example, the terms 'regional press', 'local press' and 'regional and local press' are frequently used [32]. Although regional press is the most frequent concept and includes all types of media, except for local radio stations, due to the country's size and when it comes to addressing the issue at an international level, the emphasis has been on local, that is, 'local media', 'local newspapers' and 'local journalism' [33].

To see local journalism in a digital environment, we must go back to 1996, when *Região de Leiria* started putting content online, becoming the first Portuguese medium at the local level to do so,[4] from the transposition of the contents published in the print edition (shovelware). Local radio started to make this transition from

[3] In Portugal, the most common concept is proximity journalism (*jornalismo de proximidade*). However, considering the most frequent use in international studies and in English, we will consider 'local journalism'.

[4] The pioneer of online journalism in Portugal was the daily newspaper *Jornal de Notícias*, which started its online edition on July 26, 1995 [7].

2000 [9] and the first online television of local scope appeared from 2005 [32]. Exclusively online media appeared first at a local level: on January 5, 1998, *Setúbal na Rede* appeared, dedicated to the district of Setúbal; it was published continuously until the beginning of 2017. Business models for this type of media rely mainly on two major sources of revenue: advertising and subscriptions. In the case of digital, advertising, with the support of Google and Facebook in the context of the pandemic, has essentially opened as an opportunity for this medium. The search for funding grants went from one-off to permanent possibility, as an alternative to traditional financing channels. The same can be said for crowdfunding. Regarding hyperlocal means, this is an emergent reality in the Portuguese context, as we will see below.

Local journalism in an exclusively digital environment began with *Setúbal na Rede*. The appearance of this medium was disruptive, even in legislative terms [32]. In a country accustomed to traditional formats—newspapers, radio and television—the appearance of the first online medium implied changes to fit a new media reality. In the 19-year history of this medium, we highlight the commitment to the exercise of journalism from and with the community. This way of being was transversal to the environment, since its sources of funding were not only the advertising published on the website, but also that which was generated from events that it organized. The complete preservation of the news published throughout its existence is another fact worth mentioning. Another noteworthy local online media outlet is *Tinta Fresca*, which appeared in the middle of 2000 and is still being published. Assuming a role as a "newspaper of art, culture and citizenship", it has a more local scope of intervention, namely in the municipality of Alcobaça, in central Portugal.

Seeking more recent data, we analyzed the Entidade Reguladora para a Comunicação Social (ERC) record base at the end of 2020, with the aim of identifying online media in small territories. Portugal had 1751 media outlets, of which 409 were regional, local and hyperlocal. Of these, 185 published exclusively online. Considering only those that were published regularly, we excluded all those that had a periodicity greater than one week. Thus, a total of 155 online media were selected (91.0% were published daily and 9.0% once a week).

Geographically (Fig. 1 and Table 1), Portugal's regional online media are concentrated in the North, with the district of Porto (14.2%) having the most outlets, followed by Lisbon (12.9%) and Braga (10.3%). Note also that the Beja district does not have any registered local online media.

Regarding the scope, regional online media (62.6%) predominates, followed by local (36.1%) and hyperlocal (1.3%). In relation to the nature of ownership, online media held by companies (47.7%) are almost equivalent to those held by a single person (45.2%), followed by associations (4.5%), the Catholic Church (1.9%) and cooperatives (0.6%).

When we look at the possibilities for public participation, we find that it is promoted mainly for platforms or spaces external to the online media themselves. This is true in the case of social media (79.4%), namely Facebook, Instagram and Twitter. On their own websites, participation can be seen in the possibility of commenting on news items (39.4%). With newsrooms traditionally that have few resources, there is not enough time or journalists to guarantee moderation. Perhaps

Fig. 1 Districts and autonomous regions of Portugal. Own elaboration

Table 1 Distribution of regional online media by districts and autonomous regions (A. R.). Own elaboration

District	n	%	District	n	%
Aveiro	12	7.7	Lisboa	20	12.9
Beja	0	0	Portalegre	7	4.5
Braga	16	10.3	Porto	22	14.2
Bragança	5	3.2	Santarém	6	3.9
Castelo Branco	5	3.2	Setúbal	3	1.9
Coimbra	10	6.3	Viana do Castelo	4	2.6
Évora	5	3.2	Vila Real	3	1.9
Faro	9	5.8	Viseu	5	3.2
Guarda	5	3.2	Açores (A.R.)	7	4.5
Leiria	7	4.5	Madeira (A.R.)	4	2.6

that is why most online media choose not to allow comments on their news. On the other hand, we cannot ignore studies that point to a contradictory reality experienced at a local level. While, on the one hand, journalists recognize that public participation is important, on the other, it is not widely accepted in practice [17, 33].

Regarding hyperlocal journalism, similar to other concepts already discussed, it appears to be geographically oriented and communally committed, especially at the parish, neighborhood or street level [32]. At least that is the claim. However, there is evidence that this is not always the case. Within the scope of this study and in an analysis of the Portuguese case, we found the same evidence as Baines [5], regarding the non-involvement of communities in the construction and production of hyperlocal media. An example is *Inforpaulense*, intended for the population of a small parish in central Portugal, which existed for four years. The wear and tear caused by production centered on a single person, without collaboration or even incentive, dictated its end. This is, in fact, the justification for the end of many projects supported by a single person. Sometimes the motivational factor, which leads to the creation of hyperlocal or local means and maintains them, is more decisive than economic viability. In addition, we found just one more example currently in Portugal and registered with ERC: *Jornal da Praceta*. Founded in June 2001, it presents itself as "the first electronic newspaper in a neighborhood in Lisbon". Based on the studies carried out, this is the first hyperlocal media in Portugal, although *Mais Minho* was recognized as such when it was launched in 2012 [32]. These examples provide evidence that reinforces the need for studies on journalism developed in smaller geographical contexts and in local communities.

One of the consequences of the current pandemic crisis of COVID-19 was the recognition—especially by the most resistant local media—of the importance of digital. There are even reports of local newspapers that definitively abandoned their printed version, becoming online-only. The transition is underway, with mobile devices and newsletters being the media types that have recently raised interest among the local media scene in Portugal.

5 Conclusion

This chapter examined the current state of local media in France, Spain and Portugal. Despite being countries with different territorial, administrative, social and media contexts, it is possible to find common trends in their development in the digital scenario.

After the crisis experienced in the first decade of the 2000s, local media have renewed their models and continue to occupy an essential space in the media ecosystem. On the Internet, local and hyperlocal digital media have grown in an important and generalized way in the last few years, with those journalistic initiatives that only develop their activity on the Internet gaining an increasing presence. These digital natives or pure players continue to occupy wide spaces of proximity, such as regions, but those that are limited to the most reduced geographical and social scope also grow, covering a hyperlocal area.

Local and hyperlocal digital natives are experimenting with new corporate and business models to seek sustainability in the digital scenario. However, their informative vocation means that they do not always put their economic profitability above the continuity of their activity as a service to the community.

Their orientation towards the community leads them to implement renewed strategies and forms of participation with their audiences. Users and neighbors are actively integrated into the journalistic project, either as partners of the media, as producers and collaborators in the elaboration of content, or as frequent interlocutors with journalists. In their process of integration and support of the local community, these digital media even go beyond the virtual limit of their digital editions to organize physical and participatory events with users.

The local media in the digital scenario not only grow in number and proliferate in increasingly reduced areas, but their activity continues to be fundamental for the communicative, social and democratic development of the communities they serve.

Funding This research has been developed within the research project *Digital Native Media in Spain: Storytelling Formats and Mobile Strategy* (RTI2018–093,346-B-C33), funded by the Ministry of Science, Innovation and Universities (Government of Spain) and the ERDF structural fund. This paper was also developed within the framework of *Re/media.Lab—Laboratory and Incubator of Regional Media*, a research project co financed by the Portugal 2020 Program (CENTRO-01–0145-FEDER031277; Fundação para a Ciência e a Tecnologia).

References

1. Ahva, L., Wiard, V.: Participation in local journalism. Assessing two approaches through access, dialogue and deliberation. Sur le Journalisme **7**(2), 64–79 (2018)
2. Aubert, A.: Le paradoxe du journalisme participatif. Motivations, compétences et engagements des rédacteurs des nouveaux médias. Terrains Travaux **15**, 171–190 (2009)
3. Asociación de la Prensa de Madrid: Medios lanzados por periodistas. Asociación de la Prensa de Madrid. Available at https://bit.ly/2Owmlsl (2015). Accessed 02 Jan 2021
4. Asociación de la Prensa de Madrid: Informe Anual de la Profesión Periodística 2017. Asociación de la Prensa de Madrid. Available at https://bit.ly/3bNjJ29 (2017). Accessed 02 Jan 2021
5. Baines, D.: Hyper-local: glocalised rural news. Int. J. Sociol. Soc. Policy **30**(9–10), 581–592 (2010). https://doi.org/10.1108/01443331011072316
6. Barnett, S., Townend, J.: Plurality, policy and the local. Can hyperlocals fill the gap? Journalism Pract. **9**(3), 332–349 (2015). https://doi.org/10.1080/17512786.2014.943930
7. Bastos, H.: Origens e evolução do ciberjornalismo em Portugal: Os primeiros vinte anos (1995–2015). Afrontamento, Porto (2015)
8. Boczkowski, P.: Digitizing the News: Innovation in Online Newspapers. The MIT Press, Massachusetts (2004)
9. Bonixe, L.: As rádios locais em Portugal—da génese ao online contexto e prática do jornalismo de proximidade. ICNOVA, Lisboa (2019)
10. Bousquet, F., Marty, E., Smyrnaios, N.: Les nouveaux acteurs en ligne de l'information locale vers une relation aux publics renouvelée? Sur le Journalisme **4**(2), 48–61 (2015)

11. Bousquet, F., Smyrnaios, N.: L'information en ligne et son territoire: positionnement comparé entre un pure player départemental et un quotidien régional. In: Noyer, J., Raoul, B., Paillart, I. (eds.) Médias et territoires: permanences et mutations, pp. 193–214. Presses Universitaires du Septentrion, Villeneuve d'Ascq (2013)
12. Bousquet, F., Smyrnaios, N.: Les pure players peuvent-ils renouveler l'information locale? La revue des médias. Available at https://bit.ly/3rQrPMT (2014). Accessed 21 Jan 2021
13. Bucay, Y., Elliott, V., Kamin, J., Park, A.: America's growing news deserts. Columbia Journalism Review. Available at https://bit.ly/3bPxwFw (2017). Accessed 02 Jan 2021
14. Charon, J.-M.: Les journalistes et leur public: le grand malentendu. Clemi/Vuibert/INA, Paris (2007)
15. Charon, J.-M.: Presse et numérique—L'invention d'un nouvel écosystème. Rapport à Madame la Ministre de la culture et de la communication. Available at https://bit.ly/3tjDzYx (2015). Accessed 21 Jan 2021
16. Correia, J.C., Jerónimo, P., Gradim, A.: Fake news: emotion, belief and reason in selective sharing in contexts of proximity. Br. J. Res. **15**(3), 590–613 (2019). https://doi.org/10.25200/BJR.v15n3.2019.1219
17. Correia, J.C., Jerónimo, P.: Sentimentos contraditórios: Quanto os jornalistas gostam da intervenção dos públicos? In: Antunes da Cunha, M. (ed.) Repensar a Imprensa no Ecossistema Digital, pp. 81–96. Axioma, Braga (2020)
18. Domingo, D., Quandt, T., Heinonen, A., Paulussen, S., Singer, J., Vujnovic, M.: Participatory journalism practices in the media and beyond: an international comparative study of initiatives in online newspapers. In: Future of Newspapers Conference, 12–13 Sep, Cardiff (2007)
19. Firmstone, J., Coleman, S.: The changing role of the local news media in enabling citizens to engage in local democracies. Journalism Pract. **8**(5), 596–606 (2014). https://doi.org/10.1080/17512786.2014.895516
20. Flores Vivar, J.M.: Periodismo hiperlocal, sinergia de dos entornos. Cuadernos Periodistas **29**, 38–54 (2014)
21. Franklin, B.: Local Journalism and Local Media. Making the Local News. Routlegde, London (2006)
22. Franklin, B.: The future of journalism: developments and debates. Journalism Stud. **13**(5–6), 663–681 (2012). https://doi.org/10.1080/1461670X.2012.712301
23. Guillamet, J.: Pasado y futuro de la prensa local. In: López Lita, R., et al. (eds.) La prensa local y la prensa gratuita, pp. 181–196. Universitat Jaume I, Castellón de la Plana (2002)
24. Halvorsen, L.J., Bjerke, P.: All seats taken? Hyperlocal online media in strong print newspaper surroundings: the case of Norway. Nordicom Rev. **40**(2), 115–128 (2019). https://doi.org/10.2478/nor-2019-0030
25. Harte, D.: One every two minutes: assessing the scale of hyperlocal. Jomec J. **1**(3), 1–11 (2013)
26. Harte, D., Turner, J., Williams, A.: Discourses of enterprise in hyperlocal news in the UK. Journalism Pract. **10**(2), 233–250 (2016). https://doi.org/10.1080/17512786.2015.1123109
27. Hermida, A., Thurman, N.: A clash of cultures. Journalism Pract. **2**(3), 343–356 (2008). https://doi.org/10.1080/17512780802054538
28. Horning, M.A.: In search of hyperlocal news: an examination of the organizational, technological and economic forces that shape 21st century approaches to independent online journalism (Ph.D. Thesis). The Pennsylvania University, College of Communications, Pennsylvania (2012)
29. Hujanen, J., Lehtisaari, K., Lindén, C.G., Grönlund, M.: Emerging forms of hyperlocal media. The case of Finland. Nordicom Rev. **40**(2), 101–114 (2019). https://doi.org/10.2478/nor-2019-0029
30. Jenkins, J., Nielsen, R.K.: The Digital Transition of Local News. Reuters Institute for the Study of Journalism, University of Oxford, Oxford (2018)
31. Jenkins, J., Nielsen, R.K.: Preservation and evolution: local newspapers as ambidextrous organizations. Journalism **21**(4), 472–488 (2020). https://doi.org/10.1177/1464884919886421
32. Jerónimo, P.: Ciberjornalismo de proximidade: Redações, jornalistas e notícias online. LabCom.IFP, Covilhã (2015)

33. Jerónimo, P., Correia, J.C., Gradim, A.: Are we close enough? Digital challenges to local journalists. Journalism Pract. (Online First) (2020). https://doi.org/10.1080/17512786.2020.1818607
34. Kerkhoven, M., Bakker, P.: The hyperlocal in practice: innovation, creativity and diversity. Digit. Journalism **2**(3), 296–309 (2014). https://doi.org/10.1080/21670811.2014.900236
35. Macià Mercadé, J.: La comunicación regional y local: dinámica de la estructura de la información en la España de las autonomías. Ciencia 3, Madrid (1993)
36. Metzgar, E., Kurpius, D., Rowley, K.: Defining hyperlocal media: proposing a framework for discussion. New Media Soc. **13**(5), 772–787 (2011). https://doi.org/10.1177/1461444810385095
37. Miel, P., Faris, R.: News and Information as Digital Media Come of Age. Berkman Center for Internet & Society, Cambridge (2008)
38. Negreira Rey, M.C.: Cibermedios locais e hiperlocais en España: mapa, modelos e produción informativa (PhD Thesis). Universidade de Santiago de Compostela, Santiago de Compostela (2020)
39. Negreira-Rey, M.-C., López-García, X., Vázquez-Herrero, J.: Mapa y características de los cibermedios locales e hiperlocales en España. Rev. Comunicación **19**(2), 193–214 (2020). https://doi.org/10.26441/RC19.2-2020-A11
40. Newman, N., Fletcher, R., Shulz, A., Andi, S., Nielsen, R.K.: Digital News Report 2020. Reuters Institute for the Study of Journalism. Available at https://bit.ly/3vAPxiG (2020). Accessed 02 Jan 2021
41. Nielsen, R.K., Ganter, S.A.: Dealing with digital intermediaries: a case study of the relations between publishers and platforms. New Media Soc. **20**(4), 1600–1617 (2018). https://doi.org/10.1177/1461444817701318
42. Nygren, G., Leckner, S., Tenor, C.: Hyperlocals and legacy media. Media ecologies in transition. Nordicom Rev. **39**(1), 33–49 (2018). https://doi.org/10.1515/nor-2017-0419
43. Paulussen, S., Heinonen, A., Domingo, D., Quandt, T.: Doing it together: citizen participation in the professional news making process. Observatorio **3**, 131–154 (2007). https://doi.org/10.7458/obs132007148
44. Pignard-Cheynel, N., Noblet, A.: L'encadrement des contributions "amateurs" au sein des sites d'information: entre impératifs participatifs et exigences journalistiques. In: Millerand, F., Proulx, S., Rueff, J. (eds.) Web social Mutation de la communication, pp. 265–282. Presses de l'université du Québec, Québec (2010)
45. Pignard-Cheynel N, Gerber D, Amigo L (2019) Quand les médias locaux renouent avec leurs publics: panorama des initiatives en Europe francophone. Eur. Journalism Obs.
46. Radcliffe, D.: Where are We Now? UK Hyperlocal Media and Community Journalism in 2015. Nesta, Centre for Community Journalism, Cardiff University, Cardiff (2015)
47. Rebillard, F.: Le Web 2.0 en perspective. Une analyse socio-économique de l'internet. L'Harmattan, Paris (2007)
48. Salaverría, R., Martínez-Costa, M.P., Breiner, J., Negredo Bruna, S., Negreira-Rey, M.C., Jimeno, M.A.: El mapa de los cibermedios en España. In: Toural, C., López, X. (eds.) Ecosistema de cibermedios en España. Tipologías, iniciativas, tendencias narrativas y desafíos, pp. 25–49. Comunicación Social Ediciones y Publicaciones, Salamanca (2019)
49. Salles, C.: Disrupting journalism from scratch: Outlining the figure of the entrepreneur–journalist in four French pure players. Nordic J. Media Stud. **1**, 29–46 (2019)
50. Singer, J., Domingo, D., Heinonen, A., Hermida, A., Paulussen, S., Quandt, T., et al.: Participatory Journalism: Guarding Open Gates at Online Newspapers. Wiley-Blackwell, Malden (2011)
51. Smyrnaios, N.: Quel avenir pour les pure players journalistiques en France? La revue des médias. Available at https://bit.ly/30M2fg5 (2013). Accessed 21 Jan 2021
52. Tandoc, E.C., Vos, T.P.: The journalist is marketing the news: social media in the gatekeeping process. Journalism Pract. **10**(8), 950–966 (2016). https://doi.org/10.1080/17512786.2015.1087811

53. Tenor, C.: Logic of an effectuating hyperlocal: entrepreneurial processes and passions of online news start-ups. Nordicom Rev. **40**(2), 129–145 (2019). https://doi.org/10.2478/nor-2019-0031
54. Thurman, N., Hermida, A.: Gotcha: How newsroom norms are shaping participatory journalism online. In: Tunney, S., Monaghan, G. (eds.) Web Journalism: A New Form of Citizenship?, pp. 46–62. Sussex Academic Press, Eastbourne (2010)
55. Zelizer, B.: The Changing Faces of Journalism: Tabloidization, Technology and Truthiness. Routledge, New York (2009)

María-Cruz Negreira-Rey Ph.D. in Communication, Universidade de Santiago de Compostela (USC). Lecturer of Journalism at the Department of Communication Sciences and member of Novos Medios research group (USC). Her research focuses on local journalism and the development of local and hyperlocal media in Spain, digital journalism and social media.

Laura Amigo Ph.D. candidate at the Academy of Journalism and Media of the University of Neuchâtel (UniNE). Her research focuses on the relationships between news media and audiences, editorial strategies and local journalism. She is a scientific collaborator for the LINC (Local, News, Innovation, Community) research project based at the AJM (UniNE). She teaches Cross-cultural communications at the University of Lille (France) and has previous professional experience in the journalistic and communication fields.

Pedro Jerónimo Ph.D. in Information and Communication on Digital Platforms, University of Porto and University of Aveiro. Researcher at LabCom (University of Beira Interior, Covilhã); co-editor of the journals *Estudos em Comunicação* (LabCom) – indexed to Scopus – and *Estudos de Jornalismo* (SOPCOM – Portuguese Association for Communication Sciences); founder and coordinator of SOPCOM' Local and Community Media section. His research focuses on local journalism and the development of local media in Portugal, digital journalism, and social media.

Social Implications of Paywalls in a Polarized Society: Representations, Inequalities, and Effects of Citizens' Political Knowledge

Tamás Tóth, Manuel Goyanes, Márton Demeter, and Francisco Campos-Freire

Abstract Since the emergence of the Internet, news organizations have been at crossroads. Print operations based on high advertisement rates and a robust subscription model allow media companies to bolster news workers and revenue streams. Contrarily, in online media, advertisement incomes have reduced. Interestingly, while readers' consumption patterns privilege online news, several studies empirically show that most citizens are reluctant to pay for news services, supporting the culture of free mindset. However, many news organizations implemented paywalls as a response to the general failure of free business models. With this backdrop, do citizens who pay for news, through their news repertoires, have more, less, or similar knowledge about public affairs and politics? More importantly, do paywalls engender news information inequalities among citizens who do not intend to pay for news? This chapter problematizes the presumed effects of paywalls on citizens' knowledge about public affairs and politics.

1 Introduction

The arrival of the Internet has had a major impact on journalism, and consequently, it has dramatically changed news consumption. On the one hand, online newspapers have challenged their print counterparts, and the latter suffered from a severe

T. Tóth
Corvinus University of Budapest, Budapest, Hungary
e-mail: tamas.toth3@uni-corvinus.hu

M. Goyanes
Carlos III University, Madrid, Spain
e-mail: mgoyanes@hum.uc3m.es

M. Demeter
University of Public Service, Budapest, Hungary
e-mail: demeter@komejournal.com

F. Campos-Freire (✉)
Universidade de Santiago de Compostela, Santiago de Compostela, Spain
e-mail: francisco.campos@usc.es

J. Vázquez-Herrero et al. (eds.), *Total Journalism*, Studies in Big Data 97,
https://doi.org/10.1007/978-3-030-88028-6_13

decline in copy sales as they lost their leading position in providing information and entertainment [33]. On the other hand, online newspapers are also in an economic crisis because the news is quickly available on several websites for free; therefore, the readers, who do not want to pay directly for their news consumption, can switch from one medium to another. Consequently, the fastest news provider acquires a larger audience and might have more income from advertisements. Although online newspapers form a considerable part of entertainment and supply information on the web, their total share from the digital advertisement market is meager compared to technology giants such as Facebook and Google. In other words, Facebook and Google monopolize the industry as they have more than 70% share from the advertisement market on the Internet. Therefore, online newspapers, which expect their primary incomes from the advertisements, also face economic problems [35]. Consequently, several online newspapers have started to change their strategy regarding free access to news; thus, paywalls emerged to gain extra revenue.

This chapter aims to introduce the presumed association between online newspapers' paywalls, their impact on inequalities in news consumption, and readers' willingness to pay for news services. The chapter relies on four vital elements that are parts of the arguments related to journalism and democracy: (1) the culture of free, (2) newspapers' business models regarding the different strategies of providing paid contents, (3) readers' willingness to pay for online news, and (4) the effects of paywalls on democracy. We argue that scholars should consider the elements mentioned above when analyzing the relation between (free) news consumption on politics or public affairs and democracy.

Thus, we start our introduction with the characterization of the culture of free, which is an important feature of the Internet and the essential part of the technology-utopia [17, 30]. In other words, first, we provide the features of a phenomenon that might take us a few steps closer to digital democracy, which is free online access to a public good, namely news.

2 The Culture of Free

The term 'news consumption' suggests that news appears in a commoditized context. In this light, the online media transform news into products to sell in exchange for the audience's attention, time, personal data, and money. More precisely, commoditization means that news services transform from information providers—which might fit social needs—into organizations that analyze the marketplace to discover what items could be sold for the consumers. Despite the monetarization of online news, its economic value keeps declining because a major group within the audience tends to pay less or nothing for the 'stories' (we will discuss this in detail later).

Essentially, 'the culture of free' is not equal to free culture [25] but an aspect that considers online news as a freely available public good that is part of free consumption habits and a vital element of democracy [22]. It is important to emphasize that even though the culture of free supports unrestricted access to the news services' content, the online news cannot be free of charge because the utilized electricity, electronic devices, and broadband services still have specific costs. Additionally, as mentioned above, time—especially consumers' free time when they check the updates of online newspapers—and attention are also parts of the so-called 'price'. Essentially, there is no free lunch.

News production has high fixed and low marginal costs, which means that creating content, journalists' salaries, sales, and marketing are expensive while distributing the product via the Internet is cheap [42]. In the offline world, the material products' prices will increase from the stage when the goods' planning starts to the phase when the commodity emerges in the marketplace. In contrast, in the digital sphere, the process is reversed in news production as the product's price declines and supposedly converges to zero when the consumer receives or approaches the product. In other words, the costs of providing the content are high; however, after its first production, the information will inevitably become free because many consumers tend to disregard copyrights, and they download and share content. One explanation for this phenomenon might be that the number of suppliers grows rapidly while the demand does not expand dramatically. Moreover, the generation born into the broadband services' world does not intend to pay for immaterial goods such as online news. Most of them think that paywalls are pointless if they can reach the necessary information, including news, without additional costs [22]. Specific economic examinations that focus on 'freeconomics,' 'bits economy' [1], or 'network economy' [42] support these ideas. These analyses enhance the deflationary effect of the Internet on news production. Moreover, the generation which has grown up with broadband Internet services understand, or, at least, feel that paying for online news is not that crucial.

Other analyses that scrutinize media audiences, such as reception studies, suggest that media consumption might be related to social action and identity formation [15]. In this vein, the Internet might offer the ethos of democracy because its structure might provide news consumption for free. Although people might not like news, it can still function as a public good [47] because it is useful and helps keep the audience informed and up-to-date.

In sum, as Goyanes and colleagues [22] argue, the culture of free has at least four essential features to be taken into consideration. First, there is a large gap between news production's economic value and the number of people who are willing to pay. Second, the audience knows how to download and share news; therefore, they praise freeconomics and reject paying for free access. Third, the culture of free has an additional meaning: reading news might also be a free time activity that aims to pass readers' time to catch-up with the latest trends and happenings via the pulsing news services. Finally, the audience also considers that news consumption without subscribing to paywalls is not entirely uncharged because attention, data, and time, alongside the costs of devices, broadband services, and electricity, are not for free. Although consumers must pay something to reach online news services, they regard news as a free public good like air and seawater. On the suppliers' side, as Van der Wurff [47] argues, it is pointless to exclude non-payers because the group of people who might profit from the public good would narrow, while the costs of production would remain the same. Briefly, excluding non-payers would reduce welfare, however, public goods fall under the category of government-funded commodity or service. The culture of free mindset is vital when citizens' knowledge of public affairs and politics kicks in. Taxpayers' income has funded both public affairs and politics; thus, news coverages on incidents, such as elections, legislation, frauds, etc., which are phenomena that affect ordinary people's lives, should be aligned with the culture of free because it is citizens' democratic right to be informed on these issues. At this point, one question arises: should news, which has the features of public goods, not be a part of public services provided and funded by governments? We will discuss this question in the last section of this book chapter.

In the next section, we will characterize how media organizations started to adjust to the situation wherein the news price started to reduce and implemented new business models [50], aiming to persuade the audience to pay for their product, namely, the content.

3 The Business Models of Online Newspapers

As Goyanes [19] points out, traditional newspapers have a dual concept of gaining incomes: the revenue model focuses on copy sales and advertising. In contrast, the online newspapers' revenue primarily derives from advertising, thus, they do not have enormous income from subscriptions. Indeed, there are some exceptions. For instance, at the beginning of the 2010s, *The Wall Street Journal* (WSJ) charged more than a million subscribers for approximately $100 a year [47], and the latest report showed that WSJ topped 2 million digital subscriptions until the end of 2019 [3]. We will discuss the WSJ's success in the subscription at the end of this section.

The growing popularity of online news consumption has led to a decrease in copy sales, while more readers turned to online journals to stay informed. The management of online news services is also perceived as a severe contest: the rapid flux of 'freely' available online news created brutal competition among media companies. Many traditional newspapers, which have both print and online versions, experienced the decline of copy sales, the decreasing value of online news, and the adverse effects of technology giants that created a large financial loss on the journals' side. Consequently, the digital revenue from advertising cannot compensate print losses [29]. Therefore, digital newspapers' management aims to convince readers, potential customers to pay for the news [18].

As scholars argue, despite most citizens not wanting to pay for online news, there are some tactics for persuading people to pay for news. For instance, offering personalized information for the customer, enhancing the role of high-quality journalism that needs to be supported by donors, and focusing on specific territories where individual consumers live for supplying information about local happenings [20].

Before we characterize the applicable paid content strategies of online newspapers, we also aim to briefly discuss the problems with digital giants, such as Google and Facebook, that severely affect news media. First, as we mentioned earlier, those companies monopolized the online advertisement industry. Online newspapers gain most of their income from advertisements, but as the market started to narrow, they adopted the clickbait-style, which is a way of broadcasting news that evokes sensational journalism. In other words, the tech giants undermined both the quality of online journalism and its financial support [35]. Second, misinformation and fake news might easily mislead the users, spread virally and uncontrolled on Facebook. The social harms caused by proliferating misinformation could lead to mistrust in any service that supplies news, including online newspapers. Finally, while the largest search engine and social site utilize the profit from ads to refine their surveillance methods for providing personalized bulletins, the online news services struggle to retain audiences, find donors, catch-up with other newspapers' pace, filter misinformation, and in some instances, modify their structures from free news to implement differing paywall versions.

Even so, online newspapers have an opportunity to implement the free business model based on no extra charges for the audience to have access to content. In contrast, online newspapers might fundamentally apply two types of business models: micropayments and paywalls. The former refers to limited access to the news services, while the latter is for a complete media outlet [21]. However, there are various methods to make people pay for news: (1) daily or weekly pay-per-view, (2) freemium by which users might reach out a portion of contents while the rest of the news falls under paywalls, (3) the metered model that let the audience consume a certain amount of news and then, it invites the potential customers to subscribe, (4) donations and public funding, (5) digital kiosks that simulate offline news-stands, (6) providing unique business, financial, or (hyper) local news that cannot be easily found on the Internet, (7) the speculative idea of 'experience economy' presenting news as virtual reality by following journalists in real-time to particular places, and finally, (8) line and brand extensions such as books written by journalists that are part of the same brand [18, 36, 47]. In sum, online newspapers can either provide their content for free of charge or realize income from user payments, advertising, charitable funding, transaction fees, and other extensions [32, 47].

In the subsequent section, we review the literature on readers' willingness to pay for online news to gather the most appropriate outcomes and discussions on the ties between online news consumption and democracy.

4 The Willingness to Pay for Online News

It is important to emphasize that some scholars consider the press' offline and online versions to be homogenous products [34]—which means that they are replaceable—while others claim that press is heterogeneous [12] as it consists of unique, personalized, local information, which is challenging to substitute [46]. Technically, both aspects are correct, but if we focus on the product (e.g., newspaper) rather than the industry (e.g., media and advertisement), homogenous and heterogenous newspaper titles might be an applicable distinction [19]. Specific newspapers, such as the

Financial Times and *The Wall Street Journal*, supply unique knowledge for customers interested in business news [44]. From another aspect, local or rural newspapers have an advantage based on geography [2] since they give service to smaller communities within their region and update the inhabitants on events within their territory (Ross 2007). Analyses have shown that in the United States, 72% of adult citizens follow local news [27]. Additionally, in Europe, local newspapers—whether print or online—are not inferior to their national counterparts [31].

Scholars argue that most of the people who visit media channels' websites are reluctant to pay for online news [16], but they instead tend to pay for other immaterial goods like movies, series, music, eBooks, applications, software, video, and computer games, etc. [18]. In this light, scholars argue that most people follow the 'free mentality' of online news consumption [11]. If we compare news to the products above, we might claim that the former is not purely entertaining but still useful (for instance, getting information on accidents on the highway), while the latter contains pure elements of the entertainment industry. According to specific studies, if one still aims to pay for news, the maximum price for an article could be $0.25, while users tend to pay between $0.69 to $1.29 for a song on iTunes [41, 47]. As former studies have shown, younger people, especially those who use tablets and smartphones, tend to pay for online news with higher frequency than older users from the United States [18]. Experts have previously claimed that gender is not a significant factor in predicting the willingness to pay for general online news [7, 18], and other research has shown that men are more inclined to pay for general online news [8]. However, in the United States, older people, especially women, are more likely to pay for online local newspapers because it can affect their daily routines [19]. Additionally, Fletcher and Nielsen [14] suggest that people who already paid for print news are more likely to pay for online news services.

As Graybeal and Hayes [23] claim, micropayments can be successful if the specific online news service provides general news for free and asks the audience to purchase its in-depth analyses. The presence of a prestigious journalist—which might have a positive impact on the audience's perception related to credibility, quality [10], or the exclusivity on the news, implying that the same or similar content is difficult to be found on the Internet [23]—might increase the willingness to pay for online news. Goyanes and colleagues [21] argue that Spanish readers' willingness to pay for online news is higher when a prestigious journalist publishes an article in the new media, and they penalize the mainstream news services if the condition is *exclusive*. However, in the non-exclusive context, purchasing intent also increases in the new, the mainstream, and fictitious media, if a well-known, trusted journalist writes an article [21].

After a brief overview of the readers' willingness to pay for online news in specific conditions, we discuss the intertwinement between paywalls, the willingness to pay, and possible effects on equality.

5 Paywalls, Inequalities and Possible Tensions

Hypothetically, spectacular inequalities in news consumption could quickly emerge if every online news services management decided to change from the free model to paywalls. As Goyanes [18] argues, people with a lower annual income ($75,000 or less) were less likely to subscribe to paywalls or pay for news than citizens with a high annual income ($150,000 or more) in the United States. This outcome might be a warning sign: if online newspapers modify their business models in the direction of paid content, citizens with lower incomes might have less opportunity to receive accurate news.

Being informed is crucial in public affairs and politics because people have the right to know such occurrences. As the trends suggest, the less income a person has, the higher the chance is of being uninformed on relevant issues. In other words, people with a higher income might have a better opportunity of being updated on contents that fall under the category of public affairs, which are not the type of information that fits exclusivity. In this vein, hypothetically, if news services implemented paid content strategies, online journalism cannot function as a watchdog [6] or as a gate-opener [28] power. As a watchdog, online journalism makes sense if it is a pillar of checks and balances and informs large masses. As we argued previously, the paywalls might harm the audience's size. News services on the Internet might become gate-openers, which means that online newspapers could provide the options of discussions and arguments if the broadest range of citizens could have access to the contents.

Paywalls might easily push people in the direction of Facebook or Twitter, wherein misinformation spreads rapidly. Social sites provide the information for free, and it can be a charming option to citizens who do not intend to pay for news. The questionable validity of information and fake news might challenge the citizens' right to be updated with fair and accurate information on public affairs and politics. These contents (e.g., fake news) are often outrageous [35]; thus, they might polarize the audience quickly and create filter bubbles. Sunstein [45] suggests that news consumption on the Internet might create an enormous gap between citizens who have differing political attitudes because they might get stuck into the news service that fits their preferences. Additionally, the algorithm of Facebook assesses and detects the user's preferences by analyzing their data. Therefore, the social site can offer 'relevant' news, which might be part of political propaganda. Online media, especially social sites, are fertile grounds for the proliferation of false information and conspiracy theories. Fake news might easily radicalize people and support, for instance, populist political agents, who are the masters of spreading fictitious stories, of gaining popularity or maintaining their power [4]. Populists could convince people with lower incomes and education [13] at national levels as they bypassed the gatekeepers—namely, traditional or online media—by spreading fake news on social sites. At this point, one might ask that how does the question of education appears in the issue of paywalls? The answer is plausible: news services are to inform and educate people by supplying accurate information to make the audience understand the context and flux of events.

Like several experts who support the culture of free, we argue that online news is a public good that has a constantly decreasing economic value. As a result, paid content strategies—which recently emerge with increasing frequency [43]—do not support but limit democracy because citizens with modest incomes do not have the same opportunity to be educated and informed as people with high incomes. Online newspapers must realize incomes from the market to sustain their operation, a goal that has never been so difficult to achieve. Again, the utopistic idea of government-supported news services, including every professional online newspaper, might provide a fair opportunity to give access to citizens who already paid for the broadband services and devices. Citizens' taxes, which are also included in broadband services and electronic items, could make them eligible to have rights for receiving news as a public good. Governments and states cannot function without taxes, or in other words, they would be unable to maintain public goods, make decisions, legislate, and so on, without the tax-payers contribution. Undoubtedly, governments and states greatly influence ordinary people's lives who primarily receive information about politics and public affairs via the media. Therefore, news in the online realm—especially those which highlight political events and public affairs—is part of the public good because they keep citizens informed on events that might shape their social or individual situation. The free access to online news written by trusted professionals is a common interest because the proliferation of fake news has already proven to be devastating.

6 Conclusion

Following current literature, we conclude that consequential to the digitalization of the market [49], media faces severe sustainability problems that could be fatal for those media organizations that cannot elaborate appropriate business models for obtaining financial support [9, 26]. Media platforms with a primary focus on news production are in an even more difficult situation since the high standards of professional journalism, and journalistic ethics are working against current trends by which news content should be broadcasted as fast as possible, and above all, it should be accessed freely [24, 37]. Consequential to the growing financial struggles, news providers were encouraged to develop newer business models [5]. While there are several business models offered by both scholars and media professionals [9], researchers agree that both total and hybrid paywalls might further exacerbate the problems of news media since a growing number of consumers are reluctant to pay for news consumption [16, 18]. However, alternative business models like public funding media are also problematic, since there are several competing demands from other social areas and institutions such as public funded education and the healthcare system [38]. Moreover, current financial crises make it unlikely that the public funding media model will gain a widespread political and public support [26]. We can add that the COVID-19 pandemic crisis might also encourage the public and policy makers to redirect public spending towards health care.

This chapter shows that new media business models should provide a service that is generally available and accessible to the consumers at little or no cost [48]. In news production, these features should be extended by professionalism to provide an alternative to social media where news is freely available, but news production is uncontrolled. If they are committed to democracy based on properly informed citizens and where news is a means of public good, both scholars in journalism studies and media professionals should develop sustainable business models for news media production.

Funding This work was supported by the National Program of R&D oriented to the Challenges of Society and the European Regional Development Fund (ERDF) about *New Values, Governance, Funding and Audiovisual Public Services for the Internet Society: European and Spanish Contrasts* (RTI2018-096,065-B-I00).

References

1. Anderson, C.: Free: The Future of a Radical Price. Random House, New York (2009)
2. Barnett, S., Townend, J.: Plurality, policy and the local: can hyperlocals fill the gap? J. Pract. **9**, 332–349 (2014)
3. Benton, J.: The Wall Street Journal joins The New York Times in the 2 million digital subscriber club (2020). Available at https://bit.ly/2MVSed4. Accessed 05 Jan
4. Bergmann, E.: Populism and the politics of misinformation. Safundi **21**, 251–265 (2020)
5. Brevini, B.: Public service broadcasting online: a comparative European policy study of PSB 2.0. Palgrave Macmillan, New York (2013)
6. Carpentier, N., Cammaerts, B.: Hegemony, democracy, agonism and journalism: an interview with Chantal Mouffe. J. Stud. **7**(6), 964–975 (2006). https://doi.org/10.1080/14616700600980728
7. Chyi, H.I.: Willingness to pay for online news: an empirical study on the viability of the subscription model. J. Media Econ. **18**(2), 131–142 (2005). https://doi.org/10.1207/s15327736me1802_4
8. Chyi, H.I.: Paying for what? how much? And why (not)? predictors of paying intent for multiplatform newspapers. Int. J. Media Manag. **14**(3), 227–250 (2012). https://doi.org/10.1080/14241277.2012.657284
9. Curran, J.: The future of journalism. J. Stud. **11**(4), 464–476 (2010). https://doi.org/10.1080/14616701003722444
10. Das, R., Pavlíčková, T.: Is there an author behind this text? a literary aesthetic driven approach to interactive media. New Media Soc. **16**(3), 381–397 (2014). https://doi.org/10.1177/1461444813481296
11. Dou, W.: Will internet users pay for online content? J. Advert. Res. **44**(4), 349–359 (2004). https://doi.org/10.1017/S0021849904040358
12. Doyle, G.: Understanding Media Economics. Sage, London (2002)
13. Eatwell, R., Goodwin, M.: National Populism: The Revolt Against Liberal Democracy. Penguin Books, United Kingdom (2018)
14. Fletcher, R., Nielsen, R.K.: Paying for online news: A comparative analysis of six countries. Digit. J. **5**(9), 1173–1191 (2016). https://doi.org/10.1080/21670811.2016.1246373
15. Garnham, N.: Emancipation, the Media, and Modernity: Arguments about the Media and Social Theory. Oxford University Press, Oxford (2000)

16. Geidner, N., D'Arcy, D.: The effects of micropayments on online news story selection and engagement. New Media Soc. **17**(4), 611–628 (2015). https://doi.org/10.1177/1461444813508930
17. Gilder, G.: Life After Television. WW Norton and Co., New York (1994)
18. Goyanes, M.: An empirical study of factors that influence the willingness to pay for online news. J. Pract. **8**(6), 742–757 (2014). https://doi.org/10.1080/17512786.2014.882056
19. Goyanes, M.: The value of proximity: examining the willingness to pay for online local news. Int. J. Commun. **9**, 1505–1522 (2015)
20. Goyanes, M.: Why do citizens pay for online political news and public affairs? Socio-psychological antecedents of local news paying behaviour. Journal. Stud. **21**(4), 547–563 (2020). https://doi.org/10.1080/1461670X.2019.1694429
21. Goyanes, M., Artero, J.P., Zapata, L.: The effects of news authorship, exclusiveness and media type in readers' paying intent for online news: An experimental study. Journalism (Online First) (2018). https://doi.org/10.1177/1464884918820741
22. Goyanes, M., Demeter, M., de Grado, L.: The culture of free: construct explication and democratic ramifications for readers' willingness to pay for public affairs news. Journalism (Online First) (2020). https://doi.org/10.1177/1464884920913436
23. Graybeal, G.M., Hayes, J.L.: A modified news micropayment model for newspapers on the social web. Int. J. Media Manag. **13**(2), 129–148 (2011). https://doi.org/10.1080/14241277.2011.568808
24. Larrondo, A., Domingo, D., Erdal, I.J., Masip, P., Van den Bulck, H.: Opportunities and limitations of newsroom convergence: a comparative study on European public service broadcasting organisations. J. Stud. **17**(3), 277–300 (2016). https://doi.org/10.1080/1461670X.2014.977611
25. Lessig, L.: Free Culture: How Big Media Uses Technology and the Law to Lock Down Culture and Control Creativity. The Penguin Press, New York (2004)
26. Macnamara, J.: Remodelling media: the urgent search for new media business models. Media International Australia **137**, 20–35 (2010)
27. Miller, C., Purcell, K., Rosenstiel, T.: 72% of Americans follow local news closely. Pew Research (2012). Available at https://pewrsr.ch/3cY4YdQ. Accessed 11 Dec 2020
28. Mouffe, C.: The Return of the Political. Verso, London (2006)
29. Myllylahti, M.: What content is worth locking behind a paywall? digital news commodification in leading Australasian financial newspapers. Digit. J. **5**(4), 460–471 (2017). https://doi.org/10.1080/21670811.2016.1178074
30. Negroponte, N.: Being Digital. Alfred A. Knopf, New York (1995)
31. Newman, N. (ed.): Reuters Institute Digital News Report: Tracking the Future of News. Reuters Institute for the Study of Journalism, Oxford (2012)
32. Nordenson, B.: The uncle sam solution. Columbia Journalism Review 46 (2007)
33. Picard, R.G.: Cash cows or entrecote: publishing companies and disruptive technologies. Trends Commun. **11**, 127–136 (2003)
34. Picard, R.G.: Why journalists deserve low pay (2009). Available at https://bit.ly/3cVQ3Ak. Accessed 10 December 2020
35. Pickard, V.: Restructuring democratic infrastructures: a policy approach to the journalism crisis. Digit. J. **8**(6), 704–719 (2020). https://doi.org/10.1080/21670811.2020.1733433
36. Pine, B.J., Gilmore, J.H.: Welcome to the experience economy. Harv. Bus. Rev. **76**, 97–105 (1998)
37. Połońska, E., Beckett, C.: Public Service Broadcasting and Media Systems in Troubled European Democracies. Palgrave Macmillan, Cham (2019)
38. Prahalad, C.K., Ramaswamy, V.: The Future of Competition: Co-creating Unique Value with Customers. Harvard Business Press, Boston (2004)
39. Reuters Institute (2012) Digital News Report. Available at https://bit.ly/374vu18. Accessed 11 Dec 2020
40. Ross, K.: Open source? Hearing voices in the local press. In: Franklin, B. (ed.) Local Journalism and Local Media, pp. 254–266. Routledge, London (2006)

41. Seward, Z.: Journalism online's charging clients a 20% commission. NiemanLab.org (2009). Available at https://bit.ly/2Z9hi2E. Accessed 11 Dec 2020
42. Shapiro, C., Varian, H.R.: Information Rules: A Strategic Guide to the Network Economy. Harvard Business Press, Boston (1998)
43. Sjøvaag, H.: Introducing the paywall: a case study of content changes in three online newspapers. J. Pract. **10**(3), 304–322 (2016). https://doi.org/10.1080/17512786.2015.1017595
44. Steinbock, D.: Building dynamic capabilities: the wall street journal interactive edition: a successful online subscription model (1993–2000). Int. J. Media Manag. **2**(3–4), 178–194 (2000). https://doi.org/10.1080/14241270009389936
45. Sunstein, C.: Republic.com 2.0. Princeton University Press, Princeton (2007)
46. Sylvie, G., Witherspoon, P.D.: Time, Change, and the American Newspaper. Lawrence Erlbaum Associates, Mahwah, New Jersey (2002)
47. Van der Wurff, R.: The economics of online journalism. In: Siapera, E., Andreas, V. (eds.) The Handbook of Global Online Journalism, pp. 231–250. Wiley-Blackwel, New York (2012)
48. Vukanovic, Z.: The influence of digital convergence/divergence on digital media business models. In: Clua, E. et al. (eds) International Conference on Entertainment Computing, pp. 152–163, Poznan (2018)
49. Waisbord, S.: Communication: A Post-Discipline. Polity Press, London (2019)
50. Wang, J.: New media technology and new business models: speculations on 'post-advertising'paradigms. Media Int. Australia **133**(1), 110–119 (2009). https://doi.org/10.1177/1329878X0913300115

Tamás Tóth Junior researcher at the Corvinus University of Budapest and the University of Public Service (Hungary). His research interests include populism in the United States and Hungary. Recently, he has attempted to construct a methodological and theoretical refinement in the research field above, namely explicit and implicit populism. Additionally, he also scrutinizes the center-periphery structure of globalized and internationalized science. His papers have been published in leading journals in communication and science metrics.

Manuel Goyanes Ph.D. in Journalism and Mass Communication by the University of Santiago de Compostela (Spain). Manuel Goyanes currently serves as a lecturer at Carlos III University in Madrid. His research addresses the influence of journalism, digital media, and new technologies over citizens' daily lives. He is also interested in global inequality in academic participation, the systematic biases towards global South authors, and publication trends and patterns in Communication.

Márton Demeter Ph.D. in Communication and Media Studies. He teaches media and communication at the University of Public Service (Budapest). His primary research foci are the sociology of communication research, global knowledge production and scientometrics, but he also extensively published on populist communication style. His works have been published in leading international journals. His latest monograph entitled Academic Knowledge Production and the Global South: Questioning Inequality and Underrepresentation was published in 2020 by Palgrave Macmillan.

Francisco Campos-Freire Full Professor of Journalism in the Faculty of Communication Sciences of the University of Santiago de Compostela. His main research lines are the management of media companies, the study of the impact of social networks and innovation on traditional media, and the funding, governance, accountability and transformation of Public Service Media. In the professional field, he has worked as editor and director in several Spanish newspapers and television and radio companies.

The Construction of Communicative Space: The Nature of COVID-19 Information in Italy

Laura Solito and Carlo Sorrentino

Abstract The monopolistic role played by the pandemic offers a lens through which we can observe how the digital environment is reshaping the entire communicative space. Starting from Bourdieu's concept of the "journalistic field", what we now have is a situation in which the voices of journalists and of the public sector's communicators are flanked by those of several other actors, who together are creating a polyphonic, multidimensional communicative space. Thanks to their communicative resources, they take on a unique role and position in the communicative space. Based on these premises, the present work aims to empirically explore how 10 different main actors within the sphere of public debate managed their communication strategies on Facebook during the Italian lockdown. In order to test the theoretical framework proposed here, we analyze 6 different kinds of communicative resource: reputational, institutional, professional, argumentative, performative and relational ones.

1 Introduction[1]

2020 was devastated by the COVID-19 pandemic. Global information and news focused almost exclusively on this one issue, and this was unlike anything that has happened for decades. An informational overload, produced by the ease of access to information not only in terms of the consumption thereof, but also with regard to its production, has generated a 'infodemic', word coined to underscore the difficulty of distinguishing ascertained information from less reliable, or clearly false, "rumours" [27, 34].

[1]This essay was conceived and written jointly by the authors, as were all stages of the research design, the analysis and the presentation of data and findings. More specifically, Carlo Sorrentino wrote the introduction and the first and second section, while Laura Solito the third and fourth section.

L. Solito · C. Sorrentino (✉)
Dipartimento di Scienze Politiche e Sociali, Università Degli Studi di Firenze, Firenze, Italy
e-mail: carlo.sorrentino@unifi.it

J. Vázquez-Herrero et al. (eds.), *Total Journalism*, Studies in Big Data 97,
https://doi.org/10.1007/978-3-030-88028-6_14

An information space has emerged that is highly subjective and full of questions, actions competing to be "in the public sphere". This dense public sphere [36] is the result of the expansion of communicative spaces, of the acceleration of the time it takes to exchange information, and of the shrinking of the distances and roles separating senders and receivers of information. Indeed, this transformation has resulted in the creation of numerous neologisms highlighting the existence, in each one of us, of productive and consumption practices, often occurring at the same time.

This overcrowding has meant that information is now suspended between the multiplication of opportunities and the pitfalls of excessive fragmentation [20].

The monopolization of information by the COVID-19 crisis throughout the course of 2020—only 16% of Italian journalists did not deal with the question in 2020, as per the Communications Authority's Report 2020—renders this an excellent case study by which to observe how the various actors concerned played out their parts in this dense vortex, and what communicative resources they utilized in doing so.

This on-going research project aims to illustrate the shape of the communicative space pertaining to this event. The body of data in our possession comprises messages sent via Facebook, during the first lockdown period in Italy (8 March–4 May), by 10 different social actors with clearly different roles, completely different powers, responsibilities and specific purposes, who together contributed towards filling the communicative space. Newspaper articles have been deliberately avoided, since the hypothesis we wish to work on is the central role, within digital communication, played by those sources most involved, depending on the event being recounted. In fact, the theoretical framework of our study, in keeping with the spirit of the entire volume, which focuses on the question of Total Journalism, concerns the greater relevance of those social actors most capable of establishing the newsworthiness of information, due to both the institutional role they play and to other qualities they bring to the table.

The chosen accounts, which are not intended to be exhaustive however, are the following: those of the Prime Minister's office, of Prime Minister Giuseppe Conte, of the two Regional Governments most concerned—Lombardy and the Veneto—and of the two Presidents of said Regional Governments, Attilio Fontana and Luca Zaia; those of the two institutions most involved, namely the Italian Civil Defence Corps and the Higher Institute of Health; and those of Andrea Scanzi and Renato Porro, the two most-viewed journalists in the social media sphere.[2]

First and foremost there are the institutional actors: as the latest report by Osservatorio sul Giornalismo [26] confirms, institutional sources of information have been increasingly utilized by journalists, with their share rising from around 70–86.5%, but also by members of the public, who accessed those sources to an unprecedented

[2] We used two software programmes, N-Vivo and Atlas.Ti, for the quantitative analysis of the various profiles and for the identification of key metrics, including: the total number of reactions (comments, sharing, likes and other reactions expressed by emoticons), the number of fans and their increase over the period in question; and the rate of engagement. We also conducted a qualitative analysis of the individual posts, both by establishing the presence of textual content, graphics, links or photographs in the published content, and by analyzing recurrence based on the examination of key words and semiotic markers.

extent during the period in question. The same is also true of the scientific institutions and the leading scientists who have intervened in the debate, and who have become the go-to sources of information for around one half of the population.

So, the flow of information has not been exclusively produced by journalists, but has taken on a much more magmatic, disorderly form, and derives from a variety of social actors. It is therefore interesting to examine the diverse communicative resources employed by our selected actors in order to establish their reputation, which in turn is a pre-requirement for the trust placed in them [16].

The typology we propose comprises six different types of resource:

- reputational resources;
- institutional resources;
- professional resources;
- argumentative resources;
- performative resources;
- relational resources.

They are flexible, non-mutually exclusive resources, whose importance varies depending on the changing argumentative framework and on the redefinition of the news angle from which diverse questions and events are dealt with. They are seldom clearly quantifiable, and they require each individual sender of information to possess a specific communicative capacity. The uniqueness of each resource, together with the manner in which such resource is activated, shall be described using examples taken from the results of the initial processing of our data-set. However, first of all we shall illustrate the theoretical framework of our proposed hypothesis.

The present chapter will focus above all on the theoretical framework of our working project, while the empirical data and findings we are going to report in the following sections are the result of an initial extrapolation and processing of the content gathered, designed to support the reflections proposed here.

2 A Journalistic Heresy

Bourdieu [7, 8] described the structure of the journalistic field by using two Cartesian axes, each of which having two poles: autonomy-heteronomy in one case, and orthodoxy-heresy in the other.

Although his proposal has been analyzed and criticized, above all with regard to the autonomy-heteronomy axis [2, 4, 35], it is argued that the transformations of the digital world have had a greater impact on the orthodoxy-heresy axis.

Clearly it is not easy to establish what journalistic orthodoxy consists in. We attempt to do so by making reference to the operating procedures—the "strategic rituals"—established by Tuchman [41] and Schudson [32]: facts distinguished from opinions, the completeness of information, verification using several sources, and so on. The disintermediation, hybridisation and re-mediation that characterize the digital revolution appear to indicate an inevitable leaning towards heresy. The growth of social actors who perform journalistic acts, their great public visibility and the more immediate nature of the diffusion of news by such actors—often comparable to that of professional journalists—result in the transformation of heresy into rule. Hence a paradox! Consequently, the French sociologist's conceptualisation is no longer capable of describing the dynamics of the communicative system. In fact, some years ago [37]. we suggested its place be taken by Foucault's concept of communicative dispositive:

> a dispositive is a decidedly heterogeneous ensemble that implies discourses, institutions, architectural structures, prescriptive decisions, laws, administrative measures, scientific statements, philosophical, moral or philanthropic propositions, [...] strategies between different power relationships supporting different types of knowledge and supported by those types of knowledge [15:299–300].

Agamben [1] defines the dispositive as "an encounter of power relations and knowledge relations". Foucault's expression aptly describes the web of interactions that have modified the structure of relations between senders and receivers—which Jenkins [23] calls participatory culture—that enriches cultural dialogue. Deuze [13] adds a further dimension to that of participation, namely the "reutilization" aspect: this consists in the opportunity, for both senders and receivers, to combine, reutilize and share content. Examples of this include the gradual expansion of open-source information and the joint creation of content using software applications based on Wiki.[3] The content is never completed, but is constantly a work-in-progress.

What we would like to emphasize here in regard to these processes is the significant evidence of the intrinsically relational nature of journalism; it is less clearly structured in organizational and hierarchical terms than it once was, and is now characterized by a series of practices that are often established outside of editorial offices, perhaps in more precarious, less well-paid working conditions but based on greater freedom from the restrictions of corporate control. This does not mean, however, that they centre less on collective working, but simply that this takes very different forms from that of the traditional editorial office.

Thus the heresy in question manifests itself in the form of a blurred distinction between what a journalistic product has traditionally been considered to be, and what now composes the information jigsaw.

In fact, it is this continuous process of transformation, which journalism studies have focused on for a number of years now, and with regard to which the term 'crisis' is often used to describe models challenging past certainties [42], which impacts the forms that the public sphere has taken, which for years has been substantially defined by the logic of the mass media [40]. The public sphere has been described as disrupted and disconnected [3, 12, 29, 30], which makes it difficult to establish and stabilize the processes by means of which the public agenda is defined and climates of opinion are built [11, 18, 19]. The absence of a centre further blurs the key hierarchies, to the point where Bolter [6] talks about the collapse of such hierarchies.

[3] Wiki: an application that permits the joint creation and modification of a website's pages.

This process does not necessarily lead to uncontrollable anarchy, but it certainly results in a hybridisation of models, thanks also to the increased forms of interrelationship among the various media. Bentivegna and Bocca Artieri talk of an interrelated public sphere defined by "an agenda that is the result of numerous layers of writing and reading, comprising both the individual and the collective spheres […], and that constitutes […] the current expression of the convergence of information in contemporary society" [5]. In other words, the effective metaphor of cultural chaos referred to by McNair [24] does not necessarily point to an incontrollable public sphere, but rather to the density of said public sphere in which the number of social actors in the public eye is constantly growing.

There are two particularly characteristic aspects of such developments. First of all, there is each sender's degree of awareness of the extremely strong competition from a plethora of other social actors, to get one's voice heard. Even though it does not disappear altogether, the role of the primary definer [21]—seen as the party with privileged access to the public sphere [14] and most capable of achieving hegemony over common sense, that is, of imposing its own viewpoint regarding a specific question or event, to the point where this is seen as an inescapable part of public discourse [36]—is considerably weakened nevertheless. Secondly, there is the understanding that access is favoured not only by the ability shown when negotiating with journalists and the media networks, but also by the capacity to attract a broad, constant audience on social media with whom to discuss events.[4]

3 Communicative Resources

The power to convoke those operating within the communicative space is strictly correlated to their capacity to manage the communicative resources available. As Carlson [9] rightly points out, the real difference between the authority of each sender does not depend exclusively on the ability to negotiate with journalists. What is now important is the capacity to dialogue with the many counterparties permitted by the emergence of social media. In this context, legitimation calls for transparency and sharing through the constant involvement of one's own public. This could be described as a stance based on dialogue, and thus on the relational capital that a party manages to set in motion.

Reputational resources may be defined as the series of qualities that the social actor acquires on the basis of a number of variables: for example, the social prestige acquired in the person's professional field; notoriety; recognized endorsement by institutions or by public opinion. In the case of the journalists Scanzi and Porro, for example, as with all journalists, it is clear that advantage is taken of the fame they have achieved through the TV and the press.

The importance of institutional resources is correlated to the centrality of the organization and its representatives in regard to the events and questions focused

[4] Think of the phenomenon of the 'influencers' and their importance in social media.

Table 1 The numerical and percentage increases in the number of Facebook followers of the 10 social actors concerned

Facebook account	Followers	Increase (%)
Italian Civil Defence Corps	382.6 K	238.7
Higher Institute of Health	2.8 K	32.8
Giuseppe Conte	2.0 M	177.8
Lombardy Regional Government	52.1 K	30.6
Attilio Fontana	86.6 K	114.3
Andrea Scanzi	999.6 K	170.6
Nicola Porro	189.0 K	68.8
Veneto Regional Government	14.4 K	43.7
Palazzo Chigi (Prime Minister's Official Residence)	146.0 K	119.4
Luca Zaia	173.4 K	38.9

Own elaboration

on or dealt with. Therefore, it should come as no surprise to learn that the largest increase in the number of 'followers' has been that of those institutions principally entrusted with the management of the event: the Italian Civil Defence Corps and the Prime Minister's Office (Table 1).

As far as concerns the Prime Minister's Office and the Prime Minister himself, it should be said that there is a clear division of roles. The Facebook and Twitter accounts of the Prime Minister's Office post messages relating to service matters only. The communicative field is prevalently that of the regulatory variety. The accounts of Prime Minister Giuseppe Conte, on the other hand, tend to be characterised by messages of a more emotionally-laden nature tending to involve the receivers to a greater degree.

The Prime Minister's position as the focal centre of communication is not only apparent from the intensity of posts published—over the course of the 58 days in question, the P.M.'s Office posted 126 items compared to 110 in the case of Prime Minister Conte himself—but also, and in particular, from the increase in the number of followers and of reactions on Facebook produced by the content of said posts (Fig. 1).

Continuing with our analysis of the political sphere, the profiles of the Lombardy and Veneto regional governments, and their respective presidents, Fontana and Zaia, reveal differing trends. While the Veneto Regional Government was characterized by a similar trend to the one described for Italy's national government, in the case of Lombardy the lesser degree of personalization by its President, Fontana, was offset by the more pressing rhythm of the region's institutional account (Table 1).

The institutions they represent, however, have tended to behave to the contrary. The Lombardy Regional Government has adopted tones which while remaining within the institutional context, have occasionally been of a polemical nature, which is something not commonly found for this kind of sender. This different policy is also

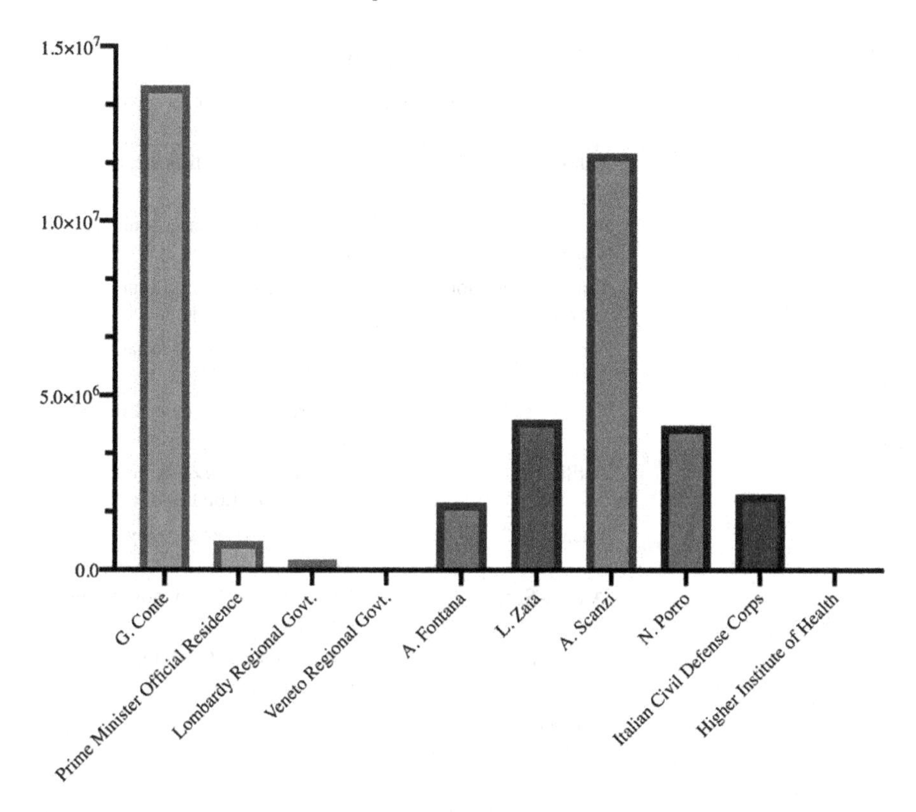

Fig. 1 Total number of reactions on Facebook. Own elaboration

borne out by the numbers: the Lombardy Regional Government's Facebook page has a lot more followers than that of the Veneto Region (170,372 compared to 32,882).[5]

Remaining within the institutional context, the Italian Civil Defence Corps mainly offers data, explanations of decisions taken and of the rules to be followed, while the Higher Institute of Health focuses mainly on healthcare information, and goes no further than denying false or misleading information. Nevertheless, it is interesting to note that the Italian Civil Defence Corps registered the greatest increase in the number of followers during the lockdown period (+238.7%), which rose from 160,226 to 542,817. This jump in numbers underscores this particular institution's importance as a benchmark for the population, due in part to its excellent reputation.[6]

Professional resources concerns both the area to which the sources of information belong—in the case of the pandemic, the voices expressed by those operating in the medical field—and the field of information. This is the reason why we have

[5] Even taking into account the smaller population of the Veneto region, this difference is nevertheless considerable.

[6] During the lockdown period, the average score given to the institution, in terms of the trust placed in it, was 8.7 out of 10. Source: ISTAT, May 2020.

Table 2 The most frequently occurring headwords

Facebook account	Headwords
Higher Institute of Health	Laws, health, disorders, know, ministry
Palazzo Chigi	Conte, measures, distance, activity, sharing
Italian Civil Defence Corps	Risk, Borrelli, green, warning, emergency
Lombardy Regional Government	*Fermiamoloinsieme*[a], Lombardy, treatment, LIS, national
Veneto Regional Government	Veneto, questions, information, presentation, *PSR*[b]
Giuseppe Conte	measures, citizens, days, European, Chigi
Attilio Fontana	Lombardy, *ForzaLombardia*[c], *Celafaremo*[d], Lombards, doctors
Luca Zaia	Veneto, thanks, emergency, home, heart
Andrea Scanzi	Salvini, Conte, Renzi, government, Meloni
Renato Porro	Conte, republic, Corriere, newspaper, freedom

Own elaboration
a. Let's stop this together
b. Rural Development Programme
c. Come on Lombardy
d. We will succeed

chosen two journalists—Scanzi and Porro—who recorded the highest number of views during the lockdown period.[7] An analysis of the most commonly used words in the accounts considered here (Table 2) confirms the coherence of the angle from which each party intervenes with regard to the question. In fact, as would be expected given the traditional Italian journalistic approach [22], they focus above all on the political argument.

The performance resources indicate the chosen themes, the language and channels used, the utilization of video clips, photos and other connotations of the published

[7] What happened in the months thereafter, in fact, clearly illustrates what we mean by "professional resource". While Scanzi and Porro enjoyed their own TV fame, the popularity on social media of Lorenzo Tosa, a former 5-Star Movement militant who subsequently became a highly-active social media journalist, although little known to the general public, in fact took over second place in the ranking of journalist-influencers from Porro, can be attributed to Tosa's journey around Italy—originally conceived as an idea for his posts, and subsequently becoming an itinerant account of the pandemic—resulted in a dramatic rise in the number of his hits, which reached the figure of 2.7 million in October. So the professional resource is attributable not only to visibility, but also to the capacity to choose a theme, to the ability to handle that theme using the right tone and language, and to the ability to tap into the interest of the most appropriate segment of the public.

content. Argumentative resources, on the other hand, concern the specific perspective from which the pandemic is spoken of, and the viewpoints from which arguments are made.

In our case—we repeat—what is really striking is the different stance adopted by the Presidents of the two Regional Governments examined here, that is, Messrs, Fontana and Zaia, whose number of posts, adopted tones and interaction reveal completely different strategies, probably as a result of the diverse characters of the persons in question, and of the different requirements of their respective regions. The performative resources of the Veneto Regional Government's President were noticeable: his constantly decisive tone as he argued with central government, who he demanded come to clear decisions, and even more so with the European Union's institutions, was met with great appreciation and congratulations. They were lacking, however, in any reference to solidarity, and any messages designed to generate emotions and a sense of belonging.[8]

As far as regards the principal communicator during the lockdown period—Prime Minister Conte—in addition to information of a more strictly institutional nature, often conveyed through live Facebook broadcasts, Conte's messages also contained information of strictly political character (for example, concerning talks with European Union representatives), together with appeals and messages of thanks addressed to the Italian public, often of an emotional tone. In other words, his communication was characterized by a greater variety of themes dealt with and of the tones adopted. Thus the significant response to Conte's messages comes as no surprise.

In the journalistic field on the other hand, Scanzi and Porro saw their 'number of views' increase dramatically, thanks to their ability to exploit the performative resources of social media. For example, the chatty tone and the clarity of their ideas, which tend to encourage polarization; their posts are always against something or someone.

Finally, there are the relational resources, that is, the capacity to trigger reactions and the engagement of those to whom one is speaking. This is the most easily measurable resource, even though it does not correspond simply to the number of followers a person or institution has, but also to the communicative approaches adopted. For example, some people adopt a top-down communicative style even on social media; others ask that their followers to participate as far as possible. The former approach comprises: the recurrent requests to follow the sender's own live broadcasts, or to refer to other broadcasts featuring the sender (their presence on other social channels or in other media, press conferences, interviews given to the mainstream media, services produced for such mainstream media); and also greater personal exposure through self-quotations, photos, videos, the disclosure of information regarding their own personal interests.

---

[8] The result of the May election confirms this. Zaia was re-elected as President of the Veneto Regional Government with over 75% of the votes cast. At the national level, in October 2020 Zaia drew level with Prime Minister in terms of public approval, at 58%. (Demos data produced for the *La Repubblica* newspaper, 2020).

Relational resources can also be considered to include the frequent suggestions made to read other colleagues' posts—very much present in Porro's case—and the reference to other posts by their chief targets, or by other figures form the political and media world.

4 Conclusion

In our analysis we have compared various different social actors in the belief that the digital revolution is transforming journalism. The communicative space has been opened up and stratified, with the increasing number of sources of information becoming more immediately accessible, and thus taking on a different role, as has the public which is now significantly involved in the production and distribution of information.

The institutional centrality of the source would account for the huge increase in followers of the Italian Civil Defence Corps, the Prime Minister's Office and Regional Governments. However, at the same time the institutions in question take on a more personalized character insofar as they identify with the persons embodying them, as shown by the number of followers of the Prime Minister and of the Presidents of the Regional Governments analysed here.[9] The use of the personal accounts of those holding important positions within those institutions differs depending on the actors concerned.

It is also interesting to note the strong political-cultural connotation of the two journalists included in our sample; it is almost as if this marks a shift away from journalism's reporting role towards more polemical terrain, which in turn favours processes of polarization often considered to be the consequence of the importance gained by social media [27, 39].

Together with institutional and professional resources, considerable importance is given to performative resources: during the lockdown period, in fact, Scanzi and Porro had more followers than other better-known journalists, thanks to the two's ability to manage the affordances of the channels they use. This gives them a considerably greater power of convocation, which is proportionately reflected in the relational dynamism found.

A communicative space in which actors, competences and resources interweave, makes it more difficult to identify the distinctive feature of each individual source of information. Rather than a collapse of hierarchies, as suggested by Bolter [6], what we have here is a multiplication of the dimensions constituting several hierarchies. We need to examine further the complex composition of such dimensions, by analysing

[9] As we all know, the debate over personalization has taken center stage in the studies conducted in recent years, in particular in regard to the relationship between communication and politics. In this case, we simply wish to refer to the fact that social media facilitate the identification of an account with a given individual.

the actors' network of influences, and by identifying the principal hubs of such networks and the specific functions that each of them performs.

Pending more detailed analyzes enabling us to substantiate what has been set out in this study, we can nevertheless outline a power of convocation that may be correlated to the diverse roots of credibility [17]. The institutional centrality of certain sources (in our case the Italian Civil Defence Corps or the Prime Minister's Office) is observed, such sources are capable of meeting the demand for official information, and of assuming the corresponding responsibility required by emergency communication, in order to meet the need for expertise on which all cognitive and moral-normative credibility is based. Nevertheless, what also emerges is a certain reshaping of the forms of convocation, involving the interweaving of professional and performative resources, as in the case of the two journalists Scanzi and Porro.

Ultimately, each social actor implements a complex communicative strategy characterized by diverse registers, in response to the levels proposed by the interrelated public sphere [5].

In other words, if Bourdieu's orthodoxy of the journalistic field is to be considered the independent, autonomous but orderly transfer of content from a source to the public through the mediation of journalists [28], or from an institution to citizens through the professional skills of institutional communicators, then we can undoubtedly claim to have entered a "heretic" space. Voices multiply and overlap, producing a density exacerbated by the immediacy of events.

Rather than counting solely on the centrality of professional abilities—those of journalists and institutional communicators—capable of performing a control and verification function, and also of acting to certify credibility and guarantee completeness and impartiality, we need to look for new, appropriate forms of legitimation and validation that foresee the possibility of combining this inevitable plurality of voices with the transparency of the ways in which information is generated and divulged [10].

This is a long and complex process that underscores the fact that the current promiscuity of communication results in a clear heretical tension underlying the radical redefinition of the communicative space. However, as the term itself indicates, heresy cannot become the norm. Consequently, we need to focus on a different way of perceiving the space established by new forms of communication.

References

1. Agamben G (2006) Cos'è un dispositivo. Nottetempo, Roma
2. Bechelloni, G.: Giornalismo e Post-Giornalismo. Liguori, Napoli (1995)
3. Bennett, W.L., Pfetsch, B.: Rethinking political communication in a time of disrupted public spheres. J. Commun. **68**(2), 243–253 (2018). https://doi.org/10.1093/joc/jqx017
4. Benson, R., Neveu, E.: Bourdieu and the Journalistic Field. The Polity Press, Cambridge (2004)
5. Bentivegna, S., Boccia Artieri, G.: Rethinking public agenda in a time of high-choice media environment. 69th Annual International Communication Association Conference, 24–28 May, Washington DC (2019)

6. Bolter, J.: The Digital Plenitude. The MIT Press, Cambridge (2019)
7. Bourdieu, P.: Champ politique, champ des sciences sociales, champ journalistique. Hermés **17**(18), 215–229 (1995)
8. Bourdieu, P.: Sur la Television. Liber-Raisons d'agir, Paris (1996)
9. Carlson, M.: Journalistic authority. Legitimating news in the digital era. Columbia University Press, New York (2017)
10. Couldry, N.: Why voice matters. Culture and Politics after Neoliberalism. Sage, London (2010)
11. Cristante, S.: Potere e comunicazione. Sociologie dell'opinione pubblica. Liguori, Napoli (2004)
12. Dahlgren, P.: The Internet, public spheres, and political communication: dispersion and deliberation. Polit. Commun. **22**(2), 147–162 (2005). https://doi.org/10.1080/10584600590933160
13. Deuze, M.: Participation, remediation, bricolage: considering principal components of a digital culture. Inf. Soc. **22**(2), 63–75 (2006)
14. Ericson R, Baranek P, Chan J (1987) Visualizing deviance. A study of news organizations. Saxon House, Farnborough Hauts
15. Foucault, M.: Dits et Écrits. Gallimard, Paris (1994)
16. Gandini, A.: L'economia della reputazione. Il lavoro della conoscenza nella società digitale. Ledizioni, Milano (2019)
17. Gili, G.: La credibilità. Quando e perché la comunicazione ha successo. Rubbettino, Soveria Mannelli (2005)
18. Grossi, G.: L'opinione pubblica. Laterza, Roma-Bari (2004)
19. Grossi, G.: Metamorfosi del politico. Rosenberg & Sellier, Torino (2020)
20. Guarnaccia, F., Barra, L.: Come si può imparare a non preoccuparsi troppo e ad amare la plenitudine digitale in Bolter J, op. cit: 5–23 (2020)
21. Hall, S.: Encoding/Decoding in the television discourse. In: Hall, S., Hobson, D., Lowe, A., Willis, P. (eds.) Culture, Media, Language, pp. 128–138. Hutchinson, London (1980)
22. Hallin, D., Mancini, P.: Comparing Media Systems. Cambridge University Press, Cambridge, Three Models of Media and Politics (2004)
23. Jenkins, H.: Convergence Culture. New York University, New York (2006)
24. McNair, B.: Cultural Chaos, Journalism, News and Power in a Globalized World. Routledge, London (2006)
25. Nielsen, R.K, Fletcher, R., Kalogeropoulos, A., Simon, F.: Communications in the coronavirus crisis: Lessons for the second wave. Reuters Institute, University of Oxford (2020). Available at https://bit.ly/3kTeCjF. Accessed 01 Dec 2020
26. Osservatorio sul Giornalismo: La professione alla prova dell'emergenza Covid-19. Autorità per le garanzie nelle comunicazioni, Rapporto di ricerca (2020). Available at https://bit.ly/3egkMJr. Accessed 01 Dec 2020
27. Pariser, E.: The Filter Bubble. Penguin Books, London, What the internet is hiding from you (2011)
28. Peters, C., Broesma, M.: Rethinking Journalism. Trust and participation in a transformed news Llandscape. Routledge, London (2013)
29. Pfetsch, B.: Dissonant and disconnected public spheres as challenge for political communication research. Javnost-The Public **25**(1–2), 59–65 (2018). https://doi.org/10.1080/13183222.2018.1423942
30. Prior, M.: Post-Broadcast Democracy. How media choice increases inequality in political involvement and polarizes elections. Cambridge University Press, Cambridge (2007)
31. Rothkopf, D.J.: When the Buzz Bites Back. Washington Post, 11 May (2003). Available at https://bit.ly/3kRP5HL. Accessed 01 Dec 2020
32. Schudson, M.: Discovering the News. Basic Books, New York (1978)
33. Solito, L.: Tra cambiamenti invisibili e immobilismi opachi. la comunicazione pubblica in Italia. Sociologia della Comunicazione **48**, 100–118 (2014). https://doi.org/10.3280/SC2014-048010

34. Sorrentino, C.: L'obiettività della competenza. Problemi dell'Informazione **28**(4), 427–435 (2003). https://doi.org/10.1445/10950
35. Sorrentino, C.: Il campo giornalistico. Carocci, Roma (2006)
36. Sorrentino, C.: La società densa. Riflessioni intorno alle nuove forme di sfera pubblica. Le Lettere, Firenze (2008)
37. Sorrentino, C., Bianda, E.: Studiare giornalismo. Carocci, Roma (2013)
38. Splendore, S.: Giornalismo ibrido. Come cambia la cultura giornalistica italiana. Carocci, Roma (2017)
39. Sunstein, C.: Echo Chambers: Bush V. Gore, Impeachment, and Beyond. Princeton University Press, Princeton (2001)
40. Thompson, J.B.: The Media and Modernity: A Social Theory of the Media. Stanford University Press, Stanford (1995)
41. Tuchman, G.: Making the News. A Study in the Construction of Reality. Free Press, New York (1978)
42. Zelizer, B.: Terms of choice: uncertainty, journalism, and crisis. J. Commun. **65**(5), 888–908 (2015). https://doi.org/10.1111/jcom.12157

Laura Solito Ph.D. in Sociology at University of Pisa and Associate Professor in Sociologia della Comunicazione since 2007 at the Department of Social and Political Sciences, University of Florence. Since 2015 she has been Vice-President of Communication and Public Engagement in the University of Florence. Her main fields of research is on the role of communication in the public administration and especially in the production and management of services and on new forms of citizenship favored by digital communication.

Carlo Sorrentino Ph.D. in Political Sociology at the University of Florence (UniFI) and Full Professor in Sociologia dei Processi Culturali since 2011 at the Department of Social and Political Sciences (UniFI). He is the Dean of the MA Course on Public and Political Communication Strategies. His main field of research is Journalism studies and in the last years he is investigating on the role of journalism in the transformation of the public sphere. He is editor of the journal Problemi dell'informazione edited by Il Mulino.

Professional Profile of the Contemporary Digital Journalist

Suzana Oliveira Barbosa, Lívia de Souza Vieira, Mariana Menezes Alcântara, and Moisés Costa Pinto

Abstract Conceptually aligned with the epistemologies of digital journalism, and in line with the definition of 'total journalism', this chapter presents a mapping of the contemporary digital journalist's professional profile, highlighting the skills for 'being' a journalist, and 'performing' journalistic activities in the new century. This is based on meta-research conducted on the Scopus and Google Scholar databases, and the Capes Catalogue, for a longitudinal study of the bibliographical framework between 2000 and 2020, combined with the application of a questionnaire with 31 legacy and local digital media editors in Brazil. Evidence indicates that contemporary digital journalistic work is experiencing a trend towards platformization and surrounded by constant adaptability to new technologies, journalism is undergoing a crisis continuum, accentuated by the COVID-19 pandemic. With insecure work, and a context of datafication and algorithmization, the journalist needs to address the need for constant innovation and qualification, without neglecting the ethos of the profession.

1 Introduction

For those aligned with the perspective of the British historian Hobsbawm [11], demarcation for the end of a century, and start of another, only occurs with a major crisis. And the COVID-19 pandemic was the effective milestone for the passing of the 20th to the twenty-first century, according to the Brazilian historian Schwarcz [34]. It is within this context of a new century that this chapter is inserted, presenting a mapping of the contemporary digital journalist's professional profile through reflection, anchored in meta-research for a longitudinal study of the bibliographical framework between 2000 and 2020, combined with the application of a questionnaire with editors from legacy and local digital news outlets in Brazil.

The objective is to indicate the skills and knowledge required from journalists, to operate in line with the definition of 'total journalism', characterized by adversity

S. O. Barbosa (✉) · L. de S. Vieira · M. M. Alcântara · M. C. Pinto
Universidade Federal da Bahia, Salvador, Brazil
e-mail: sobarbosa@ufba.br

J. Vázquez-Herrero et al. (eds.), *Total Journalism*, Studies in Big Data 97,
https://doi.org/10.1007/978-3-030-88028-6_15

of techniques, methods and tools applied by journalists who are striving for the highest possible quality in their activities [16, 17]. Therefore, it is a study based on qualitative and quantitative methodology, which is conceptually set within the scope of digital journalism epistemologies—situated as an emerging subfield in journalism studies [5–7]. The plural term is used to recognize that there are a range of different forms of knowledge, knowledge production practices and justifications for claims of knowledge in digital journalism.

The profile of journalists has featured in the research interests of digital journalism investigators since its emergence during the 1990s, being associated with the area of multimedia convergence, and the evolution of journalists' professional routines, as suggested by Salaverría [31]. Since the launch of the first journalistic sites between 1994 and 1995, researchers have discussed what journalism and journalists are in the digital era [10]. Principally in the current context, when anyone can create, edit and circulate contents on different platforms—and even mimic a news format, producing disinformation—affecting the legitimacy, credibility and authority of journalism, as a form of knowledge of the present.

In this new century, the journalists acting in a crisis continuum, accentuated by the pandemic and its consequences, job insecurity and the platformization of society and journalism [14, 39], interlaced with datafication and algorithmization [15], the distributed newsroom phenomenon [22], and need for constant innovation and qualification [24].

2 Approaches on the Digital Journalist's Role

Before presenting the different approaches on the digital journalist's role, we must conceptualize digital journalism. In the last 25 years of research, various authors have focused on this task, striving to identify the breakdowns and continuities that the internet and digital technology have represented for journalism. For Salaverría [31], it is a multifaceted, changeable and almost incomprehensible concept. Zelizer [42] affirms that the definitions of digital journalism were historically connected to expectations and optimism related to its technological possibilities. Thus, the author puts forward a concept that goes beyond the technological slant.

> Digital journalism takes its meaning from both practice and rhetoric. Its practice as news-making embodies a set of expectations, practices, capabilities and limitations relative to those associated with pre-digital and non-digital forms, reflecting a difference of degree rather than kind. Its rhetoric heralds the hopes and anxieties associated with sustaining the journalistic enterprise as worthwhile. With the digital comprising the figure to journalism's ground, digital journalism constitutes the most recent of many conduits over time that have allowed us to imagine optimum links between journalism and its publics [42:349].

Perreault and Ferrucci [28] also define the optimistic and romanticized history of digital journalism, citing studies such as those by Pavlik [25], which demonstrated how positive digital technologies could be for journalism. They distinguish a second moment of research in the area, when authors such as Ferrucci and Vos [8] identified journalists constructing their professional identity on digital breakdowns. Journalists are now starting to treat digital technologies as another part of their routines. In the view of 68 journalists from the United States interviewed by the authors, digital journalism includes the use of technologies to tell stories, it involves the dissemination of information in the quickest way possible and concentrates on the audience [28].

López-García and colleagues [17] highlight four milestones within the metamorphoses of journalism. The first, in the second half of the twentieth century, was the explosion of journalism, with the development of mass media, called 'new journalism'—narrowing the relation between journalism and literature—, precision journalism —which brings the scientific method of journalism closer—, and service, civic and solutions journalism. The digital revolution is the second moment characterized by the authors, on the turn from the second to the third millennium. Information and Communications Technologies (ICTs) began a permanent, rapid reconfiguration which changed the communication ecosystem. Hypertextuality, multimediality and interactivity emerged as distinctive features of the so-called new media, which created conditions for the appearance of digital journalism. The other major renewal, highlighted by López-García and colleagues [17], includes expressive types for storytelling, techniques based on data and virtual reality, and the growing use of interdisciplinary teams. The most recent metamorphosis is characterized as 'total journalism', a "renewed practice which expresses using multimediality, hypertextuality and interactivity, traditional journalism adapted to the present society, which uses data, immersive and transmedia techniques, among others, and which guarantees the truthfulness of information and the performance on differential values for users" [17:204].

It is within this context of a new metamorphosis and constant transformation, as the result of the entry of new digital technologies in media ecologies [30], that the digital journalist's profile emerges, and is included. Zelizer [42] highlights that, besides the 'toolbox' expected from the journalist today—made up of social networks, multimedia, big data, mobility, analyses and metrics—, the journalist's reading attributes should be discussed, such as hesitation, focus, a sense of surprise and deep understanding of reality.

Since the first decade of the 2000s, researchers such as Pierce and Miller [29] have affirmed that in the complex environment of contemporary media, basic journalistic skills continue to be important, such as writing, reporting and critical thinking. However, journalists should also recognize that they need to work on constructing other digital skills. At the time, the authors interviewed 311 North American editors, which highlighted the importance of computer-assisted reports and multimedia reporting skills.

On the other hand, in a study conducted by Weiss [41] with 444 journalists in Argentina, Brazil, Colombia, Mexico and Peru, the author confirmed that they were effectively exploring new methods and techniques in their routines, through digital media. Among the results found is the classification of the three roles performed by the digital journalist: mobilizer, disseminator and interpreter. More recently, Pereira [27] examined the careers of 11 'old media journalists' who were working in the digital media sector in Belgium, Brazil, Canada, France and Portugal. From interviews, the author observed that the first contact with the world of ICTs and entry into the digital media sector are important moments for these journalists to construct their careers. Thus, adapting to the digital ecosystem is a possibility, in order to continue working.[1]

3 Methodology

In order to present a mapping of the contemporary digital journalist's profile, initially we conducted meta-research that comprised a bibliographical review of academic work published between 2000 and 2020, which included the skills required to work in digital journalism to a certain extent. This type of study allows the construction of knowledge derived from research into one area [13], identifying approach tendencies for investigations, while providing a reflexive and critical review, involving concepts, categories, methods and research objectives [12, 19].

We used Scopus, Google Scholar and Capes Catalogue[2] databases for this research. Scopus was selected since it contains a wide selection of some of the best journals available in the area of communication and journalism. The Capes Cartalogue holds a large database of work produced at Brazilian postgraduated programs. In turn, Google Scholar was considered, since it is one of the main search tools for academic contents, with a large index of journals, and work presented at events and published in annals, dossiers, theses, dissertations, books and interviews, etc.

For greater efficiency, the search on the Scopus, Capes Catalogue and Google Scholar platforms was conducted using keywords in Portuguese and English. We looked for studies that used the terms 'digital journalism', 'online journalism', 'webjournalism' and 'cyberjournalism'. In addition, we combined these keywords with what we call 'profile variations', terms that would provide greater accuracy by producing research on the profiles and skills required to practice digital journalism: 'professional profile', 'skills', 'expertise' and 'tools'. The following are examples of some of the combinations used: "digital journalism" AND "professional profile"; "digital journalism" AND skills; "digital journalism" AND expertise, and "digital journalism" AND tools, etc.

Altogether, in the survey carried out between August and November 2020, which was refined between January and February 2021, a total of 56,246 pieces of work were found (55,966 on Google Scholar; 202 on Scopus and 78 in the Capes Catalogue),

---

[1] An example of one *old media* journalist's effort to adapt—in order to work in digital media—is that of the Bahian reporter José Raimundo, who after 30 years of working at a TV station, makes this transition. In: Demanda digital desenha novos rumos para o professional journalism. Como journalists têm se adaptado às transformações digitais e buscado no empreendedorismo uma alternativa à precarização no setor. Available at https://abi-bahia.org.br/demanda-digital-desenha-novos-rumos-para-o-jornalismo-profissional/. Accessed 25 Mar 2021.

[2] From Coordenação de Aperfeiçoamento de Pessoal de Nível Superior (foundation of the Ministry of Education (MEC, Brazil). Available at http://catalogodeteses.capes.gov.br/catalogo-teses/

which contained at least one of the keywords and/or profile variations anywhere in the text.[3] They are articles, theses, dissertations, chapters of books, communications at events, interviews, dossiers and manuals, etc., which include the above-mentioned terms in part of the text, even if the topic covered was not related to digital journalism, or any area of journalism. In the refinement stage, we limited the searches by up to ten pages in the cluster tools. To complement this, in order to obtain a corpus that could be categorized, we limited the number of pieces of work analyzed for each combination of keywords and profile variations to ten on each platform, and 'relevance' was the main criterion adopted to order the research found.[4]To finalize, we only downloaded work that contained both keywords and profile variation terms in the title, subheading, abstract, keywords, or topics. The aim was to only obtain works which addressed the changes required in the professional profile to work in digital journalism. As a result—having discarded duplicate work, and that which although containing keyword combinations and profile variations, was not related to the topic—204 texts were acquired, which form the analyzed sample. Of this total, 145 (71.0%) pieces of work were obtained on Google Scholar, 45 (22.0%) on Scopus, and 14 (6.8%) from the Capes Catalogue.

An online questionnaire was added to the bibliographical measurement and it was sent by email to editors from reference and local, Brazilian digital media. Data collection took place during the period between January 11 and 31, 2021.[5] The unrepresentative sample of 31 respondents are primarily female (51.6%), aged between 25 and 40 (61.3%), white (74.2%), have completed a lato sensu post-graduate qualification (specialization, 41.9%), and graduated in Journalism (96.7%). The editors come from seven Brazilian states, mainly São Paulo (32.3%), followed by Bahia (19.4%) and Rio de Janeiro (9.7%). They are experienced journalists (67.7% have worked in the area for more than 10 years), working at thirteen organizations: *Folha de S. Paulo*, *Globo/Extra/Revista Época*, *O Estado de S. Paulo* (*Estadão*), *Jornal Correio*, *A Tarde* and Grupo RBS—from legacy media (54.8%); *Agência Lupa*, *Aos Fatos*, *Bahia Notícias*, *Canal Reload*, *G1 Bahia*, *Alô Bahia* and Projeto Colabora—local digital media (41.9%). The majority of the editors worked over 40 h per week (51.6%) and, during the collection period, 87.1% were performing these functions at home (home office). In addition to socio-demographic questions and those on their tasks, the questionnaire included open and closed questions on a digital journalist's routine, knowledge and skills, which are listed below.

[3] The searches on the Scopus database and Capes Catalogue returned work containing just two variables, keywords and profile variables. The majority of the results on Google Scholar returned just one keyword or profile variables.

[4] Therefore, when the limiting number of ten items of work was found in a search in a given combination of terms, we moved on to research the next one.

[5] The questionnaire was structured and refined from pre-tests conducted with members of the Grupo de Pesquisa em Jornalismo On-Line (gjol.net).

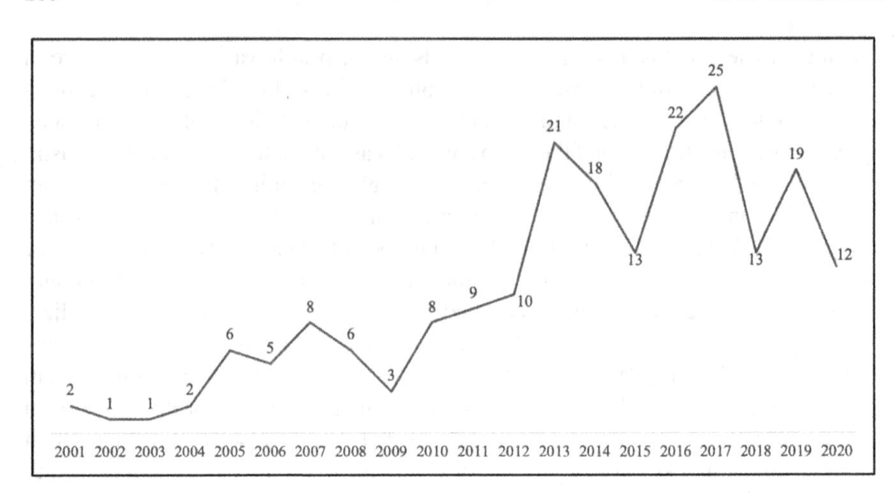

Fig. 1 Evolution in the number of categorized publications. Own elaboration

4 Professional Profile: Longitudinal Mapping and Questionnaire with Editors

The first noteworthy point in the bibliographical review is the evolution of publications in the last 20 years (Fig. 1). An increase in the number of studies that address changes in the professional profile, the emergence of new skills, expertise, and the need for digital journalists to master new tools is observed from 2012. However, growth from that year is not vertical. After that, it is important to note that, as the survey analyzed the most relevant documents, it is normal for the most recent publications to have less impact and, consequently, less relevance. Even after achieving a peak in 2017, it has been stable since then, with 2020 experiencing the highest decline since 2012, which may not represent a tendency of decline. We should take into consideration that everyone was affected in 2020—to a greater or lesser extent—by the COVID-19 pandemic, and the problems arising from it, jeopardizing work dynamics, and certainly interfering with academic production on digital journalism.

With regards to the languages in the sample, English was dominant (51.0%), and the second most relevant was Portuguese (49.0%). Although we did not use Spanish for the searches, a number of pieces of work in this language were produced, since they had titles, abstracts and keywords published in English.[6]

Scientific articles predominated (62.2%) among the types of production, followed by event communications (9.3%), dissertations (8.8%), book chapters (8.3%), theses (4.9%), monographies (4.4%), reports (0.9%), manuals and interviews (0.4% for each).

[6] Work in languages such as French, Russian and Japanese emerged in the searches, with the vast majority containing terms cited in the body of the text, but without their titles or full abstracts being translated into English.

Fig. 2 Most representative terms. Own elaboration

The majority of the work combined two or more methodologies. However, 26.2% related to case studies, followed by exploratory (21.5%) and bibliographical research (18.9%).[7] Another relevant approach to research the changes in the digital journalist's professional profile was interviews (9.18%). Survey questionnaires were used in 7.7%, while qualitative research represented 4.4%, followed by document (3.4%), observation (2.2%), experimental (2.2%), quantitative (1.9%), ethnographic (1.2%), focus group (0.6%) and descriptive research (0.3%).

In order to condense the keywords used to tag the sample work, we used a word cloud to illustrate the most representative terms (Fig. 2), namely: digital journalism (48 tags), online journalism (43 tags), journalism (39 tags), cyberjournalism (19 tags), and web journalism (19 tags). Other terms, such as convergence, internet, professional profile, university, journalism education, mobile journalism, data journalism, multimedia, metrics, and skills also appear as resonance in the cloud, indicating the thematic matrices which the research addresses.

With regards to the most favorable environments for digital journalists to develop the skills, which are essential for professional practice, universities emerge in first place (35.2%), followed by newsrooms and other workspaces (34.8%); and indicates that these skills can only be learned through a combination of theory (universities) and practice (newsrooms) in 13.2% of the publications. Only 0.4% indicate that an appropriate profile can be constructed outside of the newsrooms and academia. A

[7] It is important to highlight that some of the work was only based on theoretical references for their development, putting forward state of the art propositions and reviews on concepts and methodologies, for example.

Fig. 3 Skills highlighted between 2000 and 2010. Own elaboration

further 16.1% did not present an indication of the ideal location for learning the skills required. In respect to environments to enhance these skills, while 48.5% of the publications do not present a clear position on this, 29.4% attribute the role of the only environment that could provide new perspectives to the area to universities, and 10.78% indicate newsrooms. Other 6.8% demonstrate that neither academia or newsrooms can improve professional skills. Lastly, only 4.4% of the pieces of work argue that the academia-newsroom combination would be the best path to develop new professional skills.

Between 2000 and 2010 (Fig. 3), the aptitudes found in the sample concentrated on the development of digital skills (23.8%), mastery of media convergence (9.5%), blog tool expertise (7.1%), a multimedia profile (7.1%) and command of social networks (7.1%), in addition to skills to tackle changes (4.7%), a mastery of hypertext (4.7%) and a multi-purpose profile (4.7%). Over the period, it is noted that the digital skills which are most frequently described as compatible with career development in digital journalism are a focus on mobile journalism, a mastery of blogs, and convergent and multimedia journalism.

Between 2011 and 2020, the cloud which illustrates the main skills studied (Fig. 4) changes its configuration. Despite the continued concern with new digital skills (12.3%), the development of professional profiles able to deal with journalistic convergence (14.8%) is highlighted. Other important skills are: the development of social media expertise (9.2%), use of databases (8%) and metric and analytical tools (6.1%). Studies that indicate the following also attract attention: a multimedia profile, the capacity to generate data visualizations (3%) and infographics (3%).

The editor questionnaire allowed a more detailed view of a digital journalist's routine, and the characteristics and skills required from the professionals' perspective. In accordance with the sample data, forms of investigating the news are

Fig. 4 Skills highlighted between 2011 and 2020. Own elaboration

quite varied, with the telephone being predominant (93.5%), followed by WhatsApp (83.9%), official documents (77.4%), social networks (74.2%), email (74.2%), databases (67.7%), the street (48.4%) and geolocation tools (19.4%).

Consultation sources to create stories combine traditional means, such as rounds (74.2%), press offices (64.5%), printed newspapers (45.2%), news agencies (48.4%), anonymous source (41.9%), magazines (35.5%) and listening to the radio (12.9%), with those connected to cyberspace: social networks (83.9%), journalistic sites (77.4%), public institution sites (74.2%), databases (64.5%), private institution sites (48.4%), digital documents (45.2%), analytical data from sites and social networks (45.2%), and citizen reporters (3.2%). With regards to the formats used to produce contents, the utilization of multimedia content is common-place, with mentions of texts (100%), social network videos (71%), podcasts (64.5%), statistical infographics (64.5%), social network cards (61.3%), journalistic site videos (58.1%), photographs (58.1%), interactive infographics (45.2%), statistical maps (16.1%), animation (16.1%), webdocumentaries (16.1%), interactive maps (12.9%) and games (6.5%).

For the majority of the respondents (80.6%), content is considered collectively and distributed to different newsroom professionals. The same sample percentage totally agree that the use of technologies changes the routines and knowledge required for the journalist's work in the current digital context. Overwhelmingly, 77.4% consider metrics to be very important, or important, when defining stories, and for journalistic content formats. There are also a wide range of circulation channels: sites (96.8%), Twitter (87.1%), Facebook (83.9%), Instagram (80.6%), email and newsletters (61.3%), printed newspapers (58.1%), apps (54.8%), audio streaming platforms (41.9%), transmission lists or WhatsApp groups (38.7%), YouTube (35.5%), content aggregators (22.6%), TikTok (22.6%), radio (19.4%), free-to-air television (3.2%) and Facebook groups (3.2%).

There were differing opinions related to whether higher education in Journalism provides the knowledge and techniques to practice the profession in digital environments: 45.1% disagree or totally disagree with this statement, while 41.9% agree or totally agree. Other 12.9% neither agree or disagree. For almost all of the respondents (93.5%), the market or newsroom is the main space for training, development and further qualification, in order to obtain knowledge to practice digital media journalism. A little over half (51.6%) indicate a university degree—the same percentage that classifies specific technical courses (short duration). However 83.9% totally agree that the mastery of digital technologies is essential for journalistic work today. A little over half (58.1%) totally agree that the more skills in the use of digital technologies, the higher the quality of content produced and delivered to the public.

In relation to the technical knowledge required to execute their work, once again there is a mix: producing and editing text (100%), analytics and audience metrics (80.6%), producing and editing spreadsheets(61.3%), producing and editing photographs (45.2%), video (35.5%), audio (35.5%), data visualization (25.8%), editing software (25.8%), creating infographics (22.6%), producing and editing program code (9.7%), web scraping and data collection (9.7%), database editing (9.7%), data and document analysis (6.5%) and statistics software (3.2%). When asked about non-technical skills important to practice journalism in digital environments, the respondents cited: being creative (93.5%), being ethical (83.9%), knowing how to search for good sources (83.9%), understanding and being committed to the importance of the role of journalism in society (83.9%), being dedicated (77.4%), focused (67.7%), constructing and having good relationships with sources (54.8%), being committed to the career (38.7%), being committed to the company in which you work (32.3%), being an entrepreneur (32.3%), people management (3.2%) and good texts (3.2%).

5 Transformation and Adaptability

Evidence extracted from the meta-research indicates that contemporary digital journalistic work is undergoing a trend towards platformization [14, 39], confirming the context described by Ekström and Westlund (2020) of a general "shift" of news, to the extent that it is moving from platforms produced and controlled by traditional news media, to those outside of their jurisdiction.

Through longitudinal mapping, we found that in relation to the digital journalist's profile, convergence [1, 2, 4, 24, 32, 33] is an important conceptual framework, while multimedia and interactivity are highlighted as characteristics. More recently, the journalist has also needed to develop skills to use databases and algorithms, which invariably involve new digital skills to investigate, produce, circulate, master metrics and analytics, visualize data, manage social media, for relationships with audiences, and even new business models, without forgetting the ethos of the profession.

However, algorithms are always changing, meaning that one of the skills required is being adaptable to transformations [30]. Therefore, new technologies and digital resources require the continual development of new expertise, especially in newsrooms—as a center where everything converges, or in the new distributed newsroom model [22]—since, as a large number of case studies found in the meta-research indicate, changes are felt at first hand in these environments.

The data shows the predominant use of practical examples as the main source to understand the changes which have taken place in the professional profile over the years, which may be explained by this type of observation being able to extract which changes are taking place, both in newsrooms and universities.

The questionnaire with editors demonstrates that even in management positions, the majority of the journalists are young adults (between the age of 25 and 40), are experienced (have worked for more than 10 years in the area) and have grueling working hours, exceeding 40 h per week. Within the context of the COVID-19 pandemic, remote working is a reality for almost all of the sample. This profile is in line with research that indicates the insecurity of journalistic work in Brazil [23], related to the entry of an increasing number of younger professionals in the newsrooms, often due to their skills with technological tools [9, 21].

The diversity of journalistic organizations in this research is representative of a scenario in which national, local and niche outlets operate side by side, distributed throughout various types of media. There are national, centennial printed news outlets, such as *Folha* and *Estadão*, which has gone through a journalistic convergence process [1, 2, 4, 24, 32, 33], local media such as *Jornal Correio*, *A Tarde* and Grupo RBS; and different niches of local digital media, such as the fact-checking agencies *Aos Fatos* and *Lupa*, the *Reload* channel for young people, and Projeto Colabora, which focuses on environmental issues.

The forms of investigation, consultation sources to create stories, formats to produce contents and circulation channels, demonstrate that the use of digital technologies is present in all stages of the journalistic production process, namely: production, circulation and consumption, in accordance with classic constructionist theories [37, 38], or investigating, production and circulation, in accordance with specific digital journalism studies [18, 35]. In other words, the research findings suggest that, from the perspective of this sample's editors, there is a diversity of techniques, methods and digital tools used in their activities. Consequently, a more detailed analysis of these results is provided below.

With regards to the forms of news investigation, the telephone is predominant, followed by WhatsApp and social networks. Stacciarini [36] affirms that, by allowing text messages, audio, photos, videos and document files to be shared, WhatsApp has become a common tool for journalists who draw on groups created between sources and press professionals. However, the author shows that this mediation, which dispenses with face-to-face meetings, may have ethical consequences. Stacciarini [36] identified journalistic errors published in articles on three Brazilian news sites, which emerged following information that was posted on WhatsApp groups between sources and journalists. In terms of the context of the pandemic—and this needs to be highlighted—less than half of the respondents confirmed that they use in-person investigation (in the street) as a way of investigating, ratifying a practice that researchers identify as the "sedentary journalist" [26], or "screen workers" [3].

We can confirm that digital technologies are blended with analog consultation sources to create stories. This becomes visible when the respondents cite rounds, press offices, social networks and databases as paths to produce news. In relation to the formats, the editors mention texts, videos, audio, photos, infographics and maps. Animations, webdocumentaries and games, which are more labor-intensive in terms of production, were also quoted, demonstrating that Brazilian newsrooms are experimenting with more innovative formats. The results also indicate that metrics, which provide information on audience behavior in digital environments, are common practice in the journalistic production routine. In accordance with Vieira [40], in the current newsroom context, metrics based on clicks have started to pave the way for a more analytical and qualitative interpretation of audience data, mainly at organizations that are less dependent on publicity as their source of income.

Digital news circulation channels are predominant, with sites, social networks and emails or newsletters being responsible for the majority of mentions. WhatsApp was only cited as a circulation channel by 38.7%, indicating that it is mainly used to investigate news. The printed newspaper, a legacy media product, is also mentioned by 58.1% of the respondents.

Almost all of the sample affirm that the use of technologies changes the routines and knowledge required for the digital journalist's work. Producing and editing text, audio and video; analytics and audience metrics; producing and editing spreadsheets, code and programming and data visualization were mentioned as the technical knowledge required to do their job. However, we highlight that in response to the open question "which skills should a journalist who works in digital environments have to produce quality contents?", the understanding that the main skills did not differ greatly from those of the 'traditional journalist' was predominant. This is noticeable in the following comment by one of the respondents:

> I do not think that journalists should necessarily master all the platforms and tools, but should know what can be done with them. Therefore, I believe that the classic skills of investigating, hierarchizing and editing continue to be widespread, and teamwork makes the content more attractive for the digital environment. But I do not believe that content quality is connected with the digital environment; I think that quality comes first, independent of support. Support defines the probability of reach, and the possibilities of it being packaged in different formats, so as to satisfy different audiences.

Lastly, the editors were critical of the technical knowledge provided by undergraduate courses in Journalism, and see the market or newsroom as the main space for training, development and further qualification to practice digital journalism. Although the journalists' educational qualifications are not the object of this chapter, it draws attention to the fact that while facing an insecure market, the professionals believe that this is the best space for training, development and practice. In our view, this may have two explanations: the editors do not consider their working conditions insecure—although they responded that they work a grueling number of hours, for example—, or their own experiences as Journalism students were not satisfactory, in terms of the knowledge and skills required for digital environments.

6 Final Considerations

In this chapter, we have sought to highlight the skills that characterize the professional journalist's profile, 'being' a journalist, and 'performing' journalistic activities in the new century [11, 34]. However, while recognizing the limits of the research conducted, we understand that the study undertaken provides an important contribution, identifying what has shaped the journalist's activities in digital environments, and what is required of them for a performance in line with the concept of 'total journalism' [17]. Furthermore, this study presents quality data [20], obtained through assertive methodologies, such as meta-research and questionnaire, which enabled us to obtain an accurate scenario of the skills, expertise and knowledge required for the contemporary journalist.

In addition, the results indicate opportunities for a research agenda that seeks to close gaps and, in particular, may make progress in the epistemological challenge that the field of journalism and sub-field of digital journalism require.

References

1. Barbosa S (2013) Convergente e continuum multimídia na quinta geração do jornalismo nas redes digitais. In: Canavilhas J (ed) Notícias e mobilidade: O jornalismo na era dos dispositivos móveis. LivrosLabCom, Covilhã, p 33–54
2. Barbosa S, Silva F, Nogueira L, Almeida Y (2013) Journalistic activity on mobile platforms: A study on autochthonous products and changes to the journalist's professional profile. Brazilian Journalism Research 9:10–29. https://doi.org/10.25200/BJR.v9n2.2013.602
3. Boyer, D.: The Life Informatic: Newsmaking in the digital era. Cornell University Press, New York (2013)
4. Domingo D, Salaverría R, Aguado J, Cabrera Mª, Edo C, Masip P et al. (2007) Four dimensions of journalistic convergence: A preliminary approach to current media trends at Spain. 8th International Symposium on Online Journalism, 30–31 Mar, Austin
5. Ekström, M., Lewis, S., Westlund, O.: Epistemologies of digital journalism and the study of misinformation. New Media Soc. **22**(2), 205–212 (2020). https://doi.org/10.1177/1461444819856914
6. Ekström M, Westlund O (2019a) Epistemology and journalism. In: Örnebring H (ed) Oxford Encyclopedia of Journalism Studies. Oxford University Press, Oxford
7. Ekström M, Westlund O (2019b) The dislocation of news journalism: a conceptual framework for the study of epistemologies of digital journalism. Media and Communication 7:259–270. http://dx.doi.org/https://doi.org/10.17645/mac.v7i1.1763
8. Ferrucci, P., Vos, T.: Who's in, who's out? Constructing the identity of digital journalists. Digit. Journal. **5**(7), 868–883 (2017). https://doi.org/10.1080/21670811.2016.1208054
9. Figaro R, Nonato C, Grohmann R (2013) As mudanças no mundo do trabalho dos jornalistas. Salta/Atlas, São Paulo
10. Fulton, K., Rogers, M., Schneider, E.: What is journalism and who is a journalist when everyone can report and edit news? Nieman Reports **48**(2), 10–13 (1994)
11. Hobsbawm EJ (1995) Era dos extremos: O breve século XX (Translation: Marcos Santarrita). Companhia das Letras, São Paulo
12. Ioannidis JPA (2018) Meta-research: Why research on research matters. PLoS Biology 16(3). https://doi.org/10.1371/journal.pbio.2005468

13. Jacks N (2018) Reflexividade à vista! In: Mattos M, Barros EJM, Oliveira ME (eds) Metapesquisa em comunicação. O interacional e seu capital teórico nos textos da Compós. Sulina, Porto Alegre, p 11–14
14. Jurno A, D'Andréa C (2020) Between partnerships, infrastructures and products: Facebook Journalism Project and the platformization of journalism. Brazilian Journalism Review 16(3):502–525. https://doi.org/10.25200/BJR.v16n3.2021.1306
15. Lemos A (2020) Plataformas, dataficação e performatividade algorítmica (PDPA): Desafios atuais da cibercultura. In: Prata N, Pessoa SC (eds) Fluxos Comunicacionais e Crise da Democracia. Intercom, São Paulo, p 117–126
16. López-García X, Rodríguez-Vázquez AI (2016) Journalism in transition, on the verge of a 'Total Journalism' model. Intercom, Rev. Bras. Ciênc. Comun. 39(1):57–72. https://doi.org/10.1590/1809-5844201614
17. López-García X, Silva-Rodríguez A, Direito-Rebollal, S, Vázquez-Herrero J (2020) From meta-journalism and post-journalism to total journalism. In: Vázquez-Herrero J, Direito-Rebollal S, Silva-Rodríguez A, López-García X (eds) Journalistic Metamorphosis. Studies in Big Data 70:199–207
18. Machado E (2000) La estructura de la noticia en las redes digitales: un estudio de las consecuencias de las metamorfosis tecnológicas en el periodismo (PhD Thesis). Facultad de Ciencias de la Comunicación, Universidad Autónoma de Barcelona, Barcelona
19. Mattos MÂ, Oliveira ME (2019) Metapesquisa nos textos da Compós (2001–2010): Por uma cartografia do interacional. E-Compós 22
20. Meng, X.L.: Statistical paradises and paradoxes in big data (I): Law of large populations, big data paradox, and the 2016 US presidential election. Annals of Applied Statistics **12**(2), 685–726 (2018). https://doi.org/10.1214/18-AOAS1161SF
21. Mick J, Lima S (2013) Perfil do Jornalista Brasileiro. Insular, Florianópolis
22. Newman, N.: Journalism, Media, and Technology Trends and Predictions 2021. Reuters Institute for the Study of Journalism, Oxford (2021)
23. Nicoletti J (2019) Reflexos da precarização do trabalho dos jornalistas sobre a qualidade da informação: proposta de um modelo de análise (PhD Thesis). Communication and Expression Center, Universidade Federal de Santa Catarina, Florianópolis
24. Palacios M, Barbosa S, Firmino da Silva F, Cunha R (2019) Mobile journalism and innovation: A study on content formats of autochthonous news apps for tablets. In: Journalism and Ethics: Breakthroughs in Research and Practice. IGI Global, Hershey, p 554–578
25. Pavlik, J.: The future of on line journalism. Columbia Journalism Review **36**, 30–31 (1997)
26. Pereira, F.H.: O "jornalista sentado" e a produção da notícia online no Correio Web. Questão **10**(1), 95–108 (2004)
27. Pereira, F.H.: "Old journalists" in a "new media" environment? A study on the career choices of digital journalists. The Journal of International Communication **26**(1), 36–58 (2019). https://doi.org/10.1080/13216597.2019.1677261
28. Perreault, G., Ferrucci, P.: What is digital journalism? Defining the practice and role of the digital journalist. Digit. Journal. **8**(10), 1298–1316 (2020). https://doi.org/10.1080/21670811.2020.1848442
29. Pierce, T., Miller, T.: Basic journalism skills remain important in hiring. Newsp. Res. J. **28**(4), 51–61 (2007). https://doi.org/10.1177/073953290702800405
30. Robinson, S., Lewis, S.C., Carlson, M.: Locating the "digital" in digital journalism studies: Transformations in research. Digit. Journal. **7**(3), 368–377 (2019). https://doi.org/10.1080/21670811.2018.1557537
31. Salaverría R (2019) Digital journalism: 25 years of research. Review article. El Profesional de la Información 28(1). https://doi.org/10.3145/epi.2019.ene.01
32. Salaverría R, García Avilés JA, Masip P (2008) Convergencia periodística en los medios de comunicación. Propuesta de definición conceptual y operativa. In: Actas I Congreso AE-IC. AE-IC, Santiago de Compostela
33. Salaverría, R., García Avilés, J.A., Masip, P.: Concepto de convergencia periodística. In: López García, X., Pereira Fariña, X. (eds.) Convergencia digital, pp. 41–64. Universidade de Santiago

de Compostela, Santiago de Compostela, Reconfiguração de los medios de comunicación en España (2010)
34. Schwarcz LM (2020) Quando acaba o século XX. Companhia das Letras, São Paulo
35. Schwingel C (2012) Ciberjornalismo. Paulinas, São Paulo
36. Stacciarini I (2019) O WhatsApp como ferramenta de apuração: Erros jornalísticos originados em grupos restritos a repórteres e fontes na área de segurança pública do DF (PhD Thesis). Faculty of Communication, Universidade de Brasília, Brasília
37. Traquina N (2005) Teorias do Jornalismo: porque as notícias são como são. Insular, Florianópolis
38. Tuchman, G.: Making news: a study in the construction of reality. The Free Press, New York (1978)
39. Van Dijck, J., Poell, T., de Waal, M.: The Platform Society: Public values in a connective world. Oxford University Press, Oxford (2018)
40. Vieira L (2018) Métricas editoriais no jornalismo online: ética e cultura profissional na relação com audiências ativas (PhD Thesis). Communication and Expression Center, Universidade Federal de Santa Catarina, Florianópolis
41. Weiss, A.S.: The digital and social media journalist: A comparative analysis of journalists in Argentina, Brazil, Colombia, Mexico, and Peru. Int. Commun. Gaz. **77**(1), 74–101 (2015). https://doi.org/10.1177/1748048514556985
42. Zelizer, B.: Why journalism is about more than digital technology. Digit. Journal. **7**(3), 343–350 (2019). https://doi.org/10.1080/21670811.2019.1571932

Suzana Oliveira Barbosa Associate Professor in the Department of Communication, and a permanent professor on the Post-Graduate Program in Contemporary Culture and Communication at the Faculty of Communication, Universidade Federal da Bahia. She is in her second term as the director of FACOM, is the current coordinator of the Grupo de Pesquisa em Jornalismo On-Line (GJOL), and vice-coordinator of the Compós Journalism Studies Working Group. Her research areas include the following topics: digital journalism, data journalism, convergent journalism, and innovation journalism.

Lívia de Souza Vieira Assistant Professor in the Department of Communication at the Faculty of Communication, Universidade Federal da Bahia. Member of the Grupo de Pesquisa em Jornalismo On-Line (GJOL). PhD in Journalism at Universidade Federal de Santa Catarina (UFSC). Visiting academic at Birmingham City University (2016–2017). Research focuses on transformation in the news production digital environment, with emphasis on the ethical issues.

Mariana Menezes Alcântara Ph.D. student on the Post-Graduate Program in Contemporary Culture and Communication at the Faculty of Communication, Universidade Federal da Bahia. Member of the Grupo de Pesquisa em Jornalismo On-Line (GJOL), in which studies fact-checking platforms and innovations in journalism. Has a Master's Degree in Scientific Culture and Communication and Popularization of Science by the Multidisciplinary Graduate Program in Culture and Society (Pós-Cultura, UFBA). Bachelor in Journalism and specialist in Scientific and Technological Journalism at UFBA.

Moisés Costa Pinto Ph.D. student on the Post-Graduate Program in Contemporary Culture and Communication at the Faculty of Communication, Universidade Federal da Bahia (UFBA). Member of the Research Group on Online Journalism (GJOL) in which studies the relationship between digital journalism and algorithms. Bachelor in Journalism at UFBA.

Audiences First: Professional Profiles, Tools and Strategies of Digital Newsrooms to Connect with the Public

Ana-Isabel Rodríguez-Vázquez, Marius Dragomir, and Noelia Francisco-Lens

Abstract For the media, and therefore journalism, to be sustainable in today's digital ecosystem, significant changes must occur in the relationship between journalistic companies and their audiences to generate the trust and added value necessary for the audience to feel that the worth paying for news content. This article summarizes the key findings of an exploratory research over a five-year period (2015–2020) on an initial sample of 100 digital native media, as well as legacy media from Spain, the rest of Europe, and the United States. The data collection had a double objective: to study the emerging professional profiles and the audience measurement tools that strengthen editorial teams in their attempt to improve their relationships with the audience, and to explore strategies aimed at improving the connection with the public through through participation and co-creation, both online and offline.

1 Introduction

Linked from its origins to technological innovation, journalism is undergoing today a profound restructuring process [3, 44] that affects the journalist's professional role, which is transitioning from a rather romantic profile where the journalist's mission was to chase stories of interest, to an interdisciplinary function where news literacy and computer skills are indispensable [4]. Forced by rapid technological advances, a rapidly changing network society and new media consumption patterns, during the past decade or so, journalism has been seeking new directions and forging new alliances, including interaction with citizens as part of the news gathering process through participatory services and tools as well as new ways of doing journalism [22].

A.-I. Rodríguez-Vázquez (✉) · N. Francisco-Lens
Universidade de Santiago de Compostela, Santiago, Spain
e-mail: anaisabel.rodriguez.vazquez@usc.es

M. Dragomir
Center for Media, Data and Society, Central European University, Vienna, Austria
e-mail: dragomirm@ceu.edu

J. Vázquez-Herrero et al. (eds.), *Total Journalism*, Studies in Big Data 97,
https://doi.org/10.1007/978-3-030-88028-6_16

In their attempt to adjust to the age of audience-centred news [36], digital newsrooms have been adapting their operational structures and routines to achieve a double goal: to create stories that would stir such a high interest that they would persuade audiences to buy subscriptions or memberships, make donations or micro-payments, or participate in crowdfunding to support journalism. By doing that, news media have made the public an integral part of their own sustainability planning process.

The quality of the journalistic act combined with technologies that allow for an improved user experience is the recipe that news media outlets generally use today as a way to create and cultivate a new kind of relationship with their audience. This transition process, which implies making better use of social research techniques and of technologies and automation models to produce high-quality journalism with a public service value and socially useful, can be seen as a path towards Total Journalism, a journalism concept fit for our age that preserves the core essence of journalism and is performed in a network society defined by mobility, connectivity and information ubiquity [22].

In this context, the process of searching for and producing news requires, for example, knowing the advantages of algorithms as identifying sources of digital traffic [10], big data and data processing [1], audience analysis [50], or content system management [40]. To work in different sectors of today's media system, journalists need to be part of integrated teams in which professionals with different profiles (engineers, designers, statisticians, photographers, videographers, audience editors and others) provide insights helping news media to add a high level of value to the news product that will motivate the user to pay for.

A certain level of technological preparedness is expected from journalists today [37]. This preparedness does not consist of the incorporation of technologies, as a set of tools, into the journalistic practice, but of equipping journalism professionals with more technology-related skills and competences that permit them to take advantage in their work of the opportunities brought about by the computational model, in which the *software* has taken precedence [24].

2 The State of the Question

Along with the 'demassification' of the mass media and the advance of the prosumer predicted by Alvin Toffler in the late 1970s [51], we have seen the emergence of the personalization of the news offering and, with it, the audience participation in co-creating content in multiple spaces and platforms.

The growing role of the audience in news and information production gained an important acknowledgement during the 18th International Symposium on Online Journalism (ISOJ), held in Austin-Texas, USA, in 2017. Journalists, media outlets and academics who participated in the event recognized that placing the audience at the center of the journalistic universe by prioritizing the consumer interests on all platforms has become the biggest challenge that publishers have to face.

Since then, various international events have delved into discussing the challenges of audience measurement and the newsroom roles that have emerged in recent years to track the digital footprint of the audiences and to establish more direct relationships with the users. These concerns were also the topic of various studies published in the past five years (2015–2020) by authors such as: Lamot and Paulussen [20], Nelson and Tandoc [28], Carlson [2], Rodríguez-Vázquez et al. [42], Groot and Costera [15], Hernández-Serrano et al. [17], Cherubini and Nielsen [6], Peña et al. [35], Ruddock [43], Zalmanson and Oestreicher [57], Welbers et al. [53], Harcup [16], López and Silva [23], Masip et al. [25], Meso et al. [26], Ortega et al. [33], O'Brien and Toms [32], Patriarche et al. [34], and Jönson and Örnebring [18].

In digital media, 'audience-first' is not just a slogan, but a concept that encapsulates a fundamental change in how media operate. With the audience put at the heart of the media operations in the digital age, everything changes: the design of workflows, the roles and skills required of new professionals, the tools and technologies used to measure audiences, and the decisions-making process.

As part of this transformation, the media have revisited the concept of civic engagement, refocusing it 'emotional engagement', which manifests itself through the interaction of users with a digital medium. The media have been focusing on identifying and quantifying the level of this engagement, which reinforces again the increased appeal that the audience has for the media in the digital age [29].

In the new digital engagement ecosystem (characterised by the time spent by users on specific content multiplied by the return frequency and divided among access devices—cross-platform UVs), the news media tend to reinforce and increasingly value the engagement metric as a key indicator of their users' predefined behavior.

Interactions on social media, exposure to the news interface, the time spent on consuming content, the perceived ease of use, the endurability, novelty and the participation [32], all these have become operational elements that are used on a daily basis by analysts working on user engagement. They are reinforced by the Mutteres formula [27] that consists of a combination of the number of clicks (CI), the duration of the visit (Di), the rate at which the visitor returns to the site over time (Ri), the brand awareness (Bi), the willingness to contribute with comments (FI) and the likelihood of participation in specific activities on the site, designed to increase awareness and create a lasting impression (II).

Smith and colleagues [47] propose a typology of audience-centric digital metrics for content customization based on the following criteria:

- Audience location: Where's the audience located? Place of origin (measurement of traffic from the domestic market or abroad).
- Audience size and composition: How is the audience? The reach determines the audience reached and data such as the number of visitors, the average number of unique users and their percentage value in the context of the market. Demographic data make it possible to know the profile of the audience and to collect information segmented by age, gender and income level.
- Audience behavior: data about the consumption rate and the levels of attention and loyalty of the users. These metrics give information related to the content

consumed, the duration of visits, or return rates (return frequency in percentage terms).
- Audience source: number and percentage of direct access to the media content, accessing by search, accessing derived from other media or social networks, among others.
- Audience by device: distinguishes the public access by screen views, helping obtain the share of users accessing content through desktops, mobile devices or tablets.

These forms of measurement gather data about news consumption behaviors, allowing journalists to better respond to their audiences, but there are problems, both old and new, that appear in measuring digital news audiences [2]. 'Measurable journalism' is a term that encompasses the cultural and material change triggered by and related to the digital platforms, which are able to provide individualized, real-time quantitative data on consumers' behavior. Eight dimensions of measurable journalism are emerging, around which we structured part of our study [2]:

- Material: digital analytics and metrics software.
- Organizational: new newsroom professionals who monitor and/or react to metrics; new types of marketeers interacting with journalists and news managers.
- Practice: use of data in story assignment and placement decisions; new forms of public participation.
- Professional: acceptance or resistance to audience metrics used in decisions of journalistic interest; concern for professional autonomy; trust in improving the connection with the public.
- Economic: changes in content monetization; basing resources and staff-related decisions on data.
- Consumption: personalized content or content-related recommendations for news audiences.
- Cultural: concern about the popularity of decisions about journalistic interest; debates about individualized news versus collective news.
- Public Policy: concerns about audience privacy and data tracking.

3 Objectives and Method

Our methodological approach is based on a mixed method using a combination of case studies and exploratory research, the two together, rather than analyzing realities through de-contextualized segments, permitting a holistic understanding of these realities [5, 56].

More flexible in its methodology than the descriptive or explanatory research, the exploratory research seems to be the most appropriate for this study as it is designed to take stock of the unexpected. We chose this approach because it helps us to become more familiar with relatively unknown phenomena [7] and discover perspectives that have not been identified thus far.

On the other hand, the case study research, which emerged in the social sciences as an attempt to understand different problems in their social context [11], allows us to understand how the profile of journalist has evolved towards a more technological one as a professional who works in the current media with renewed narratives, and also enables us to delve into the most recent main innovation trends in the industry.

The research draws on two lines of work developed in the last five years in the framework of the following two projects led by the research group New Media (Novos Medios) of the USC: *Uses and Information Preferences in Spain's New Media Landscape: Journalism Models for Mobile Devices* (CSO2015-64662-C4-4-R) and *Digital Native Media in Spain: Storytelling Formats and Mobile Strategy* (RTI2018-093,346-B-C33), both funded by the Ministry of Science, Innovation and Universities and co-financed by the European Regional Development Fund (ERDF). The two lines of work focus on the following key objectives:

O1. To study the new professional profiles that present in today's newsrooms whose function is to improve their relationship with the audience. These profiles are studied in both legacy media and digital native media from the moment of incorporating the audience editor and the engagement editor in the newsroom, and at the same time the tools used for tracking the user's digital footprint in real time (measurement of traffic-consumption and social audience) are identified.

O2. To explore the strategies that the media companies implement in order to improve their connection with the public, with particular attention paid to the content that brings value to the audience engagement (participation and co-creation), which would lead to the identification of a combination of strategies adjusted for both *online* and *offline* contexts.

The research is part of a regular process of monitoring and analysis carried out on a sample of more than 100 digital native media and legacy media from Spain, the rest of Europe and the United States, in their online versions (web and mobile apps), aimed at understanding the evolution and adjustment of their newsrooms and narratives to the demands of the journalistic profession at a time when news is increasingly centered on audiences.

Following on the work done by Ragin and Becker [38], this paper gathers cases from a sub-sample of media outlets whose main selection criterion was the presence of innovation elements related to at least one of the two objectives set out in the study. From the sample, the media outlets selected for analysis in this article are the following: *The Washington Post*, *The New York Times*, *The Guardian*, *CNN*, *NBC News*, *HuffPost*, *BuzzFeed*, *Quartz*, *The Texas Tribune*, *The Economist*, *Vox*, *El País*, *La Vanguardia*, *elDiario.es*, the *Financial Times*, *Dublin Inquirer*, *Illinois Newsroom*, the *Financial Times*, *Público* and *ProPublica*, among others.

4 Findings

4.1 Professional Profiles and Measurement Tools

In today's digital ecosystem, the media no longer play only the role of informing their audiences, but they also commit to them. Knowledge about and experience with audience participation become a priority when professionals are recruited to work in the newsrooms. They are also a strategic goal that news media aim to achieve.

As a result, there is an increase in investments in resources and infrastructures that allow media to design new strategies aimed at improving their interaction with the public and to achieve the highest level of audience loyalty without compromising the quality of the news product. Functions such as social media editors, analytics editors, audience editors, growth editors or engagement editors have been incorporated into the professional operations of media outlets to fulfil a proactive mediating role between content creators and the audience [12, 55].

Hence, in job adverts posted by publications such as *The Guardian, The Washington Post, HuffPost, BuzzFeed, The Philadelphia Inquirer* or *Quartz*, the position of audience editor foresees a data-analyzing professional who is capable of placing the data in the right context and interpret such data in the decision-making process. Such a professional is expected to have a special instinct to identify new trends, an insatiable curiosity, the desire to defend the community, marketing skills, the ability to adjust to the changes in consumer behavior, the talent of finding creative and compelling editorial approaches for paying readers, the skill of developing content distribution and messaging strategies, and, ultimately, the ability to look for creative ways to deliver content to loyal as well as potential consumers. Other outlets such as *The Texas Tribune, ProPublica, Vox and The Economist* introduced the audience editor function with new added roles such as engagement or growth editor. In both groups of publications, the main objective of the professionals filling the newly created positions is to maintain an open and direct dialogue with the audience, beyond the metrics analysis. The weight of developing strategies for user loyalty through the personalization of the news product (curation) and encouraging public participation in the news production (co-creation) lies on the shoulders of these engagement professionals.

In the Spanish media environment, we find references to such professionals in newspapers like *elDiario.es and LaVanguardia.com.* The Catalan newspaper *La Vanguardia* published in 2019 a job advert for the position of audience analyst. During the two weeks that the role remained open on the professional social network LinkedIn, a total of 36 candidates submitted their applications. The position requirements included knowledge and skills in cross-platform product and performance analysis, audience analysis (user behaviour, audience qualification and competitive benchmarking), high analytical capabilities, the capacity of comprehensively processing large amounts of data, teamwork, openness to innovation and capacity to handle different sources of traffic data analysis (Comscore, Google Analytics, Google Tag Manager, Google Search Console, Facebook Insights or Power BI).

For news providers, another objective they seek to achieve by using these tools is to be able to analyze audiences from a quantitative viewpoint. In addition to the data provided monthly and annually by specialized audience measurement companies such as Comscore or Nielsen, in most of the analyzed countries, the media outlets incorporate into their newsrooms other tools that allow them to understand, internally and in real time, the digital footprint of their audiences. Among others, the majority of them use tools such as Adobe Analytics, Chartbeat, Parse.ly, Amplitude, Crazy Egg, SimilarWeb, Sistrix, Tableau Software, Welovroi, DogTrack, SocialFlow, EzyInsights, CrowdTangle, Semrush, Trendinalia, Elephant, BuzzSumo and others.

Measuring news consumption in a cross-media, hyperconnected environment characterized by high mobility can be extremely difficult. That is why content teams in the newsrooms are increasingly spending more time and creative effort designing strategies and approaches that are expected to enable them to reach the right audience: the one that is the most valuable in the long term, willing to participate in funding the media as a way to contribute to their sustainability [46].

As such, the teams of audience-centred professionals (including the editors themselves) look at the news lifecycle in a holistic way. Starting from the stage of story selection and even during the follow-up phase after the publication of the piece, these professionals work on promoting the news content, adapting it to specific audiences and platforms, and analyzing the interest generated by it in order to discover whether the content was valuable for the audience and the story could be considered successful.

This has been the main objective of the Knight-Lenfest Newsroom project. Started in 2015 as Knight-Temple Table Stakes, the project, which brought together more than 50 professionals from four leading metropolitan media outlets in the USA, and further expanded in 2017, placed the audience at the centrepiece of the news production cycle, a strategy meant to add value to the media product [14]. The project led to the identification of seven key strategic aspects organized in the so-called 'Table Stakes' described by Rodríguez and colleagues [41:87] as follows:

- "Personalised content for targeted audiences: focusing on particular audiences with needs, interests and problems to provide information coverage superior to competitors. Using behaviour and usage patterns, four audience segments were identified: grazers, test drivers, intenders and subscribers [46, 47].
- To publish on the platforms used by your targeted audiences: media have a responsibility to identify where their audiences are, instead of expecting their audience to reach them. To do so, they must make an effort to publish on the platforms used by the chosen target audiences.
- Always active, always there: to offer digital content adapted to the rhythms and habits of target audiences, to their availability of time and attention, to the problems they need to solve across the platforms they use. Changing routines to accommodate the production of contents to suit the audience rather than the newsroom or the print schedule. Digital content, first.
- To convert the audience: to transform sporadic users into loyal, regular, valuable and paying audiences. This involves engaging the audience until they are ready

to pay for content, products and services, which aids in valuing the journalistic brand sufficiently to recommend it to other users.

- Through innovation, and the testing and development of different sources of income from the audience, long-lasting relationships are built with new products, services or businesses of value to the target audience and the community.
- Expansion: through partnerships, third-party services, content creation, marketing, distribution… with the purpose of risk sharing.
- Interfunctional mini-publishers to drive growth in target audience segments".

From a qualitative point of view, taking into account the high value of the engagement metrics in the newsrooms, worth mentioning is also a project carried out by Impact Architects in 2019 in collaboration with the Graduate School of Journalism at the University of New York whose aim was to establish whether there is a relation between the trust generated by the journalism focused on an engaged audience and the resulting income. To this end, the media participating in the project allowed the public to become part of the news production process, providing them with details about how the journalistic work is done in newsrooms and subsequently reporting on how the audience's contributions had been incorporated into the process, all being an attempt at user involvement and engagement with the goal of improving trust [13]. As part of this experiment, a total of 130 international editors shared the belief that, although journalistic engagement has a commercial value, journalism must also help build a community through which journalists can collaborate with the public to improve their work, not simply to promote it only [39].

4.2 Strategies to Connect the Media with the Audience

The media sector has been experiencing increasing concerns about the rise in the number of users who report actively avoiding news (32%) because news has a negative effect on their mood (58%) or because they feel powerless in intervening in the events, both in their nearest environment and in a global context [31].

In addition, *paywalls* introduced by some media drive many users away from quality news. According to numerous studies, faced with a limited budget to invest in subscriptions, a part of the audience expresses a clear preference to pay for entertainment and not for news [30, 31].

Media strategies over the past two years, clearly aimed at driving audiences towards subscriptions, have been disrupted by the COVID-19 pandemic. The uncertainty triggered by the crisis undermines media sustainability in a context where free and paid-content models co-exist, illustrating the different perspectives of the news providers present on the market: those clearly in favor of the *paywall* like *The Wall Street Journal*; the reticent ones like *The Guardian*; or the moderately exploratory ones like *The Times*.

Qualitative research on the motivations of the public to pay for consuming news online shows that users value the suppliers who offer them a plurality of views or perspectives in their news content, a variety of content through which they can make fortuitous discoveries, or the ability to customize the news media interface by prioritizing the content according to their interest [19]. Seducing the audience is becoming more and more complicated and, to meet this challenge, the strategies implemented by the media go in different directions:

- Strategies for approaching/capturing audiences (newsletters, alerts, user service centers, etc.).
- Strategies for public/community involvement and participation in news production (closed communities that allow dialogue between journalists and users, debates and discussions of issues, integration of explanatory journalism or solutions, meetings with the audience in online and offline spaces).
- Strategies for using video and audio in information products.
- Strategies of gamification.
- Innovation strategies in social media formats.
- Subscription strategies (clubs offering VIP products and services).

Below are some examples in different categories selected through the research using the aforementioned sample of media that was carried out during the last five years.

High levels of content consumption through the use of email, an effective means of attracting the oldest and most highly committed news consumers [31] prompted media outlets like *The Washington Post* to design a strategy of developing specialized news bulletins sent via email to increase the direct traffic to their website. Younger audiences are persuaded to do the same through mobile notifications.

In March 2017, *The Washington Post* created the Facebook Group PostThis, a closed group for newspaper readers interested in stories about journalism accountability where they could ask questions about how the journalistic work was done. This *Post* space did not become a subscriber platform like the one created by *The Boston Globe* or a community focused on a topic of interest supported by chatbots, but a space for people interested in the responsibilities that journalists have where journalists can also explain to their audiences how they produced their news stories. More recently, building on evidence that e-mail newsletters prompt users to consume news content and in line with its strategy to generate more direct traffic to its website, *The Washington Post* launched around 80 different newsletters.

The New York Times strives to customize digital news delivery with the goal of reaching 10 million subscribers by 2025 [48]. In June 2017, it announced a new strategy to boost the connection with its audience through the creation of the Reader Center that is aimed to capitalize on the knowledge and experience of its users by improving the response of the outlet to the opinions, comments and advice received from the audience, boosting the transparency of its news coverage; experimenting with new formats that impact and attract the public; incorporating the voices of readers into digital platforms; and helping journalists build communities of readers interested in covering different topics.

Following the withdrawal of *The Guardian* from the Facebook Instant Articles and the Apple News platform in April 2017, the newspaper expressed its commitment to keep readers on its portal as long as possible and strengthen the paid subscription model to achieve the one-million target by 2019. *The Guardian* then redesigned its strategy moving away from the paywall and embracing instead the registration wall. It launched Hope is Power, a campaign whose goal is to reach two million paying followers before 2022 by focusing on deepening the relation of the newspaper with existing subscribers while encouraging new readers to spend more time on *The Guardian* website.

The international news channel CNN has been implementing a vertically designed innovation strategy centered on mobile devices and focused, on the one hand, on the creation of specialized content designed for *CNNMoney* (on economic and business issues) and *CNN Politics* (content that has been gaining increased visibility thanks to the intensity generated by the U.S. election campaign of 2020), and, on the other hand, on strengthening the Health, Technology and Travel sections. Content design is combined with a research strategy geared on identifying and analyzing news consumption habits that are monitored through a heat map built into mobile phone applications. For example, knowing and taking advantage of the location of the app buttons, which must be reachable by the thumb to allow people to easily use the platform on public transport (when the user has only one free hand to operate the mobile), led to an increase in the number of *CNN*'s active users along with the time spent by them on the platform's news apps. *CNN*'s strategy is anchored in content that is presented in a more interactive and personalized style.

NBC News has been focusing its audience engagement strategy on 'vertical sites' that are designed as public participation labs and are also distributed as newsletters that can attract a combined monthly audience of some 67 million unique users, according to 2018 data. With a content publishing rate of 200 pieces per month, *NBC* created Think, a space for op-eds, in-depth analyses and essays about news and current events created to experiment with audience groups and specific content [54].

El País created Join the Conversation, a space open to direct public participation in the process of news creation. Through this space, the outlet aims to use the knowledge of the audience and to listen to their ideas and requests to build a community of readers. One of the strategic novelties introduced by the newspaper is the audiovisual rubric El País Video, which is available both on the website and through the Facebook Live/Watch feed.

The research has also identified two examples of community approach strategy, with events open to audience participation, both online and offline. One is the News & Brews project initiated by the Illinois Newsroom as a discussion group held in public places such as coffee shops where citizens can discuss with journalists about issues affecting the community. The other one is a 'citizens agenda' approach applied as part of the election coverage by *Dublin Inquirer*, which, inspired by publications such as *The Angeles Times* and *The Tyee*, used a standard Google form to collect suggestions for questions and issues (almost 2,000 answers were received) that the citizens wanted to raise with politicians ahead of the 2020 elections [58].

In addition to audience engagement initiatives spearheaded by well-known international media outlets, our research also identified such projects run by local media. Based on the Audience-first strategy and experimenting with tactics learned in the Knight-Lenfest Table Stakes project, the local newspaper *Atlanta Journal-Constitution* launched in March 2020, in the midst of the COVID-19 pandemic, a Facebook campaign aimed at engaging the children of their readers. To counter the wave of dispiriting news about the coronavirus, the newspaper involved its community of readers (many of whom had been already sharing personal stories) into the Art of the Heart, a campaign asking its readers to send any drawings made by their children that featured frontline workers such as postmen, police officers, firefighters, supermarket employees or dispatchers doing their job in the times of the pandemic.

The initiative, which has expanded both in the newspaper print edition and on its online platforms, generated 56,000 page views, the interaction of nearly 8000 users, total income of US$ 35,000 over the course of eight weeks and positive comments praising the initiative (such as, "I really like the efforts you're taking to humanize our hometown newspaper") [52].

Trying to attract young subscribers, the *Financial Times* designed FT Schools, a project through which the newspaper creates online accounts for students aged 16 to 19 giving them access to the publication's media content. Some 40,000 pupils from 2,800 schools in 101 countries signed up for an *FT* account, which allows them to receive two weekly emails with content related to their fields of study as well as videos and articles on how to prepare for an exam or for work interviews [9]. The *Wall Street Journal* and *Politico* also use events to maintain physical contact with their readers through the Journal House [21] and the Politico Live projects, respectively.

Combining the offline and online spaces to better interact with the public, *The New York Times* has created events for subscribers that gives readers, regardless of their location, access to the Times newsroom. Online, on the telephone or in person, *The NYT* initiates conversations and organizes panels, conferences or live experiences to foster a direct dialogue between its journalists and the audience as part of subscribers-only spaces such as All Access Plus and Home Delivery.

In the search for youth audiences, news providers increasingly turn to social media. *The Economist* enhances the dialog of its editors with young audiences through Instagram [45], *The Telegraph* does so through Snapchat Discover; and *The Dallas Morning News*, *NBC News*, *Bloomberg*, *BuzzFeed*, *Vice* and *The Washington Post* (with Dave Jorgenson) use TikTok in their attempt to connect with the GenZ.

As part of its video and audio (podcast) strategy, Facebook has shifted its focus onto longer-form programs designed for Facebook Watch, prompting Netflix, Amazon and HBO to follow suit, investing significant amounts of money into expanding their portfolio of longer-form content. Examples of this video format expansion include the documentary series Explained, produced by Vox and broadcast on Netflix; videos from *The Washington Post*'s satire/humor department released via Amazon; the documentary series The Weekly produced by *The New York Times* and broadcast by Hulu, Odisea and Movistar+. They also include original series from the *Post*'s satire and humor department such as those conducted by Dave Jorgenson ('Short Takes') on different platforms and social networks, which are especially designed to attract young audiences on TikTok.

Podcasts have existed for many years, but these episodic digital audio files seem to be reaching nowadays a critical mass as a result of improved content and easier ways of distribution. Various publications (like the *Financial Times*) believe that podcasts help attract audiences, especially young followers and new subscribers, reinforcing the widespread belief that audio is the new gold standard of journalism [49].

The Guardian, The Washington Post, Politiken, Aftenposten, The Economist and the *Financial Times* are among dozens of publications that have released daily podcasts since 2019. This follows the unbridled success of *The New York Times* Daily podcast program with about five million listeners a day, which is being broadcast on public radio and is expected to become a video series as well. Meanwhile, the *BBC* has renamed its on-demand radio app BBC Sounds to better reflect the shift to on-demand consumption and the growing interest of the podcast generation. Some news media also began to follow the example of established audio platforms, such as Spotify, which have begun to pay providers of popular premium content. Thus, some news organizations like *Politiken* began to restrict access to some of their daily news products to subscribers only. These new sources of income help media produce more professional content and increase the value of their content, but some fear that the purity and authenticity of the podcast experience will be lost in this process.

In this respect, following a suggestion, from a #LaRepúblicaDePúblico subscriber, to share songs that inspired strength during the confinement imposed by the COVID-19, the newspaper Público decided to collect suggestions from the audience to create a playlist with songs that were likely to make the quarantine more bearable. It distributed the playlist on Spotify under the title of *Quarantine Days and 500 Nights*.

An example of gamification used to engage the audience is *The Waiting Game* created by *ProPublica* that is based on real-world case files of five asylum seekers from five different countries and interviews with the medical and legal professionals who evaluate and represent them. This is an experimental news game that allows the user to take on the role of an asylum seeker from the moment he chooses to travel to the United States to the final decision of an immigration judge.

5 Conclusions

When media, through technological development, become more ubiquitous, intrusive and hyperconnected; and when the participation of the public as an indispensable actor increases, the focus shifts to increasingly complex audience research, which presents great challenges, but also great opportunities, especially in the context of the advances made recently in the IoT (Internet of Things) [8].

If the goal is to move towards new financing models that place the audience at the epicenter of the business model, by focusing on the metrics of engagement to give the public what they want, the media outlet will strengthen its utility until it becomes indispensable for its audience.

Our research confirms that staff in the newsrooms take on new challenges as professionals in newly emerged roles, including data, video, audience and analytics editors, work together to better understand and listen to their audiences, to experiment, to act quickly, to better integrate data and to better tell stories through their platforms. As Ismael Nafría, of the Knight Center put it, "the Journalism with capital letter is back" [36]. New professional roles converge on a common goal: achieving greater engagement, trust and loyalty between media and users with a view to enable stronger relationships in the future and strengthen strategic business lines whose aim is to generate revenues from direct contributions made by the audiences.

However, in spite of the large amounts of audience data that are being circulated in the newsrooms, establishing benchmarks for news impact is still a very complex process. Therefore, numerous media tend to turn use of digital metrics into a goal.

In the context of Total Journalism, the relationship of the media with the audience is fundamental and the audience-first strategy seeks to reach journalistic goals by employing format innovation strategies and improving the connection of the media with the public through audience participation in the journalistic work as well as the emotional and economic involvement of the audience in the media outlet.

Acknowledgements Special thanks to Norina Solomon (Media Power Monitor) for the linguistic revision of this text.

References

1. Bruns, A.: Big data analysis. In: Witschge, T., Anderson, C.W., Domingo, D., et al. (eds.) The SAGE Handbook of Digital Journalism, pp. 509–527. Sage, London (2016)
2. Carlson, M.: Confronting measurable journalism. Digital J **6**(4), 406–417 (2018). https://doi.org/10.1080/21670811.2018.1445003
3. Casero-Ripollés, A.: Contenidos periodísticos y nuevos modelos de negocio: Evaluación de servicios digitales. El Profesional de la Información **21**(4), 341–346 (2012). https://doi.org/10.3145/epi.2012.jul.02
4. Codina, L.L.: Tres dimensiones del periodismo computacional. Intersecciones con las ciencias de la documentación. Anuario ThinkEPI **10**, 200–202 (2016). https://doi.org/10.3145/thinkepi.2016.41
5. Creswell, J.W.: Qualitative Inquiry and Research Design: Choosing Among Five Approaches. Sage, Thousand Oaks (2007)
6. Cherubini, F., Nielsen, R.K.: Editorial analytics: How news media are developing and using audience data and metrics. In: Digital News Project 2016. Reuters Institute for the Study of Journalism, Oxford (2016). Available at https://bit.ly/3qVFe5i. Accessed 01 Apr 2020
7. Dankhe, G.L.: Investigación y comunicación. In: Fernández-Collado, C., Dankhe, G.L. (eds.) La comunicación humana: ciencia social, pp. 385–454. McGraw-Hill, México (1989)
8. Das, R., Ytre-Arne, B.: Critical, agentic and trans-media: frameworks and findings from a foresight analysis exercise on audiences. Eur. J. Commun. **32**(6), 535–551 (2017). https://doi.org/10.1177/0267323117737954
9. Davies, J.: How the Financial Times is building brand loyalty among young readers. Digiday. (2019). Available at https://bit.ly/30HPxiC. Accessed 01 Sep 2020
10. Diakopoulos, N.: Algorithmic accountability: Journalistic investigation of computational power structures. Digital J. **3**(3), 398–415 (2015). https://doi.org/10.1080/21670811.2014.976411
11. Eisenhardt, K.M.: Building theories from case study research. Acad. Manage. Rev. **14**, 532–550 (1989). https://doi.org/10.5465/AMR.1989.4308385
12. Ferrer-Conill, R., Tandoc, E.C.: The audience-oriented Editor. Digital J. **6**(4), 436–453 (2018). https://doi.org/10.1080/21670811.2018.1440972

13. García-McKinley, E.: Engaging for trust: what news organizations can (and should) do right now. Impact Architects (2019). Available at https://bit.ly/3cCPJFe. Accessed 01 Sep 2020
14. Griggs, T.: What are the seven "Table Stakes" essentials? Better News (2017). Available at https://bit.ly/2OAC2P5. Accessed 28 Sep 2020
15. Groot, T., Costera, I.: What clicks actually mean: exploring digital news user practices. Journalism **19**(5), 668–683 (2017). https://doi.org/10.1177/1464884916688290
16. Harcup, T.: Asking the readers: Audience research into alternative journalism. J. Practice **10**(6), 680–696 (2015). https://doi.org/10.1080/17512786.2015.1054416
17. Hernández-Serrano, M.J., Renés-Arellano, P., Graham, G., Greenhill, A.: Del prosumidor al prodiseñador: el consumo participativo de noticias. Comunicar **50**, 77–88 (2017). https://doi.org/10.3916/C50-2017-07
18. Jönson, A.M., Örnebring, H.: User-generated content and the news: Empowerment of citizens or interactive illusion? J. Practice **5**(2), 127–144 (2011). https://doi.org/10.1080/17512786.2010.501155
19. Kantar: Attitudes to Paying for Online News (2017). Available at https://bit.ly/3cDbOU6. Accessed 28 Jun 2020
20. Lamot, K., Paulussen, S.: Six uses of analytics: digital editor's perceptions of audience analytics in the newsroom. J. Practice **14**(3), 358–373 (2020). https://doi.org/10.1080/17512786.2019.1617043
21. Lim, S.: How The Wall Street Journal is using its event business to plug the print gap (2019). Available at https://bit.ly/2Orgq85. Accessed 8 Feb 2020
22. López-García, X., Rodríguez-Vázquez, A.I.: Journalism in transition, on the verge of a "Total Journalism" model. Intercom **39**(1), 57–58 (2016). https://doi.org/10.1590/1809-5844201614
23. López, X., Silva, A.: Estrategias para la participación de los usuarios en la producción de contenidos de tres cibermedios de referencia: BBC.com.uk, NYT.com y TheGuardian.com. Estudios sobre el Mensaje Periodístico **21**, 145–164 (2015). https://doi.org/10.5209/rev_ESMP.2015.v21.50669
24. Manovich, L.: Elsewhere (2019). Available at https://bit.ly/2Q4scp8. Accessed 01 Mar 2020
25. Masip, P., Guallar, J., Suau, J., Ruiz-Caballero, C., Peralta, M.: News and social networks audience behavior. El Profesional de la Información **24**(4), 363–370 (2015). https://doi.org/10.3145/epi.2015.jul.02
26. Meso, K., Agirreazkuegana, I., Larrondo, A.: Active Audiences and Journalism. Analysis of the Quality and Rregulation of the User Generated Contents. Universidad del País Vasco, Bilbao (2015)
27. Mutteres, A.: Engagement: the new digital metric. Reflections of a newsosaur (2011). Available at https://bit.ly/2OAG5Lh. Accessed 01 Mar 2020
28. Nelson, J.L., Tandoc, E.C.: Doing 'well' or doing 'good': what audience analytics reveal about journalism's competing goals. J Stud. **20**(13), 1960–1976 (2019). https://doi.org/10.1080/1461670X.2018.1547122
29. Nelson, J.L., Webster, J.G.: The myth of partisan selective exposure: a portrait of the online political news audience. Social Media + Society **3**(3) (2017). https://doi.org/10.1177/2056305117729314
30. Newman, N.: Journalism, Media and Technology Trends and Predictions 2019. Reuters Institute for the Study of Journalism, University of Oxford (2019a). Available at https://bit.ly/3lhtUiB. Accessed 08 Mar 2020
31. Newman, N.: Digital News Report 2019—Executive Summary and Key Findings. Reuters Institute and University of Oxford (2019b) Available at https://bit.ly/3rWf9Ew. Accessed 08 Mar 2020
32. O'Brien, H.L., Toms, E.: The development and evaluation of a survey to measure user engagement. J. Am. Soc. Inf. Sci. **61**(1), 50–69 (2010). https://doi.org/10.1002/asi.21229
33. Ortega, F., González, B., Pérez, M.E.: Audiences in revolution. use and consumption of mass media groups' apps for tablets and smartphones. Rev. Lat. Comun. Soc. **70**, 627–651 (2015). https://doi.org/10.4185/RLCS-2015-1063en

34. Patriarche, G., Bilandzic, H., Jensen, J.L., Jurisic, J.: Audience Research Methodologies: Between Innovation and Consolidation. Routledge, London (2015)
35. Peña, S., Lazkano, I., García, D.: European newspapers' digital transition: new products and new audiences. Comunicar **46**, 27–36 (2016). https://doi.org/10.3916/C46-2016-03
36. Piccato, F.: La era de las noticias centradas en la audiencia. La Voz (2017). Available at https://bit.ly/3vuDenI. Accessed 08 Mar 2020
37. Powers, M.: "In forms that are familiar and yet-to-be invented": American journalism and the discourse of technologically specific work. J. Commun. Inq. **36**(1), 24–43 (2012). https://doi.org/10.1177/0196859911426009
38. Ragin, C.C., Becker, H.S.: What is a Case? Exploring the Foundations of Social Inquiry. Cambridge University Press, Cambridge (1992)
39. Reuters Institute: Media, Journalism and Technology Predictions 2016 (2016). Available at https://bit.ly/2NoRttk. Accessed 18 May 2020
40. Rodgers, S.: Foreign objects? Web content management systems, journalistic cultures and the ontology of software. Journalism **16**(1), 10–26 (2015). https://doi.org/10.1177/1464884914545729
41. Rodríguez-Vázquez, A.I., Costa-Sánchez, C., García-Ruiz, R.: New information consumptions. The impact of audiences on journalistic roles. In: Toural-Bran, C., Vizoso, Á., Pérez-Seijo, S., Rodríguez-Castro, M., Negreira-Rey, M.-C. (eds.) Information Visualization in the Era of Innovative Journalism, pp. 82–93. Routledge, New York (2020)
42. Rodríguez-Vázquez, A.I., Direito-Rebollal, S., Silva-Rodríguez, A.: Audiencias crossmedia: nuevas métricas y perfiles profesionales en los medios españoles. El Profesional de la Información **7**(4), 793–800 (2018). https://doi.org/10.3145/epi.2018.jul.08
43. Ruddock, A.: Investigating Audiences. Sage, London (2016)
44. Salaverría, R.: Estructura de la convergencia. In: López, X., Pereira, X. (eds.) Convergencia Digital: Reconfiguración de los medios de comunicación en España. Universidade de Santiago de Compostela, Santiago de Compostela (2010)
45. Scott, C.: "It's not just an audience, it's a community": How the economist is engaging with young people on Instagram (2018). Journalism.co.uk. Available at https://bit.ly/30YRYgZ. Accessed 15 Mar 2020
46. Smith, D., Hope, Q., Griggs, T.: Customize the funnel for your own context, strategy and goals for different audience segments. Knight-Lenfest Newsroom Initiative (2017a). Available at https://bit.ly/3bMzx51. Accessed 15 Mar 2020
47. Smith, D., Hope, Q., Griggs, T.: Target audiences: measures of success and tracking progress in closing the gaps. Knight Foundation (2017b). Available at https://bit.ly/3vpyj7H. Accessed 15 Mar 2020
48. Spayd, L.: A Community of One: The Times Gets Tailored. New York Times. The New York Times (2017). Available at https://nyti.ms/3tngtAt. Accessed 08 Apr 2020
49. Sternik, I.: Tendencias de audio en medios de comunicación (webinar). Audio: el nuevo oro del periodismo. Laboratorio de Periodismo (2019). Available at https://bit.ly/3vCuoVk. Accessed 19 Sep 2020
50. Tandoc, E.C.: Journalism is twerking? How web analytics is changing the process of gatekeeping. New Media Soc. **16**(4), 559–575 (2014). https://doi.org/10.1177/1461444814530541
51. Toffler, A.: The Third Wave. Bantam Books, New York (1980)
52. Waligore, M.A.: How the Atlanta Journal-Constitution engaged the community to thank front-line workers and generated ¢35,000 in advertising revenue along the way. Better News (2020). Available at https://bit.ly/3vuNbS8. Accessed 20 Sep 2020
53. Welbers, K., Van Atteveldt, W., Kleinnijenhuis, E., Ruigrok, N., Schaper, J.: News selection criteria in the digital age: professional norms versus online audience metrics. Journalism **17**(8):1037–1053 (2015). https://doi.org/10.1177/1464884915595474
54. Willems, M.: How NBC News' vertical sites became audience engagement laboratories. Digiday (2019). Available at https://bit.ly/3tnROvs. Accessed 19 Sep 2020

55. Willens, M.: The latest key newsroom job: membership editor. Digiday (2019). Available at https://digiday.com/media/latest-key-newsroom-job-membership-editor/. Accessed 19 Sep 2020
56. Yin, R.K.: Case Study Research: Design and Methods. Sage, Thousand Oaks (2009)
57. Zalmanson, L., Oestreicher, S.: Turning content viewers into subscribers. MIT Sloan Manage. Rev. **57**(3), 11–13 (2016)
58. Zirulnick, A.: Case study: How the Dublin Inquirer set a citizens agenda. Medium (2019). Available at https://bit.ly/3qYcYPv. Accessed 19 Sep 2020

Ana-Isabel Rodríguez-Vázquez Professor of Audiovisual Communication and member of the Audiovisual Studies research group from Universidade de Santiago de Compostela (USC). Her lines of research are focused on the study of audiovisual information, broadcasting evolution, circulation of audiovisual products and social audience. She lectures on Television Genres, Programming and Audience, and Circulation of Cultural Products. At present, she is the Secretary of the Faculty of Communication Sciences (USC).

Marius Dragomir Director of the Center for Media, Data & Society (CMDS) and Visiting Professor teaching research practice at Central European University (CEU) in Vienna. His work focuses on mapping media ownership structures, study of state media and public media models, journalism business models, analysis of disinformation platforms, study of social media and mapping of regulatory mechanisms for media and social media.

Noelia Francisco-Lens Graduate in Audiovisual Communication from Universidade de Santiago de Compostela (USC) and holds a Master's Degree in Journalism and Communication (New Trends on the Producion, Management and Dissemination of Knowledge), also from the USC. Ph.D. student in Contemporary Communication and Information at the Faculty of Communication Sciences (USC). The main lines of investigation are the innovation of the media through multimedia and interactive online products, and the new forms of engagement with the connected audience.

Co-creation and Curation of Contents: An Indissoluble Relationship?

José Sixto-García, Pablo Escandón Montenegro, and Lila Luchessi

Abstract From the popularization of co-creation as a marketing strategy that enables the generation of joint activities between organizations and users, this phenomenon is explored in the media, especially in digital native ones. Co-creation allows the participation of the public in the ideation, development and marketing of journalistic products, but for this situation to continue guaranteeing the right of citizens to receive truthful information, it is essential to cure content. How the media filters information produced by audiences, how journalistic and non-journalistic products differ and how co-creation could affect misinformation are some of the questions answered in this chapter.

1 Co-creation of Journalistic Products

Co-creation has gone from marketing to journalism. Just as marketing complemented the purely economic sceneries with others linked to social environments or the transmission of ideas from the 70s of the past century, companies started, especially from the 2010 decade [30], to implement strategies based on the generation of joint activities with their clients. Journalism companies are no exception to these types of practices and have involved the public in the creation of the products they offer, especially digital native media.

Ideation, design and production of services and contents are increasingly entrusted in sectors to the collaboration between the producer and the consumer [22], so that consumers constitute nowadays a fundamental part of the organizations and the distinction of their roles is considered obsolete in production and consumption [16, 17]. Products have a greater added value if those who consume them participate in its

J. Sixto-García (✉)
Universidade de Santiago de Compostela, Santiago de Compostela, Spain
e-mail: jose.sixto@usc.es

P. Escandón Montenegro
Simón Bolívar Andean University of Ecuador, Quito, Ecuador

L. Luchessi
National University of Río Negro, Río Negro, Argentina

J. Vázquez-Herrero et al. (eds.), *Total Journalism*, Studies in Big Data 97,
https://doi.org/10.1007/978-3-030-88028-6_17

creation process [26, 29] and, thus, the development of individualized experiences based on co-creation suppose a great competitive advantage for the coming years [33].

Co-creation implies, then, that users become collaborators of the companies, so that organizations make value proposals instead of creating them, and consumers themselves are the ones who co-create [21]. This implies that marketing is not directed to the market, as it was traditionally, and it is directed towards users more than ever, because it is involved in the marketing activities to co-create value [27], so that there is a mutual benefit. Some investigators are even talking about a new stage in marketing, which they call collaborative marketing, where any stakeholder is welcomed provided that it proposes a collaboration that could be successful for the organization [31].

Co-creation allows consumers to co-build a self-experience in a personalized context, but also different for each one of them, which supposes, on the other hand, that they could help each other when it comes to solving problems, because participative forums are promoted [1, 29]. In previous investigations [33] it has been proven that in the specific case of digital native media, several fundamental elements are needed so the public can co-create products. First of all, co-creation could not have been possible without technology [8, 10, 15, 23, 31, 36] which allows users to comment, recommend and participate in the elaboration of products. Technology enables consumers to empower themselves [28] to the point of creating content, but for this to happen a web architecture is needed to favor the visibility of participative spaces and to guarantee its accessibility, because the public produce content motivated for its self-satisfaction and knowing that they count with a place where they can share data they have [34], which favors the existence of co-creation practices in corporative spaces instead of other external spaces such as social media.

Prahalad and Ramaswamy [29] synthesize that any co-creative process must be based on dialogue, access with its users and knowledge of the risk that co-creating must imply and the transparency between media and co-creators. From these considerations we can deduct that co-creation must be understood as a voluntary process and that in any case it should compel or force the consumer to participate if he or she does not wish to do so. Some recent investigations [35] conclude precisely that public demands more adequate spaces for public debate and interaction with journalists or the editorial department, and not so much production options or personalization of contents. Current co-creation processes must transcend the initial stage of co-production [3, 9] and give the possibility of co-creation in different areas of the company, in development (ideation + renovation) as well as in production (planning + execution) and marketing (promotion + distribution) [22].

2 Does Co-creation of Journalistic Products Imply Content Curation?

During the 90s, when Internet is no longer limited and the web becomes massive, users have access to an unmanageable volume of contents. Rapidly, the figure of curators goes from art to digital contents. GeoCities constitutes a fundamental antecedent for the curation of contents starting from the easiness to manage them according to the interests and personal selections [2]. In this case, despite not being curated by professionals or experts in information, contents were administered and selected by users with thematic interests and amateur knowledge.

Journalistic products present a particularity with respect to other types of companies or sectors. Journalistic co-creation does not suppose that any citizen can become a journalist or that audiences are necessarily capable to produce journalistic pieces according to the technical demands of the profession and the ethical and deontological regulations. Therefore, the public can co-create products facilitating graphic or audiovisual material, data or testimonies, etc. which provide value to the story created by the reporter—even an online survey can determine the configuration of the topics of a media company [12]—and which compensates the impossibility of the media to have a reporter in each place in which something is going on.

Nevertheless, when there is total co-creation, digital media habilitate spaces in its corporate sites—fundamentally on the Web—so users can publish and spread information they know. In this case there could be several alternatives, from media which grants that space to the public so they can submit their own story to others which habilitate their spaces under editorial supervision. In the first case there is greater creative liberty for audiences, but a questionable journalistic accuracy, because that information has not been submitted to journalistic filters and the media accepts no liability for the contents published in those spaces as happens, for example, in the French digital native *Mediapart*.

Another alternative consists in editorial supervision prior to publishing the co-created contents. In 2000 Yeonho Oh creates the Korean information website *OhmyNews*. The experience suggests that every citizen is a reporter. However, the collaboration is published under the edition of professionals that the media refers to as *news guerilla*. *OhmyNews* consumes and creates its information jointly with users if they are registered. That allows it to send news which are published in Korean, English and Japanese. The creators of *OhmyNews* present a website as a 3.0 journalistic experience or citizen journalism, even though they actively participate in the care of quality of the information.

This is where curation of contents has a fundamental role. It is a concept that is introduced in academic studies from 2010 after having initiated its entrance in the professional sector shortly before [11]. The first definitions over curation refer to a professional profile in charge of selecting and sharing the best and most relevant content over a specific issue, completing thus the function which is done by algorithms [4]. Before this McAdams [24] had alerted over the need to make curation in Internet journalism, so he recommended the example of museum curators, who selected the most representative based on their experience and knowledge.

In the concrete case of media, curation, along with reporting and the traditional edition, configurate a new journalism [7], but the curator who works in journalism is required to contribute with judgment to become more than just a collector of contents [13]. When we refer to curation in co-created journalistic contents, among the practices which identify López-Meri and Casero-Ripollés [18] for content curators—recommendation of their own contents, of reporters from the competition, non-mediatic actors, alternative contents to those spread by the media, verifying data and information, and debunk rumors—, recommendation of products created by non-mediatic actors or alternative to those traditionally spread by the media would fit among their roles, even though, without doubt, verification of data and information, as well as debunking rumors, result in essential tasks to guarantee the right to an accurate information. Thus, it is also necessary that the curation task is made by a reporter or a communicator [6]. The value provided by co-creators dwells in the role of the sources, but they cannot be requested the same accuracy and professionalism required to reporters because they are not.

2.1 What Does Curation in the Informative Process Mean?

Professionals who interact with citizens who collaborate in different journalistic websites were displaced from their traditional duties. In front of the great volume of circulating data, the management of the informative process is more and more mediated by second-hand sources which publicly intervene through digital platforms. Reporters, facing the new configuration of their jobs, are secluded in their offices, away from first sources [19].

Due to budget matters, from team formation or from proximity with events, it is almost impossible to give an account of the incidents that rapidly gain platforms and social agenda. Online collaboration and the ability of life transmissions allow instant access to almost all events. Coverage, the main ingredient for news construction, remains greatly in the hands of those who are in the place, with a mobile device and being part of some digital network.

Journalists' contacts with traditional sources are made by guidelines and with specific means. In the same way, professional coverage of events is made in cases where it has been planned with enough time. For the rest of events, which feed breaking news and last-minute coverage, users are who contribute with information, photos, videos or testimonies. The journalistic task is not anymore to access to the source in search of information but to verify the data submitted by users, as primary sources for the informative system.

The news, aesthetic and editorial criteria of users of networks not always coincide with those professionally agreed or the with the corporately chosen for a particular product. The debate over the abilities of citizens to capture events and their conditions for publication seem to impede the circulation of their products, are shared through personal accounts and force traditional media to negotiate with the users' interests so as not to lose audience [20].

Information users cease to be loyal consumers of the traditional mediatic brands. In their journey they access to the same mediatized sources which arrive at the same time as journalists. They arrive at elaborated news from integrating nodes of thematic interest. In those nodes, information is shared that afterwards circulates through networks in which their members participate. Informative consumption is not given any more exclusively navigating through journalistic websites. Access is given by interactions made with peers within the same network.

Chaotic navigation and intentional diffusion of some information [14] brings closer to an unexpected discovery process. Serendipity is rescued by a good part of studies that work over digital coverage events. Consumers' attitude is inclined not to search for information, but to find it. In the field of theories over journalism this is called incidental consumption of news [5].

While COVID-19 pandemic took back in time some consumptions to traditional ways of information and increased the number of users who pay for professional information [25], the change of habits in searching news became first screen consumption of information from the search engine, or in the best of cases, it is shared by acquaintances and friends who receive it. Digital platforms and messenger services are the main spaces to access information. Confidence is not expressed towards media or reporters but, in the majority of cases, to whom circulates among their network of contacts.

3 Methodology

To understand the co-creation and curation processes in digital media, seven cases of digital native media in Spanish in Spain and Latin America were considered (Table 1). At first, an exploratory work was made which took us to the selection of ten representative media in the Ibero-american culture and curation of online contents.

An analysis of content was made over them which allowed us to adjust the sample to seven cases which constituted the second phase of the work: the instance of in-depth interviews with who produce other media. A questionnaire was designed to make this with open questions that were answered by the responsible person of *Cerosetenta*, *El Pitazo*, *Curarnos*, *Lado B*, *GK*, *Mínimos*, *elDiario.es* and *El Surtidor*.

Those answers were compared with the analysis of content to organize a matrix of analysis that allowed to establish the relations between the producers and users. Starting from the study of these relations the following variables can be established for the interpretation of data. Citizen journalism, cooperation of users, relation with the sources, checking and metadata, traffic and algorithms are the variables established for the analysis of contents and the collaboration of producers.

Table 1 Sample

Digital media and contact	Country	Platform
Lado B Mely Arellano (co-director)	Mexico	Web: ladobe.com.mx
El Pitazo César Batiz (director)	Venezuela	Web: elpitazo.net
Cerosetenta Alejandro Giraldo (director)	Colombia	Web: cerosetenta.uniandes.edu.co
El Surtidor Alejandro Valdez (director)	Paraguay	Web: elsurti.com
GK Isabella Ponce (co-founder)	Ecuador	Web and newsletter: gk.city
elDiario.es Ander Oliden (chief of information)	Spain	Web: eldiario.es
Mínimos Fernando Casella (founder)	Argentina	Newsletter: minimos.com.ar
Curarnos Pere Ortín (founder)	Spain, Colombia, Mexico	WhatsApp

Studied digital native media platforms answered to diverse users and realities, as well as submission formats and varied support (websites, newsletters, WhatsApp or blogs) and constitute the opening doors for active participation and generation of journalistic contents.

4 Results

Co-creation from citizen's journalism, understood as cooperation of users, is a process of contact between the editorial department of the media and the sources and this is the way in which *GK*, *Lado B* and *Cerosetenta* work, because they do not have an explicit section for publishing works of the readers-users, but they maintain a strict curation for external participations through applying regulations so the texts comply with the minimum quality standards required. In this way, there is a two-way process of editorial work: to accompany from the beginning the construction of the story or to send a staff reporter or associated to make the verification work, survey of data and writing in different formats.

From the seven digital media studied, only three, *elDiario.es*, *El Pitazo* and *El Surtidor* have sections explicitly dedicated to the participation of users with their journalistic texts, even though they recognize that they have a long way to go for integrating formats and diversifying them, but the process is common: a professional reporter reviews the information, makes a follow-up with the creator and verifies that all data to be published is real. In the case of blogs from associates to *elDiario.es*, the team has an opinion column that oversees verifying the compliance with the regulations and editorial policies of the digital media.

The ways to participate with themes, initiatives and submission of productions are varied. These include from the contact and submission by email, to writing from the platform in blogs and WhatsApp contact or via social media to the integration in collaborative platforms, such as the so-called *la minga* in *El Surtidor*, which is a brainstorming method to establish a theme and where all the offers are valid and integrated in the creation of the journalistic product.

Participation of those who do not belong to the editorial department team is established through the platform and the creation of an account as collaborator or member, who contributes and is a part of the business model of the digital media. In relation to the active participation of citizens and the generation of stories and information, the representatives of the media consulted, coincide in that with a proper formation by the editor, any citizen can contribute with quality texts. Therefore, they are opened to publishing or participating with information that will be valued and validated for the construction of the informative agenda of each media company, that has a specific focus and has moved away from generalism and daily news.

In this way, the relation with sources is a continuous verification and contrasting, because in many cases, the issues are suggested from the needs of the communities through one of its representatives, who suggest the stories and the editorial team or an associate writes according to the editorial regulations of the media. Checking of information is an indispensable process for all digital media, even for those who have participative spaces such as blogs and newsletters, formats in which the contribution of collaborators are constantly verified, because in their editorial policies express it as a deontological principle and that strengthens the credibility between the community which consumes and follows the diffused information.

Collaborators of media, except those who participate in blogs of *elDiario.es*, are a part of and wide network of professionals linked with the specific themes of the journalistic agendas of each media, that is the directive teams establish the relations of assignment of themes and formats with who do not belong to the editorial department. Thus, digital media are based on the direct collaboration of experts, much more than the spontaneous citizen participation, because from the editorial and information departments, the quality of content is fundamental and for the construction of stories, reports and specials, the work shared and with multiple visions is the most adequate to understand the analyzed reality. An example of this is *GK*'s initiative, having a section of feminine sources validated in various issues, as a network of experts to be consulted and thus open a spectrum of expert voices that are taken as valid and different sources, to equilibrate the informative access.

While journalistic projects in digital platforms, metadata and the relation with other algorithms is necessary, they do not depend directly on them. For all media, its growth and exposure facing searches and search engines is completely organic, because the relationship with their readers-users is the main source of traffic and not clickbaits or paid advertising. In all cases the interviewee considers that in doing so they would betray the essence of the journalistic project and that its users and collaborators would become aware of how the flow and traffic is managed. They are not behind an accelerated growth of followers in networks, but they want an in-depth reading in their digital spaces where stories are anchored, because they consider that their members, subscribers and collaborators are citizens that do not fall into disinformation or manipulation or information bias.

Active participation of citizenship is indispensable from the various forms of inclusion. Mainly, in generating agendas and submitting texts, especially opinion and analysis, because all the ones we consulted consider that their users combat misinformation with their suggestions and productions since in the digital spaces there is a greater critical reflection facing informative consumption.

Curarnos is a particularly exceptional case, it had an ephemeral life and in its diffusion they adopted the most usual consumption of citizens: mobility and WhatsApp ubiquity associated to a telephone number in Spain, through which cultural contents were submitted and users were in charge of the diffusion and replication of information within their network and contacts.

During this initiative's functioning, users had no direct or indirect relationship with the creation of contents, but they were indispensable in the way of circulating. There resided its active participation, considered by one of its founders, as essential now of choosing who were included in the beginning to form part of the circulation network of content. The same producers and writers of the projects were the ones to define who would be their first contacts for submission. That was their only link with readers, that afterwards they replicated and maintained the contact through the telephone number, as their only way.

The experience, that could be understood as a case of diffusion or viralizing contents, places the users in an indispensable role so that the material can circulate, be shared and disseminated. Even so, the most usual formats for participation of the users are texts, expressed in blogs and, in the case of *El Surtidor*, are focused on audios and graphics, with memes and illustrations, due to its origin and aesthetic style as digital native media. Also, professional intervention is indispensable for arranging the information that socially circulates through digital networks. In the case of *Mínimos*, the publication by subscription sets out the rules of the game in the section.

Mínimos occupies the curation space from which interacts with users organizing the issues over which they look for information, through submitting electronic bulletins, in which the readers-users participate. *El Pitazo* makes a network from incorporating allies that allow to reach the whole territory with professional coverage. Its objective is to "guarantee the rights of the information to the sectors of the population economically less favored in Venezuela" and for that, it generates associations of professionals engaged in networks. Nevertheless, it does not include the underprivileged as sources or collaborators, but it guarantees the professional quality of the information it considers relevant for them.

All the interviewees recognize that co-creation and citizen participation must be greater in digital media, accompanied by a continuous formation so that citizens can have the journalistic point of view and not the private complaint or to look for a personal benefit.

5 Conclusions

While curation appears as a loan from the artistic and marketing systems to the information industry, there are not many cases found with respect to co-creation. Mostly, information volume that circulates in digital platforms is so great that it complicates its checking and organizing and that constitutes a preoccupation for whoever tries to establish interactions with users.

For the moment, while everybody thinks that they can learn to communicate and that citizens' participation is important, they tend to prefer being them those who control the quality and scope of what is published. At some point, the relation between asymmetry that journalists establish in their traditional forms of relationship with their audiences is sustained in the digital ecosystem. To the extent they allow collaboration for co-creation, they reserve curation for themselves.

The contribution of audiences for the generation of contents usually is resumed as their own in the majority of digital media. In those cases, in which these contributions are considered as less professional, they are dismissed. In general, there is an agreement with regards to the value of citizens as information sources. Nonetheless, there is clear reticence with regards to citizens as possible managers to establish the dynamics, editorial lines and criteria over the socially necessary information [32] which is intended to circulate.

Anyway, spokespersons from the community are included who request the inclusion of subjects to be contained in agendas with respect to what could be considered as the general collection of themes. In an environment in which a sense of infinity is presented, there are immeasurable volumes of information without hierarchies. These productions are unapproachable for users and producers. In a first step, curation establishes the parameters of ordering the information and a hierarchy which responds to sectorial interests, group positioning and professional valorizations.

From a well-intentioned place, it tends to question non-professional actors but with the vocation of adapting to conventional suggestions. Where it is possible—even if it flags co-creation—it is supposed for users not to have the pretension of managing for themselves their communication modes. Even if in the context of over-information, it is indispensable a professional management of data, the interactions with users tend to be utilitarian. They are asked for collaboration, but not always their recognition is visible.

In some cases, the intention of curators is to open a space for co-creation. Nonetheless, there is also reticence of the citizens to get involved in information systems. The need for training, for reporters as well as for citizens in a condition of co-creation, is an effort that, in many cases, there is no one who wants to do it.

With respect to the actions for spreading and sharing they entail less effort. However, in systems in which the confidence is in a critical stage, it is very difficult to make a network that collaborates with the circulation of information. It is feasible to understand that first it is necessary to generate community and from there look for collaboration and co-creation. Even so, if it is possible to accomplish a loyalty mailing of users who replicate contents, the alliance of confidence between peers guarantees the construction of a community which relates in two ways. On the one hand, a vertical communication between a digital media and a small community. On the other hand, the ties between users and their own networks, with which it establishes horizonal links for distribution of those products.

Even if incipient, the co-creation process is thought on three ordered forms. On the one hand, in relation to the contribution of ideas and issues of coverage; on the other, for the concrete provision of elaborated contents in the form of images, audios and videos or data cloud; finally, as indispensable nodes for the circulation and diffusion of contents on the Web.

Funding This research has been developed within the research project *Digital Native Media in Spain: Storytelling Formats and Mobile Strategy* (RTI2018-093346-B-C33), funded by the Ministry of Science, Innovation and Universities (Government of Spain) and the ERDF structural fund.

References

1. Aitamurto, T.: Balancing between open and closed: co–creation in magazine journalism. Digital J. **1**(2), 229–251 (2013). https://doi.org/10.1080/21670811.2012.750150
2. Bashkar, M.: Curaduría. El poder de la selección en un mundo de excesos. Fondo de Cultura Económica, Ciudad de México (2017)
3. Bendapudi, N., Leone, R.: Psychological implications of customer participation co-production. J. Mark. **67**(1), 14–28 (2003). https://doi.org/10.1509/jmkg.67.1.14.18592
4. Bhargava, R.: Manifesto for the content curator: The next big social media job of the future? (2009) Available at https://bit.ly/3uSWURY Accessed 30 July 2020
5. Boczkowski, P., Mitchelstein, E., Matassi, M.: Incidental news: How young people consume news on social media. In: 50th Hawaii International Conference on System Sciences, 4–7 Jan, Hawai, pp. 1785–1792 (2017)
6. Codina, L.: Curación de contenidos para periodistas: definición, esquema básico y recursos. Universitat Pompeu Fabra, Barcelona (2018). Available at https://bit.ly/3c4Ach4 Accessed 21 Sept 2020
7. Díaz-Arias, R.: Curaduría periodística, una forma de reconstruir el espacio público. Estudios sobre el Mensaje Periodístico **21**, 61–80 (2015). https://doi.org/10.5209/rev_ESMP.2015.v21.51129
8. Fiore, A., Kim, J., Lee, H.: Effect of image interactivity technology on consumer responses toward the online retailer. J. Interact. Mark. **19**(3), 38–53 (2005). https://doi.org/10.1002/dir.20042
9. Frow, P., Nenonen, S., Payne, A., Storbacka, K.: Managing co-creation design: a strategic approach to innovation. Br. J. Manag. **26**, 463–483 (2015). https://doi.org/10.1111/1467-8551.12087
10. Gruner, K., Homburg, C.: Does customer interaction enhance new product success? J. Bus. Res. **49**(1), 1–14 (2000). https://doi.org/10.1016/S0148-2963(99)00013-2
11. Guallar, J., Codina, L.: Journalistic content curation and news librarianship: differential characteristics and necessary convergence. El Profesional de la Información **27**(4), 778–790 (2018). https://doi.org/10.3145/epi.2018.jul.07
12. Holmes, S.: But this time you choose! approaching the interactive audience in reality TV. Int. J. Cult. Stud. **7**(2), 213–231 (2004). https://doi.org/10.1177/1367877904043238
13. Jarvis, J.: El fin de los medios de comunicación de masas. Gestión 2000, Barcelona (2015)

14. Jenkins, H.: “Cultural acupuncture”: Fan activism and the Harry Potter alliance. In: Geraghty, L. (ed.) Popular Media Cultures, pp. 206–229. Palgrave Macmillan, London, (2015). https://doi.org/10.1057/9781137350374_11
15. Kim, J., Ju, H., Johnson, K.: Sales associate’s appearance: links to consumers’ emotions, store image, and purchases. J. Retail. Consumer Sci. **16**(5), 407–413 (2009). https://doi.org/10.1016/j.jretconser.2009.06.001
16. Kotler, P.: Marketing Management. Prentice Hall, New Jersey (2002)
17. Kotler, P.: Marketing 4.0: Transforma tu estrategia para atraer al consumidor final. LID Editorial Empresarial, Madrid (2018)
18. López-Meri, A., Casero-Ripollés, A.: Las estrategias de los periodistas para la construcción de marca personal en Twitter: posicionamiento, curación de contenidos, personalización y especialización. Revista Mediterránea de Comunicación **8**(1), 59–73 (2017). https://doi.org/10.14198/MEDCOM2017.8.1.5
19. Luchessi, L.: Descentramientos, influencias y reacomodamientos en el ejercicio periodístico. In: Amado, A. (ed.) Periodismos argentinos: modelos y tensiones del siglo XXI, pp. 37–50. Konrad Adenauer Stiftung—Infociudadana, Buenos Aires (2016)
20. Luchessi, L.: Viral news content, instantaneity, and newsworthiness criteria. In: Rampazzo, R., Alzamora, G. (eds.) Exploring Transmedia Journalism in the Digital Age, pp. 31–48. IGI Global, Hershey (2018)
21. Lusch, R., Vargo, S., O’Brien, M.: Competing through service: insights from service-dominant (SD) logic. J. Retail. **83**, 5–18 (2006). https://doi.org/10.1016/j.jretai.2006.10.002
22. Malmelin, N., Villi, M.: Co-creation of what? Modes of audience community collaboration in media work. Convergence **23**(2), 182–196 (2015). https://doi.org/10.1177/1354856515592511
23. Malthouse, E., Hofacker, C.: Looking back and looking forward with interactive marketing. J. Interact. Mark. **24**(3), 181–184 (2010). https://doi.org/10.1016/j.intmar.2010.04.005
24. McAdams, M.: Curation, and journalists as curators. Teaching Online Journalism (2008)
25. Newman, N., Fletcher, R., Schulz, A., Andı, S., Nielsen, R.K.: Digital News Report 2020. Reuters Institute for the Study of Journalism (2020). Available at https://bit.ly/3cS6PQU. Accessed 23 Nov 2020
26. Ostrom, A., Bitner, M., Brown, S., Burkhard, K.A., Goul, M., Smith-Daniels, V., et al.: Moving forward and making a difference: research priorities for the science of service. J. Serv. Res. **13**(1), 4–36 (2010). https://doi.org/10.1177/1094670509357611
27. Petri, J., Jacob, F.: The customer as enabler of value (co)-creation in the solution business. J. Ind. Mark. Manag. **56**, 63–72 (2016). https://doi.org/10.1016/j.indmarman.2016.03.009
28. Piller, F., Ihl, C., Vossen, A.: A typology of customer co-creation in the innovation process. In: Hanekop, H., Wittke, V. (eds.) New Forms of Collaborative Production and Innovation: Economic, Social, Legal and Technical Characteristics and Conditions, pp. 1–26. University of Göttingen, Göttingen (2011)
29. Prahalad, C., Ramaswamy, V.: Co-creation experiences: the next practice in value creation. J. Interact. Mark. **18**(3), 5–14 (2004). https://doi.org/10.1002/dir.20015
30. Ramaswamy, V.: Co-creating value through customers’ experiences: the Nike case. Strategy Leadership **36**(5), 9–14 (2008). https://doi.org/10.1108/10878570810902068
31. Roser, T., DeFillippi, R., Samson, A.: Managing your co-creation mix: co-creation ventures in distinctive contexts. Eur. Bus. Rev. **25**(1), 20–41 (2013). https://doi.org/10.1108/09555341311287727
32. Schiller, H.: Information Inequality. Routledge, New York (1996)
33. Sixto-García, J., López-García, X., Toural-Bran, C.: Oportunidades para la cocreación de contenidos en los diarios nativos digitales. El Profesional de la Información **29**(4) (2020). https://doi.org/10.3145/epi.2020.jul.26
34. Shao, G.: Understanding the appeal of user-generated media: a uses and gratification perspective. Internet Res. **19**(1), 7–25 (2009). https://doi.org/10.1108/10662240910927795

35. Suau, J., Masip, P., Ruiz, C.: Missing the Big Wave: Citiziens' discourses against the participatory formats adopted by news media. J. Practice **13**(10), 1316–1332 (2019). https://doi.org/10.1080/17512786.2019.1591928
36. Wang, L., Baker, J., Wagner, J., Wakefield, K.: Can a retail web site be social? J. Mark. **71**(3), 143–157 (2007). https://doi.org/10.1509/jmkg.71.3.143

José Sixto-García Assistant Professor of Journalism at the Department of Communication Sciences of the University of Santiago de Compostela (USC). He was the director of the Social Media Institute (2013-2019) and his research is focused on new media, audience and social networking. He belongs to the research group Novos Medios (USC). His publications include books like *Digital Marketing Fundamentals or Professional Management of Social Networks.*

Pablo Escandón Montenegro Ph.D. in Communication and Contemporary Information, Master in Information and Knowledge Society, Master in Digital Journalism. Professor and Academic Coordinator in the postgraduate program of Digital Communication at Simón Bolívar Andean University of Ecuador (UASB-E), editor of URU Academic Journal of Culture and Communication from the Academic Area of Communication at UASB-E. External member of Novos Medios Research Group (USC, Spain) and collaborating member of the Museum I+D+C Group.

Lila Luchessi Director of the Institute for Public Policy and Government Research of the National University of Río Negro. From 2009 she is a Tenured Professor at the UNRN and UBA (1997). She is the author of many books and articles, and she is a member of several academic and editorial committees in the field of social communication. She holds a Ph.D. in Political Sciences (UB) and a Bachelor's Degree in Communication Sciences (UBA).

The Challenges of Artificial Intelligence

Platforms in Journalism 4.0: The Impact of the Fourth Industrial Revolution on the News Industry

Josep-Lluís Micó, Andreu Casero-Ripollés, and Berta García-Orosa

Abstract Robotics, artificial intelligence, machine learning, big data, blockchain, virtual and augmented reality are becoming widespread in our society. These technologies have triggered the fourth industrial revolution and will bring about profound changes in social relations, working life, politics, the education system, the economy and journalism. In this context, digital platforms play a key role. Given that these devices are reorganizing cultural and social processes on a large scale, we can speak of the emergence of a platformization of society. This transformation will require new skills, new products, new environments and new business models. In particular, it will affect the relationship between the public and journalism, its financing, its revenue-earning channels and its news production processes. This chapter will analyze how the fourth industrial revolution will affect journalism from a business and structural point of view.

1 Introduction

Current society is immersed in Industry 4.0, which involves not only significant changes in production systems, but also important modifications in the lives of citizens and in social organization. The omnipresence of technological innovations in all areas accounts for a substantial transformation in the relationships between individuals and social groups [4]. As in previous revolutions, multiple factors come together to explain a complex social, economic, political, cultural and, in this case, global

J.-L. Micó (✉)
Universitat Ramón Llull, Barcelona, Spain
e-mail: josepllluisms@blanquerna.url.edu

A. Casero-Ripollés
Universitat Jaume I de Castelló, Castellón, Spain
e-mail: casero@uji.es

B. García-Orosa
Universidade de Santiago de Compostela, Santiago de Compostela, Spain
e-mail: berta.garcia@usc.es

J. Vázquez-Herrero et al. (eds.), *Total Journalism*, Studies in Big Data 97,
https://doi.org/10.1007/978-3-030-88028-6_18

change. This chapter offers a contextual evaluation of Industry 4.0 in journalism from a business and structural point of view.

2 The Fourth Revolution: Culture 4.0

Although scientific and technological innovation play a relevant role in all revolutions, these are not due exclusively to the emergence of new technologies. Rather, they are a process where a different use of existing resources makes the modification of social relations and the power structure possible [4]. In each of these complex phenomena, there are some notable and lasting changes that have modified the productive and social systems of the time with transformations that have shaped our current society. The first industrial revolution, at the end of the eighteenth century, was brought about by the steam engine, which lead to factory mechanization. During the second industrial revolution, between the end of the nineteenth century and the 1950s, electricity favored the division of labor and mass production. With the third revolution, which began in the late 1950s, the automation of these actions was completed thanks to technological advances [5]. The fourth revolution or Industry 4.0 is still current today and has popularized drones, driverless vehicles, smart homes, smart cities and especially all kinds of robots and technology based on artificial intelligence [44].

Robotics, artificial intelligence, 5G, machine learning, big data, blockchain and virtual and augmented reality are starting to play a very prominent role in several areas such as education, health, security, transportation, logistics, communication, entertainment, corporate customer care and home care and the importance of their role will increase in the coming years [13, 28]. As in previous revolutions, progress may destroy some jobs, especially the most basic ones or those requiring less training, however, progress has historically had the opposite effect by helping to create jobs [37]. Profound changes are expected in social and working relationships with significant challenges for countries that will have to adapt quickly and effectively to these innovations in order to avoid becoming isolated in the new political, social and economic configuration. Governments, businesses and civil society will need to design global strategies to integrate robots and the rest of the digital agents into their operating dynamics [53].

Within this context, culture has undergone the greatest changes and has faced the greatest challenges. The demanding and participatory culture resulting from this phenomenon is evident in all organized settings surrounding technology, where any user can become a sender and receiver of all kinds of messages containing text, audio, graphics, photography, video and multiple combinations of these elements. In addition, people can interact with dispersed and distant audiences with devices that are permanently connected, with artificial devices that have the ability to learn and through smart buildings and cities [7].

This cultural change requires unprecedented skills that allow people to engage with conventional media, such as Web 2.0, as content creators and distributors, as well as allowing them to navigate areas where the real and the virtual are intertwined. Human groups and communities of practice are formed using these tools and platforms, for example the use of smartphones, wearables and social networks. These groups grow out of curiosity, interest, passion or engagement in a variety of human activities including politics, sports and art. Institutions, companies and private players currently compete simultaneously on different fronts, channels and supports to attract consumers, customers and even robots.

The paradigm shift is so radical and far-reaching that it has not yet been accurately defined by the academic community. Prospects are not favorable for digital transformation theorists. New developments such as the Internet of Things, artificial intelligence, machine learning, big data, cloud computing or blockchain complicate their daily work [40], which must be done almost in real time.

Since 2004, when Tim O'Reilly coined the term Web 2.0 to refer to a universe dominated by applications that facilitate collaboration and information sharing, the transformations have been very significant. One of the most notable milestones occurred some years later when Facebook and Twitter came onto the scene and achieved great popularity. These media, as well as the recent addition of YouTube, Instagram, WhatsApp, Twitch and TikTok, have experienced exponential growth. The social network site created by Mark Zuckerberg reached 100 million users in just 9 months while radio needed almost four decades to gather 50 million listeners in the United States and television required 13 years.

As in previous revolutions, the incursion of a new element, the Internet in this case, has broken the monopoly of communication companies and political actors when it comes to producing and distributing content [8]. Classical calculations about who, how and when such data is generated and consumed are no longer valid [26].

During the age of the fourth industrial revolution, a genuine blending of abilities rather than a potpourri of skills is needed in order to enrich communication and foster an understanding of the whole. Consequently, the idea of paradigm is no longer appropriate. Unity does not derive from a shared basic knowledge; it is conveyed through the will to solve complex and common conflicts. Within this framework, digital media and digital tools play a key role [51]. These systems are the privileged platforms that citizens use to communicate [16] using their mobile devices, their home and work computers, tablets, consoles and connected devices such as watches and cars. Consequently, platforms are not just applications. They are ecosystems because they go beyond their role as intermediaries by shaping our lives [17]. Therefore, we can speak of a platformization of society, as there is an insertion of infrastructures, economic processes and governmental frameworks linked to these technologies in different economic sectors and areas of life [38]. As a result, digital media technologies have become inextricably intertwined with our daily lives to the point that it is difficult to do without them. Hence, a deep mediatization has been generated that has positioned digital platforms as a vector of social transformation in our societies [25].

People need months to learn how to drive but autonomous cars, equipped with artificial intelligence systems, can do so in a few days [42]. Radiologists take an average of 13 years to become experts in medical imaging yet, it only takes a computer a few hours to achieve this. Following a full training course, a human user will have learned how to navigate complex environments while a robot can teach itself to operate in just a few minutes. Automation is spreading across all kinds of activities and sectors: industry, finance, commerce, healthcare, social care, administration and also journalism.

One of the most significant facts in this context is that, in recent years, technological companies have gone from being service providers and content distribution channels to political and media players with great influence on public opinion [8]. The reshaping and erosion of informational and cultural plurality [6, 47] are risks that all social and political actors will have to undertake in the coming years.

In recent years, the role of digital platforms has been growing not only as transmitters of content and cognitive frameworks but also as gatekeepers, a function that is traditionally reserved for journalists. These companies have evolved beyond the role of distribution channels and have become players that control what the public sees and even what format and type of journalism and culture is provided [3]. Using different strategies, the platforms steer media companies towards certain content that is of interest to their business strategies, such as live videos. Platforms also dictate the publisher's activity through design standards and they provide training with tools that are supposedly free. Therefore, digital platforms are actually publishers. Google, Apple, Facebook, Snapchat and Twitter, among others, are publishers with teams and strategies that focus on promoting journalism.

The relationship between public actors and digital platforms is increasingly relevant and asymmetric [46]. Specifically in journalism, technology companies can also incorporate their logic and become a transforming force that challenges the traditional professional concept [52]. Facebook with Instant Articles, Apple with News App, Google with Accelerated Mobile Pages and Snapchat with Discover have designed formats where they can place their branded content.

No longer are only consumers incidentally exposed to content through algorithmically controlled recommendations on third-party platforms, but platforms promote specific strategies for public opinion. The audience is large and captivated. According to the *Digital News Report*, one-third of the population segment between the ages of 18 and 24 claims that social networks are their main source of information [35].

The role of Cambridge Analytica with Brexit or the latest actions of Facebook and Twitter with Donald Trump's messages, fake news and proliferation of bots bring up the debate concerning who is responsible. Faced with this scenario, the European Union issued a warning concerning the threat to democracy that this situation poses and other countries have drafted laws regulating artificial intelligence (Italy in 2014, France in 2016, United Kingdom in 2017). Recently, in November [19], the European Commission adopted the Action Plan Against Disinformation in order to minimize disinformation in the European elections. Specifically, the EU insisted on addressing the overall surveillance system that these platforms have put in place, by implementing more protection rather than more data. The European Economic and Social Committee (EESC) held a public hearing on February 5, 2020, to explore the impact of campaigns aimed at increasing participation in the last European elections and to analyze ways in which to combat misinformation [20]. In addition, it aimed at ensuring increased public participation in political decision-making in the EU in the coming years.

Platformization is a significant challenge and its solution is still in its early stages. This could result in serious consequences for journalism as the platforms' ecosystem encourages the dissemination of easy-to-consume, low-quality content that is distributed virally.

3 Journalistic Companies 4.0

Public opinion is strongly influenced by the media. This opinion does not respond directly to reality, but to a pseudo-environment created by journalists through their mediation activity [11, 27]. Even today, in the midst of the digital boom, companies in this sector provide much of the material that society uses in its dialogues. According to Maxwell McCombs [29], the issues that the media emphasizes become the most important in public consideration just as issues that are not featured are sidelined.

Unlike those who act sporadically and without adjusting to a regular work system on platforms such as Facebook, Instagram, YouTube, Twitter, TikTok, Telegram or WhatsApp, professional mediators, as journalists, have the responsibility to rigorously choose the topics of debate. Concerning the press, liberals in the eighteenth century stated that this task is linked to the necessity to favor social and political dialogue in order to strengthen democracy [34]. This was clearly highlighted by sociologist Gaye [48] when she said that the quality of civic debate depends on the information available.

Media companies have traditionally been very focused on increasing audience numbers in order to obtain advertisements and earn revenue to survive. Nonetheless, users have gradually gained power to the point that they have an impact on the agenda. This is due to the quantitative factors of shares, visits, page views, likes and retweets, which are as important as qualitative factors. The impact of the news on the public is thus placed above journalistic quality. The media has increasingly focused on what is interesting to the detriment of what is important. The former attracts audiences and generates clicks and comments. The latter, on the other hand, triggers more far-reaching outcomes. Both elements can be combined, although this is not easy.

Biologists call the behavior of persistent or opportunistic species the R-strategy. These species have many offspring but pay little attention to them, which is why their mortality rate is high. In the communication sector, this pattern would be equivalent to the production of a large number of news items in the hope that, due to their repetition, they will reach some recipients. This is especially true for the Internet, social media, mobile devices and gamified content [45]. It is not relevant whether some information, such as Facebook posts or Twitter messages, gets lost along the way because the media brand is not adversely affected.

In contrast, the K-strategy is characteristic of species with sparse offspring that require a lot of attention. In journalism, this would imply circulating complex productions based on depth and quality in order to attract an audience [15]. In other words, it implies wagering on creativity and accuracy. Failure of a news item prepared in accordance with the precepts established by the R-strategy is not a serious matter, since the probabilities of failure are high. However, should this occur with journalistic products prepared according to the K-strategy, it would be a disaster because the circulation of this content is only possible after complex, slow and costly work.

Architects, town planners, institutions and ordinary neighbors often regret the aesthetic and environmental deterioration of their cities, towns and landscapes. All in all, it is strange that journalists and communication entrepreneurs complain about information pollution present, both in quantity and quality, given the predominance of the R-strategy [24]. Overabundance is also affecting advertising, which is moving away from short ads and promotions without interruption to more substantial advertising. Advertisers and journalists understand that their audiences are always ready to go elsewhere, therefore, they do not want to lose the battle for their attention. No matter how relevant the topic is that they want to explain, if they do not capture the interest of the viewer, their task will be fallow.

Unless journalistic companies establish ties of mutual interest and reciprocal benefit with their audiences, they will have a difficult time in the future [31]. In other words, to survive in the digital world, between artificial intelligence and big data as well as the Internet of things and augmented reality, it is not enough to be a hunter in the classic style [23, 43]. Communication hunters tend to reduce diverse information into a basic, standard package in an effort to define it to different people and groups who are in situations and contexts that have little to do with each other. However, nowadays, polyphony is taking precedence over the solo voice as are good manners over bad manners. It is a mistake to address millions of people from a position of hegemonic broadcasting. If content continues to be cloned on TV, radio, newspapers, Instagram, Twitter or Telegram, it suggests savings in production costs. Ultimately, however, this strategy will lead to a reduction in revenue as users will end up searching for tailored news material.

Explicit or implicit consensus among journalistic companies concerning the publication of certain sensitive issues is more difficult to attain today than it was in the past. Concepts such as 'citizen journalism' or 'participatory journalism', phenomena such as cyberactivism, tools such as social media and equipment such as mobile devices have confirmed this. The audience has experienced a role change. In a short period of time, the monologue of the news industry has been replaced by the prominence of the audience [14]. Industry 4.0 poses the need to redefine the relationship between journalists and audiences in order to move towards a model that establishes a fluid conversation between the two. This transformation would require updating media operations, starting with the news selection process. In facilitating dialogue between the media and the public, technology makes journalism more efficient in its mission as a social mediator. It would seem sensible to go back to the origins of the profession and understand what was proposed to this profession then and what technology makes possible today that was impossible before [36]. With this approach, progress would be made towards the establishment of a real community without naively forgetting the rules and guidelines of journalistic work.

All types of media have articulated participation and personalization mechanisms through various platforms. Thanks to methods such as web analytics, it is possible to know in a detailed and exhaustive way the preferences of the audience; minute by minute, 24 h a day, 365 days a year [10]. They are channels that promote participation, but they are also data providers that have emerged as a marketing tool unprecedented in the history of journalism and are vital in a sector that has not found a solid business model within the network [21]. Over two hundred years after its birth, the media continues to be the main intermediary of general interest in modern societies. In the digital age, newspaper companies are less effective but they continue to generate most of the information circulated [9].

For decades, journalism has been seen, heard and touched with printed newspapers. Digital devices contribute to refining the sense of touch. Today's media is managed interactively, unlike the way it was done before with newspapers and magazines [2]. Innovation can transport users to a dimension yet to be discovered through virtual, augmented and mixed reality, with the Internet of things and wearables, with artificial intelligence and machine learning, among others. Oftentimes, this qualitative leap is not reflected in quantitative plans because sales, audiences, revenues and profits do not increase. Nonetheless, the transformation can earn political, cultural, social and moral rewards. These returns should not be disregarded, although they would not be capitalized in the income statements. We refer to hypothetical situations because economic limitations prevent us from implementing many of these advances at the present time.

Digital journalism, defined by tensions between continuity and change and between digital and legacy [18], has led to the birth of digital natives in order to respond to various needs and challenges. Examples such as *Il Post*, *Mediapart* and *Observador* stand out, all of which were created with a focus on their audience and seek to be differentiated from other media through subject specialization. These modifications overlap the essence of journalism under a new umbrella called 'constructive journalism' or 'solutions journalism' [1]. Technology acts as a driver [56] or, at times, as an enhancer, which still has major challenges to overcome.

Journalistic companies are still learning how to take advantage of the inclination of millions of people to promote, distribute and advertise other people's content on a large scale. Technological advances available to the media and its audience help narrow the gap between desire and its materialization. However, without minimal financial investment, the public will be unable to own 100% of the news. Unless the companies within this sector find funding, people will not be able to access, touch and experience information with 4.0 technologies.

Journalistic platforms and their professionals build their personality through intentional speech and planned or spontaneous behavior. Therefore, it can be said that these groups are not what they say they are, but what they transmit through their actions in the different areas of media. The Web, smartphones and smart TVs or virtual reality and 3D glasses should not be fabulous, modern packages to contain the same old material. They are gateways to the totality of information where physical barriers dissipate when journalistic ideas are coupled with money [34].

In the book *Mass Communication Theory*, published in 1987, Denis McQuail established the characteristics of conventional press, radio and television companies that can be applied to social media. According to McQuail [30], digital platforms are a growing and evolving industry that creates jobs, produces goods and services and strengthens related sectors. In addition, by developing their own rules linked with society, digital platforms create an independent institution. They operate as a very powerful resource and simultaneously they are a mechanism for control, manipulation and innovation. Digital media provides a place where public affairs are discussed, both nationally and internationally. They are a dominant source of definitions and images of reality for individuals, but equally so on a collective scale because they express values and standards of judgment that are combined with news and entertainment [55].

Advertising is another key aspect effected by the redefinition of journalism imposed by Industry 4.0. Advertisements finance a large amount of conventional content beyond digital business, in newspapers, magazines, radio stations and television channels [39]. Those with commercial interests expect advertising to continue funding much of the material streaming on the Web, computers, smartphones, tablets and any other Internet of Things device such as vehicles, household appliances, clothing, accessories, etc. [54]. The formulas are becoming increasingly sophisticated. Concept and promotion methods have changed because of the Web and its multiple derivatives. Brands are aware that it is only worth paying for what draws the users' attention to their products or for what creates a virtual space where they can relate to a multitude of potential consumers.

Marketers and customers have understood that, in this day and age, simply advertising on the Web is the wrong strategy. The key is to offer real value to people, thus obtaining more data and turning the information into a useful, practical, tailor-made service. For example, a watch that measures the calories we consume or sensors in every room of a home.

The low costs of the Web mean that, under normal circumstances, individuals and organizations would not be able to release certain content into circulation but now they are able to publish what they and their audiences want for free [40]. New actors are emerging in this new environment as well. Among them, digital influencers stand out. Daily, they show how they raise themselves to brand status and are essential agents in this commercial system [12]. Their power has grown tremendously in a short time [26]. These opinion leaders manage to earn a lot of money thanks to companies and institutions that hire them to promote products and services on their private social networks and blogs.

Institutions behave like commercial brands and promotion has become the dominant social language [31]. For this reason, advertising is omnipresent. There is no environment or scenario, whether public or private, that is unaffected by the influence of advertising and marketing campaigns, offers, discounts, sales and exchanges. As a result of this domination, the reaction of the recipients is becoming more complex. This is especially true in the virtual world, where the invasion is more diffuse, although the users' capacity to complain and suggest alternatives is also greater [15]. Companies, political parties and administrations base their positive image on advertising and public relations. Besides legitimizing those who pay for this, it transcends its natural environment to become a cultural product on the street, in the media, on the Internet and on technological devices.

Within these changes, journalism is also undergoing revolutionary transformations in the 4.0 environment. These changes are not only occurring in its business scope and relationship with its audience, but also its productive routines [49, 50]. Renewed practices and identities that respond to the current context are already beginning to be apparent [22] during this time dominated by the influence of platforms [3, 41]. In 2014, *Los Angeles Times* published a story about an earthquake just three minutes after it happened. This informative feat was possible because an employee had developed a robot, Quakebot, to automatically write text from material generated by the US government, more precisely, by its geological department. This program has its own author profile on the newspaper's website [32].

Companies within the digital industry, such as Narrative Science, designed most of these tools until recently. However, in recent years, news organizations have started creating their own bots such as Juicer at the BBC, Heliogra at *The Washington Post* and Cyborg at *Bloomberg*. From *The Economist,* editor Kenn Cukier states: "We should not think about what is best for the staff, but for the public. We did not cling to the pen when the typewriter arrived, and neither should we do so now..." [32]. Companies are making progress in automatically tailoring information for specific groups and even individuals. For example, in an informative piece about road safety measures, the examples and testimonials would change depending on the recipients' vehicle and personal circumstances. It is a matter of materializing the concept of The Daily Me, popularized by the founder of the MIT Media Lab, Nicholas Negroponte, or to make the current formula of Google News and other similar services more sophisticated.

Machines are also detecting trends, for example, in national markets and economies. The *Financial Times* is one of the newspapers making use of this innovation and its managers are also considering using other programs. Prestigious journalists such as Cait O'Riordan, who has worked for leading media outlets such as the BBC and the *Financial Times*, and Calum Chace, *Forbes* columnist and author of the book *The Economic Singularity*, believe that there is "too much opinion being offered". Like many of their colleagues, they miss having a greater volume of quality information and rigorous analysis by experts. Machines also introduce a new player into this complex task: artificial intelligence [33].

"Human audiences want informed interpretations, not just data structured by an algorithm" says O'Riordan. However, he recalls that there are systems such as IBM's Project Debater that are capable of expressing their own reasoning. For this reason, the journalist quickly clarifies that opinions of this nature "could become significant" in a short amount of time [32].

4 Conclusions

The impact of Industry 4.0 on journalism is a challenge for both the industry and society as a whole. In this context, digital platforms and artificial intelligence are the main drivers of change. Their importance is causing the emergence of a platformization of society, which places these devices at the center of the reorganization of cultural and social practices. The use of bots, virtual and augmented reality, blockchain, big data and machine learning, among other technologies, offer new opportunities for journalism, but also create threats. In this chapter we have examined how the new 4.0 environment particularly affects the relationship with the audience, funding channels such as advertising and news production. Currently, several phenomena and dynamics are profoundly redefining these essential aspects of journalism. The outcome of this process will determine not only what journalism will be like in the coming decades by establishing its limits and potential, but also what its democratic and social contribution will be. So far, the news media has played an essential role in the articulation of public opinion. The great challenge is to know whether it will continue to do so in the future.

Funding The work of Josep-Lluís Micó-Sanz in this chapter is part of the grant *Industry 4.0, Communication and Religion*, funded by the Catalan Government (Generalitat de Catalunya). The work of Andreu Casero-Ripollés is part of grant UJI-B2020-14, funded by Universitat Jaume I under the Research Promotion Plan 2020. The work of Berta García-Orosa is part of the research project *Digital Native Media in Spain: Storytelling Formats and Mobile Strategy* (RTI2018-093346-B-C33), funded by the Ministry of Science, Innovation and Universities (Government of Spain) and the ERDF structural fund.

References

1. Aitamurto, T., Varma, A.: The constructive role of journalism. Journal. Pract. **12**(6), 695–713 (2018). https://doi.org/10.1080/17512786.2018.1473041
2. Anderson, C.: Towards a sociology of computational and algorithmic journalism. New Media Soc. **15**(7), 1005–1021 (2013). https://doi.org/10.1177/1461444812465137
3. Bell, E.J., Owen, T., Brown, P.D., Hauka, C., Rashidian, N.: The platform press: how silicon valley reengineered journalism. Tow Center Digit. Journal. (2017). https://doi.org/10.7916/D8R216ZZ
4. Bianchi, P.: 4.0: La nueva revolución industrial. Alianza, Madrid (2020)
5. Brynjolfsson, E., McAfee, A.: The Second Machine Age: Work, Progress, and Prosperity in a Time of Brilliant Technologies. Norton, New York (2014)
6. Cardenal, A.S., Galais, C., Majó-Vázquez, S.: Is Facebook eroding the public agenda? Evidence from survey and web-tracking data. Int. J. Public Opin. Res. **31**(4), 589–608 (2019). https://doi.org/10.1093/ijpor/edy025
7. Carpentier, N.: Media and Participation. Intellect, Bristol (2011)
8. Casero-Ripollés, A.: Research on political information and social media: key points and challenges for the future. El Profesional de la Información **27**(5), 964–974 (2018). https://doi.org/10.3145/epi.2018.sep.01

9. Casero-Ripollés, A.: Influence of media on the political conversation on Twitter: activity, popularity, and authority in the digital debate in Spain. ICONO 14, Revista de comunicación y tecnologías emergentes **18**(1), 33–57 (2020). https://doi.org/10.7195/ri14.v18i1.1527
10. Cherubini, F., Nielsen, R.K.: Editorial Analytics. Reuters Institute for the Study of Journalism, Oxford (2016)
11. Childs, H.L.: Public Opinion. Van Nostrand, Princeton (1965)
12. Coll, P., Micó, J.L.: Influencer marketing in the growth hacking strategy of digital brands. Observatorio **13**(2), 87–105 (2019). https://doi.org/10.15847/obsOBS13220191409
13. Condoluci, M., Araniti, G., Mahmoodi, T., Dohler, M.: Enabling the IoT machine age with 5G: machine-type multicast services for innovative real-time applications. Access IEEE **4**, 5555–5569 (2016)
14. Costera Meijer, I.: Understanding the audience turn in journalism: from quality discourse to innovation discourse as anchoring practices 1995–2020. Journal. Stud. **21**(16), 2326–2342 (2020). https://doi.org/10.1080/1461670X.2020.1847681
15. Deuze, M.: What is multimedia journalism? Journal. Stud. **5**(2), 139–152 (2004). https://doi.org/10.1080/1461670042000211131
16. van Dijck, J., Poell, T., De Waal, M.: The Platform Society: Public Values in a Connective World. Oxford University Press, Oxford (2018)
17. Dubravac, S.: Digital Destiny: How the New Age of Data Will Transform the Way We Work, Live, and Communicate. Regnery, New Jersey (2015)
18. Eldridge, S.A., Hess, K., Tandoc, E.C., Westlund, O.: Navigating scholarly terrain: introducing the digital journalism studies compass. Digit. Journal. **7**(3), 386–403 (2019). https://doi.org/10.1080/21670811.2019.1599724
19. European Commission.: Communication from the Commission to the European Parliament, the Council, the European Economic and Social Committee and the Committee of the Regions. Tackling Online Disinformation: A European Approach (2018). Available at https://eur-lex.europa.eu/legal-content/EN/TXT/?uri=CELEX%3A52018DC0236. Accessed 31 Jan 2021
20. European Economic and Social Committee.: La UE debería endurecer su postura sobre las noticias falsas (2020). Available at https://www.eesc.europa.eu/es/news-media/news/la-ue-deberia-endurecer-su-postura-sobre-las-noticias-falsas. Accessed 31 Jan 2021
21. Galli, B.J.: Role of big data in continuous improvement environments. Int. J. Appl. Logistics **9**(1), 53–72 (2019). https://doi.org/10.4018/IJAL.2019010104
22. García-Orosa, B., López-García, X., Vázquez-Herrero, J.: Journalism in digital native media: Beyond technological determinism. Media Commun. **8**(2), 5–15 (2020). https://doi.org/10.17645/mac.v8i2.2702
23. Geraci, R.: Apocalyptic AI: Visions of heaven in robotics, artificial intelligence, and virtual reality. Oxford University Press, Oxford (2010)
24. Hammi, B., Khatoun, R., Zeadally, S., Fayad, A., Khoukhi, L.: IoT technologies for smart cities. Networks IET **7**(1), 1–13 (2018)
25. Hepp, A.: Deep Mediatization. Routledge, New York (2019)
26. Holton, A.E., Lewis, S.C., Coddington, M.: Interacting with audiences. Journal. Stud. **17**(7), 849–859 (2016). https://doi.org/10.1080/1461670X.2016.1165139
27. Lippmann, W.: Public Opinion. Free Press Paperbacks, New York (1997)
28. Lowenstein-Barkai, H., Lev-On, A.: Complementing or substituting? Int. J. Human-Comput. Interact. **34**(10), 922–931 (2018). https://doi.org/10.1080/10447318.2018.1471571
29. McCombs, M.: Setting the Agenda. Polity Press, Cambridge (2004)
30. McQuail, D.: Mass Communication Theory: An Introduction. Sage, Thousand Oaks (1987)
31. Micó, J.L.: Periodismo y comunicación corporativa: Desafíos y tendencias en la cuarta revolución industrial. Revista de la Asociación Española de Investigación de la Comunicación **6**(11), 5–16 (2019). https://doi.org/10.24137/raeic.6.11.1
32. Micó, J.L.: Cuando los robots discutan con los tertulianos… y les ganen. La Vanguardia (2021). Available at https://bit.ly/3vCXmUY. Accessed 31 Jan 2021
33. Napoli, P.: Automated media: an institutional theory perspective on algorithmic media production and consumption. Commun. Theory **24**, 340–360 (2014). https://doi.org/10.1111/comt.12039

34. Nechushtai, E., Lewis, S.C.: What kind of news gatekeepers do we want machines to be? Comput. Human Behav. **90**, 289–307 (2019). https://doi.org/10.1016/j.chb.2018.07.043
35. Newman, N., Fletcher, R., Kalogeropoulos, A., Levy, D., Nielsen, R.K.: Digital News Report. Reuters Institute for the Study of Journalism, Oxford (2017)
36. Örnebring, H.: Technology and journalism-as-labour. Journalism **11**(1), 57–74 (2010). https://doi.org/10.1177/1464884909350644
37. Pavlik, J.V., Bridges, F.: The emergence of augmented reality (AR) as a storytelling medium in journalism. Journal. Commun. Monogr. **15**(1), 4–59 (2013). https://doi.org/10.1177/1522637912470819
38. Poell, T., Nieborg, D., van Dijck, J.: Platformization. Internet Policy Rev. **8**(4), 1–13 (2019). https://doi.org/10.14763/2019.4.1425
39. Pérez-Seijo, S., Vizoso, Á., López-García, X.: Accepting the digital challenge: business models and audience participation in online native media. Journal. Media **1**(1), 78–91 (2020). https://doi.org/10.3390/journalmedia1010006
40. Ragnedda, M., Destefanis, G.: Blockchain and Web 3.0. Routledge, New York (2019)
41. Rashidian, N., Tsiveriotis, G., Brown, P.D., Bell, E.J., Hartstone, A.: Platforms and publishers: the end of an era. Tow Center Digit. Journal. (2020). https://doi.org/10.7916/d8-sc1s-2j58
42. Russell, S., Norvig, P.: Artificial Intelligence: A Modern Approach. Prentice Hall, New Jersey (2010)
43. Schmidt, E., Cohen, J.: The New Digital Age: Reshaping the future of people, nations, and businesses. John Murray, London (2013)
44. Schwab, K.: The Fourth Industrial Revolution. World Economic Forum, Geneva (2016)
45. Shpakova, A., Dörfler, V., MacBryde, J.C.: Gamifying the process of innovating. Innov. Organ. Manage. **22**(4), 488–502 (2020). https://doi.org/10.1080/14479338.2019.1642763
46. Smyrnaios, N., Rebillard, F.: How infomediation platforms took over the news: a longitudinal perspective. Polit. Econ. Commun. **7**(1), 30–50 (2019)
47. Tamir, E., Davidson, R.: The good despot: technology firms' interventions in the public sphere. Public Underst. Sci. **29**(1), 21–36 (2020). https://doi.org/10.1177/0963662519879368
48. Tuchman, G.: Making News. TheFreePress, New York (1978)
49. Vázquez-Herrero, J., Negreira-Rey, M.-C., López-García, X.: Let's dance the news! How news media is adapting to the logic of TikTok. Journalism (Online First) (2020). https://doi.org/10.1177/1464884920969092
50. Vázquez-Herrero, J., Negreira-Rey, M.-C., Rodríguez-Vázquez, A.-I.: Intersections between TikTok and TV: channels and programmes thinking outside the box. Journal. Media **2**(1), 1–13 (2021). https://doi.org/10.3390/journalmedia2010001
51. Westlund, O.: Mobile news. Digital. Digit. Journal. **1**(1), 6–26 (2013). https://doi.org/10.1080/21670811.2012.740273
52. Wu, S., Tandoc, E.C., Salmon, C.T.: When journalism and automation intersect: assessing the influence of the technological field on contemporary newsrooms. Journal. Pract. **13**(10), 1238–1254 (2019). https://doi.org/10.1080/17512786.2019.1585198
53. Yang, G.-Z., Bellingham, J., Dupont, P.E., et al.: Grand challenges of science robotics. Sci. Robot. **3**(14), 1–14 (2018). https://doi.org/10.1080/17512786.2019.1585198
54. Zanella, A., Bui, N., Castellani, A., Vangelista, L., Zorzi, M.: Internet of Things for smart cities. IEEE Internet Things J. **1**(1), 22–32 (2014). https://doi.org/10.1109/JIOT.2014.2306328
55. Zelizer, B.: What Journalism Could Be. Polity Press, Cambridge (2017)
56. Zelizer, B.: Why journalism is about more than digital technology. Digit. Journal. **7**(3), 343–350 (2019). https://doi.org/10.1080/21670811.2019.1571932

Josep-Lluís Micó Professor, Chair of Journalism at the Ramon Llull University. He is the academic Vice Dean of the Blanquerna School of Communication and International Relations, where he directed the Degree in Journalism and the Master in Advanced Journalism Blanquerna–Grupo Godó. He works as a technology analyst in media such as *La Vanguardia* and *Radio Nacional de España–Ràdio 4.*

Andreu Casero-Ripollés Professor of Journalism and Dean of the School of Humanities and Social Sciences at Universitat Jaume I de Castelló. Previously, he was Head of the Department of Communication Sciences and Director of Journalism Studies. He has been a visiting researcher at the Columbia University (United States) and the University of Westminster (United Kingdom), among others. He studies political communication and the transformation of journalism in the digital environment.

Berta García-Orosa Full Professor at the University of Santiago de Compostela (USC). BA in Communication Sciences, B.A. in Political and Administration Sciences and Ph.D. in Communication Sciences (USC). She has studied communication and politics for more than 20 years. She has collaborated in more than 50 research projects and international research networks, the results of which have been published over 100 times in databases such as Scopus and JCR and publishing houses.

Apocalypse or Redemption: How the Portuguese Media Cover Artificial Intelligence

João Canavilhas and Renato Essenfelder

Abstract Artificial intelligence (AI) occupies an increasingly important place in the contemporary societies. One of the most visible aspects is the personalization of online content, news or advertising, but the emergence of personal assistants and autonomous vehicles are phenomena that also attracts the attention of the media and, by this way, reach the population. This article studies the Portuguese reality by analyzing the coverage of artificial intelligence by five national newspapers. The conclusions allow us to say that, although it was considered huge, the coverage proved to be superficial, with a high rate of reuse of sources from other texts, presents little diversity in opinions, and does not stimulate debate in society. Among the most common approaches are economics and political. The coverage was positive in most cases, especially in the health area, where the media never criticized the effects of AI.

1 Introduction

Since the 1950s, the field of artificial intelligence (AI) has gone through alternate moments of euphoria and disappointment, but in the last two decades, the world has experienced a new "golden age" [13]. The present moment is highly favorable, a situation made possible by several factors, among which should be highlighted the increase big data available for machine learning, a term coined in 1959 by Arthur Samuel to define the field of study that seeks to give computers the ability to learn without being explicitly programmed [14].

In parallel, the processing capacity of the machines has been exponentially increasing in recent years: "Prior to 2012, AI results closely tracked Moore's Law,

J. Canavilhas (✉)
Universidade da Beira Interior, Covilhã, Portugal
e-mail: jc@ubi.pt

R. Essenfelder
Universidade Fernando Pessoa, Porto, Portugal
e-mail: ressenfelder@ufp.edu.pt

J. Vázquez-Herrero et al. (eds.), *Total Journalism*, Studies in Big Data 97,
https://doi.org/10.1007/978-3-030-88028-6_19

with compute doubling every two years. Post-2012, compute has been doubling every 3, 4 months" [26:5].

Today, AI is present in applications as varied as self-driving vehicles, voice recognition, autonomous planning and scheduling, gaming, anti-spam software's, logistical planning, robotics and machine translation (Russel and Norvig 2016). AI resources are incorporated in computers and smartphones, but also in televisions, cars and even household appliances. AI is now among the most modern services provided by small technological startups and by mega-corporations that shape the contemporary worldview such as Google, Facebook, Instagram, Netflix, Spotify, Uber, Airbnb, Amazon, among others.

2 Artificial Intelligence and Society

Although artificial intelligence is increasingly present in our lives, it is interesting to note that there is no clear and consensual definition [3, 11, 25].

The term first appeared in 1956, during a seminar at Dartmouth College, organized by a pioneer in the field, the scientist John McCarthy [30], who had the objective of discussing aspects of human intelligence that could be simulated by machines. At the time, McCarthy resisted to use more specific terms such as 'computational intelligence' because he wanted to consider analog device research as well [20].

More than 60 years after that inaugural seminar, experts still propose different answers to the question "What is artificial intelligence?" (Martinez-Plumed et al. 2018).

AI can be analyzed as a field of computer science research, but also as a vast set of engineering practices [30].

Conceptually, definitions can be divided into four major groups: AI as a system that thinks rationally; as a system that imitates human thought; as a system that acts rationally; or as a system that mimics human action.

For Bellman [1], AI is the automation of activities associated with human thought, such as learning, making decisions and solving problems. Winston [35] discards the human paradigm: for him, AI is the field that is dedicated to study computations that make it possible to perceive reason and act.

Nowadays, the predominant line among engineers is that of the rational agent, i.e., artificial intelligence that solves problems without worrying about imitating human thought or behavior. This paradigmatic option has also helped to boost the growth of the area in recent decades, because it is easier to build machines that fly when we do not try to imitate birds perfectly, but only solve problems .

In the last 50 years, there has been a greater trend of investment in the areas of machine learning, health and medicine, and personal assistants [19].

2.1 AI and Public Interest

General interest in the topic has grown exponentially. Between 1998 and 2018, the volume of peer-reviewed AI papers has grown by more than 300%, "accounting for 3% of peer-reviewed journal publications and 9% of published conference papers" [26:5]. According to the *AI Index 2019*, published by the Stanford University, at the graduate level, AI is now by far the most popular specialization among computer science PhD students in North America, with over twice as many students as the second most popular specialization (security/information assurance). "In 2018, over 21% of graduating Computer Science PhDs specialize in Artificial Intelligence/Machine Learning" [26:6].

The interest in the subject goes beyond university walls: a global study conducted by McKinsey & Company [5] states that in 2017, 20% of companies of various sizes and sectors of the economy adopted at least one AI solution in their businesses, a figure that in 2018 reached 47%. In December 2019, a Stanford study revealed that this number rose to 58% [26]. This popularization of AI has been accompanied by the growth of journalistic coverage on the subject and in recent years the expression 'artificial intelligence', as well as others associated with it (such as 'machine learning', 'deep learning' and 'Tensor Flow') have been increasingly mentioned in Google searches and in journalistic stories from all over the planet.

A survey conducted in the massive database of the GDELT Project, which monitors the internet news, printed and broadcast in 65 languages, shows that about 0.5% of all news in the world quoted the expression 'artificial intelligence' in June 2019, which compares with about 0.15% in early 2017 [26].

However, what do the media say about artificial intelligence? At a global level, there is a clear prevalence of one issue: the impact of this technology on the human work. Of all the news that cites AI, 17.7% are associated with the words 'job', 'jobs', 'employment' or 'unemployment', according to Perrault and colleagues [26]. In contrast, the percentage that contained either 'biases' or 'biased', one of the most relevant discussions in the development of AIs today, was only 2.4%.

As the interest and relevance of artificial intelligence in the world grows, it becomes more important to understand the construction of public debate on the subject and to stimulate discussion. There is a consensus that much remains to be done in this field because, in general, the subject is unknown to the public, and the debates between specialists and public authorities are not properly reflected in the media.

To some scholars, the issue in the press is often divided between two opposing poles: (a) the utopia of a better future, without diseases and with abundance for all; (b) the dystopia of the robotic apocalypse, when machines revolt against men [7], or when technology deepens inequalities and provide the emergence of a caste of superhumans, an elite of men and women who will be able to afford the latest technology implants to become smarter, stronger and healthier in the face of other mortals, trapped in the frailties of the human condition [12].

A content analysis carried out in six British media that looked at the expressions 'artificial intelligence', 'machine learning', 'deep learning' or 'neural networks' found that almost 60% of them were linked to market agendas, such as the launch of products or services [3]. Interestingly, almost 12% of all articles containing one of these expressions referred to Tesla's billionaire founder, Elon Musk. In other words, the debate on AI remains predominantly restricted to the business world. From this perspective, AI is essentially a way to optimize processes, maximize profits and increase the comforts of modern life.

The clear predominance of economic agendas impoverishes the debate on this complex and multifaceted topic. Lack to the British coverage "a wider range of voices in discussions of AI. Academics, activists, politicians, civilians, and civil servants, amongst others, can all contribute to a rich and sophisticated public debate around AI" [3:9].

When coverage is shallow, it disregards the important differences that AI scholars have about "what AI is, what it will be able to do, and how it can be designed, regulated, and integrated into society" [3:1].

Contributing to the research that seeks to know what the media are talking about AI, in this study we analyzed 123 articles published in five Portuguese media (*Expresso*, *Correio da Manhã*, *Jornal de Notícias*, *Público* and *Observador*) in the first two months of 2020. In addition to trying to understand how the Portuguese press covers the topic of 'artificial intelligence' today, we compare the results obtained with British research and data from the *AI Index 2019*.

It should be noted that in June 2020 Portugal received for the first time the classification of "highly innovative country" by the European Innovation Scoreboard (EIS), which highlights the relevance of broadening the debates on what kind of technological innovation the society is stimulating and how the media reflect this.

3 Empirical Study

This exploratory study uses a quantitative approach to facilitate the description of the complexity of a given hypothesis or problem and the crossing of multiple variables [28].

In view of the scenario presented above, the research problem that arose was the following: In what way do the main Portuguese journalistic websites cover the topic of artificial intelligence? From this central question derive three more: What are the main themes associated with AI? Is the value attributed to AI in journalistic coverage positive, negative or neutral? Is there a difference in media approach?

For the study, 123 articles were collected in five Portuguese media, which make up a representative panorama of the national production of online journalism in the country: the weekly with the largest circulation (*Expresso*), the three largest daily newspapers (*Correio da Manhã*, *Jornal de Notícias* and *Público*) and the largest digital native (*Observador*).

3.1 Sample and Observation

The sample is composed by texts containing the expression 'artificial intelligence' and published between January 1 and February 29, 2020. We also searched texts that include expressions belonging to the same semantic universe of AI, namely 'deep learning', 'machine learning', 'neural networks' and 'Tensor Flow'.

Throughout the analyzed period, only five texts contained one of these expressions without being associated with the expression 'artificial intelligence', a first sign that the coverage of the subject still tends to be more generic.

It should be noted that in this period there was no mention of 'neural networks' or 'Tensor Flow'—the name of the most popular system for creating and training neural networks for AI applications.

After the collection of the 123 texts, separated by newspaper and by date of publication, each one was analyzed and coded individually to form a set of variables designed to meet the objectives of the research, enabling the realization of multiple cross-references. The variables were:

1. Textual genre

 1.1. News: shorter texts, associated with daily facts, such as products or service launches, political speeches, financial results, events.
 1.2. Report: longer texts, generally with more sources and with analytical ambitions to deepen the discussion.
 1.3. Opinion: editorials or texts signed by guests or specialists.
 1.4. Advertising report: also known as 'native advertising', are texts sponsored by companies.

2. Origin

 2.1. Local: article produced by the newspaper itself.
 2.2. News agency: article signed by a news agency and republished.
 2.3. Expert or authority: stories signed by guests (experts on the subject, politicians or authorities).

3. News Peg—following Brennen's et al. [3] categories.

 3.1. Market: the pretext for the publication is a market announcement (company results, product launches, service reviews, corporate events, etc.).
 3.2. Academy: the pretext is an academic issue, such as scientific publications, interviews with researchers, scientific events, etc.
 3.3. Government: government actions, discussions in Parliament, laws, promotion programs with public money.
 3.4. Civil society: related to behaviors, focused on characters, actions of organized society, such as protests, petitions or NGO actions.
 3.5. Others: for texts without a clear pretext.

4. Source—quantity (0–5) and type, following Brennen and colleagues [3], the principles are, like the 'news peg', variable

4.1. Market: companies.
4.2. Academy: universities or research centers.
4.3. Public power: Executive, Legislative or Judiciary.
4.4. Civil society: non-specialized sources or NGO members.
4.5. Writing: reproduction of quotations aired in other media.

5. Theme: main subject of the content. After successive filtering and grouping processes, the following thematic nuclei of each analyzed text were defined: economy, politics, health, curiosities and others.
6. Value: approach to the theme (positive, negative or neutral in relation to artificial intelligence).
7. Weight: a measure of the relevance of mentioning artificial intelligence within the text. This variable is important because it was perceived that in some cases the expression 'artificial intelligence' was only a rhetorical resource. Here it was measured if the text treated the AI as something relevant to the story or if the expression was lateral. The classification has a scale 1–3, being 1 "not very important", 2 "moderately important" and 3 "very important" in cases where the AI is the central theme of the text.

3.2 Results and Discussion

Once the data is collected, we analyze the texts according to the group of variables stipulated. In the case of the variable sources, we remind that only the sources with direct citation were considered, that is, between quotation marks. Figure 1 shows the total distribution of the sample by media.

The newspaper *Público* dedicates more space to the topic, followed by *Expresso.* They are the two most important media in the country, aimed at urban readers with a high level of education [29] and generally considered the most credible Portuguese newspapers [27]. It is followed by the only digital native in the sample, the *Observador*, launched in 2014, which is part of the sample for comparison with other

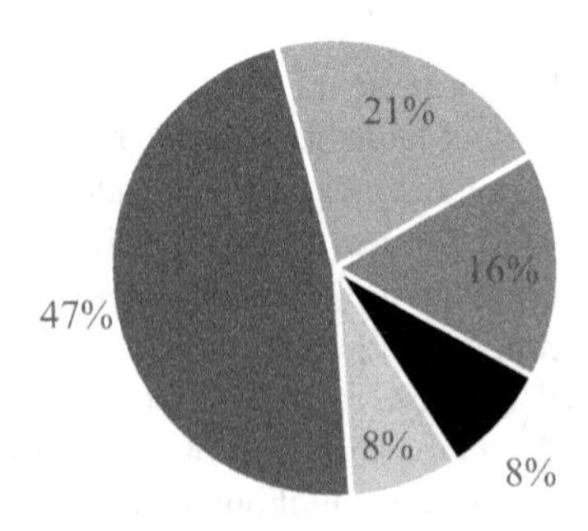

Fig. 1 Texts containing the expression 'artificial intelligence'. Own elaboration

traditional media and also because it is the first Portuguese media with a political position—in this case, European liberal [9]. *Correio da Manhã* and *Jornal de Notícias*, influential periodicals of a more popular nature, are tied with ten texts each containing the expression 'artificial intelligence'. It can already be pointed out that the subject is markedly more present in the reference newspapers.

In general, the coverage was considered voluminous, with an average of about 60 articles/month referring to AI. In other words, if these technologies are incorporated into the daily lives of citizens, their existence seems to be more and more incorporated into Portuguese journalism as well. However, this expressive quantity is not synonymous with quality, since most texts have zero or one source of information cited in direct discourse, and even in articles that record more than one source, there is no diversity of points of view.

Sources, Plurality and Diversity

When evaluating the quality and depth of a journalistic text, one of the important aspects is to verify the presence of diversified and reliable sources of information, carefully selected [18]. Nowadays, the ease of access to online content is accompanied by the potential threat of growth and professionalization of malicious sources of misinformation, which can confuse even professional journalists [22].

In the case of online journalism, the situation seems to be even more complicate due to the intense and exhausting pace of work [2], a situation that leads reporters to depend on the sources that are most available to respond in an agile and minimally reliable manner. In this scenario, professional sources are gaining strength, such as press releases produced by public relations, information retrieved from other news sites and news agencies [6], convenient and institutionalized sources.

Unfavorable working conditions among digital journalism producers "give reason to assume that online journalists are at a heightened risk of using sources that are fraudulent, distorted or inaccurate" as Manninen [18:215] notes in a study conducted with online journalists in Finland.

Numerous studies on journalism deal with the role of sources in journalistic processes, namely the impact of sources in defining what is news [17], political, ideological and ethical issues [23], and how technological transformations have altered the dynamics in the search of information [8]. In addition, classic concepts such as gatekeeper [4, 33] have been reviewed in the light of hyper-connected post-modernity [10, 24].

In Michael Schudson's sense, the sources are "the deep, dark secret of the power of the press. Much of this power is exercised not by the news institutions themselves but by the sources that feed them information" [32:134].

In the online environment, this relationship becomes even more complex. As Sundar and Nass [34] point out, there are multiple layers of sources on the web confusing the public's perception of the original source of information: the news sites linked to newspaper companies dispute space with social-bookmarking sites (e.g., digg.com), social networking sites (e.g., Facebook) and microblogs (e.g., Twitter) and each element in the communication chain can be perceived and defined as a source. Especially problematic, therefore, are the frequent news articles entirely

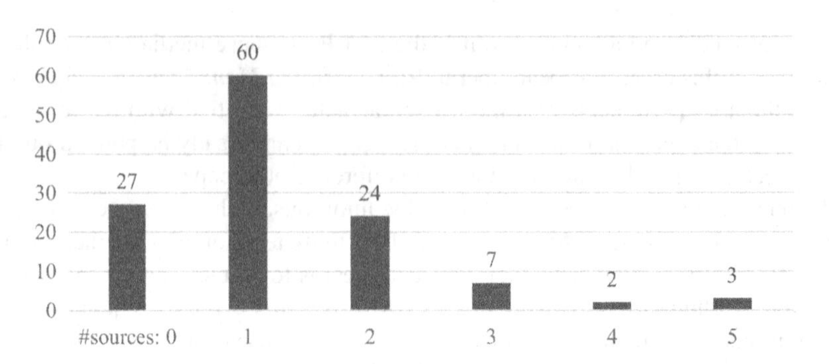

Fig. 2 Quoted sources by article (n = 123). Own elaboration

written based on quotations from other media. Inattentive readers will perceive the source of that information as the newspaper that compiled it, not the original sources, whose reputation may be questionable [15].

In the case of this study, most articles have zero or only one source, which is in line with the above prognosis. Figure 2 shows the distribution of the texts according to the number of sources cited in direct speech, i.e., using quotations in quotation marks.

The data allow us to verify that 71% of the sample is composed by the sum of texts without any source (22%) and texts with only one source (49%) cited in quotation marks, which causes some concern regarding what could be expected from reference vehicles.

As Manninen [18] states, the ease of access to sources and the reduction in the costs of circulating content did not revolutionize journalism's production methods, but only simplified and accelerated already consolidated processes. "The increase of productivity has been realized as more products, instead of better products. [...] With no temporal or spatial limits to content, there is never time to waste—any excess could be put toward creating more content" [18:215].

Brennen and colleagues [3] notes that in the case of special press coverage, including science and technology journalism, the situation is even more dramatic, with successive cuts made over the last few years by the economic crisis that plagues most media groups. "These changes mean that some outlets cover these stories less frequently, task non-specialist reporters with reporting these stories, give their reporters less time and fewer resources to cover them, or encourage more reliance on press releases or wire articles" [3:2].

When we analyze the textual types codified in this study (advertising report, news, opinion and report), the prognosis is partially better, since 17 (63%) of the 27 texts without quotes are opinion articles, signed by experts or authorities in their fields of work. Another 9 (33%) are news, and the remaining case is an advertising report sponsored by a private company (the only case of custom-made text in the sample). Not quoting statements from third parties is relatively common in opinion texts, but we consider relevant the fact that about one third of the occurrences of articles

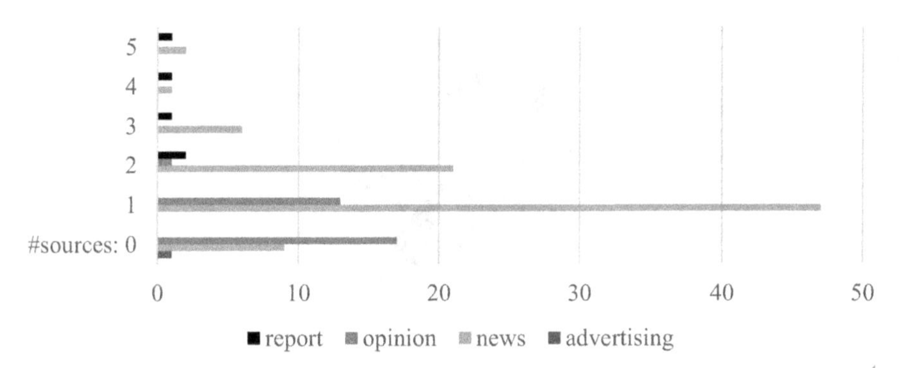

Fig. 3 Quoted sources by textual type. Own elaboration

without any source are framed as 'news'. Of this one third, half were produced by the same media, the *Observador* site, which seems to reinforce the hypothesis already discussed previously that, although exclusively digital vehicles have not created the processes that lead to the pressure for journalists' productivity, they have intensified them.

If in the category 'texts without any cited source' the opinion-active gender predominates, in all the other extracts there is a prevalence of news materials, as illustrated in Fig. 3.

Considering the total sample, 29% of the texts have more than one source expressly cited. But what are these sources? Will there be plurality and diversity of voices in the articles dealing with artificial intelligence in the Portuguese press? Kischinhevsky and Chagas point out that "the diversity of voices is advocated by journalism theorists as a criterion to ensure quality in an informative coverage" [16, 21] reinforce that the concepts of plurality and diversity are not only of interest to the students of mass communication, but also fundamental to the maintenance of the public sphere and the health of democracy itself.

Here, we consider plurality as a quantitative aspect in the selected sample: the more sources, the more plural will the media discourse be. However, diversity is a qualitative order and presupposes different perspectives to enrich the journalistic text. Thus, there is no diversity when a reporter interviews several fonts from the same field or perspective, reinforcing the weight of a single argument. As Kischinhevsky and Chagas state, "knowing where the news comes from is not only a rhetorical question, but a decisive factor in maintaining journalism as a social institution and part of the set of public interests in a democratic society" [16:114].

The plurality of texts is compromised by the low number of sources heard in the sample. In the total of 123 texts collected, 152 fonts were mapped (average of 1.24 per text). When we exclude from this total the 'written' sources (references to speeches, press releases and other previously published reports), the total of citations drops to only 75, that is, only these sources were effectively interviewed by the authors of the articles. In this case, the average of sources duly interviewed per text drops to only 0.61, which weakens the principle of plurality of voices.

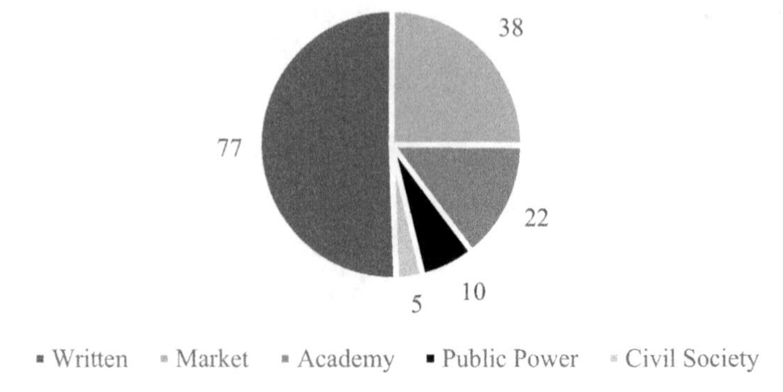

Fig. 4 Quoted sources origin. Own elaboration

To analyze the aspect of diversity in the sample, we classified the sources according to their origin. The result is what appears in Fig. 4.

Most (51%) of the sources with direct citation come from other texts, mainly from reports published in prestigious international newspapers such as *The New York Times* and *The Guardian.* In the sequence, market sources appear (25%), academics (14%), interviews with members of one of the three Powers (7%), and members of civil society, whether in the form of representatives of non-governmental organizations or ordinary citizens (3%).

The data confirm Brennen's et al. [3] study on AI coverage in the UK press. The British press, however, seems to be less dependent on written sources (about 40% of the total, compared to 51% in Portugal), but more dependent on the market (33% in the UK and 25% in Portugal). The other proportions were very similar.

When we analyze in detail how the news combines sources from different backgrounds, we find that only 9 articles in the sample (7% of the total) mix sources from different backgrounds. Still, in most of these few cases the combination is made between one or more written source (mainly newspaper articles) and one or more market sources. Qualitatively, therefore, it is possible to state that the diversity of sources is also a point to improve in Portuguese coverage of artificial intelligence.

Themes, Euphoria and Fear

Considering the topics addressed in the texts of the sample, there was a predominance of economy (36%), followed by political issues (33%) and health (14%). An expressive set of news about sparse themes, sometimes of humorous tone, was grouped under the heading of 'curiosities' (e.g.: "The creator of copy and paste dies"), making up 15% of the total. The remaining subjects, with a more fragmented theme, dealt with themes such as environment and education and appear under the heading 'others', as shown in Fig. 5.

The trend of a predominantly economic AI coverage was observed in the UK [3], where 60% of the stories had as news peg the announcement of a product or service. In the Portuguese case, it is interesting to note that although the economic agenda predominates, there is also a notable emphasis on the social-political discussion of

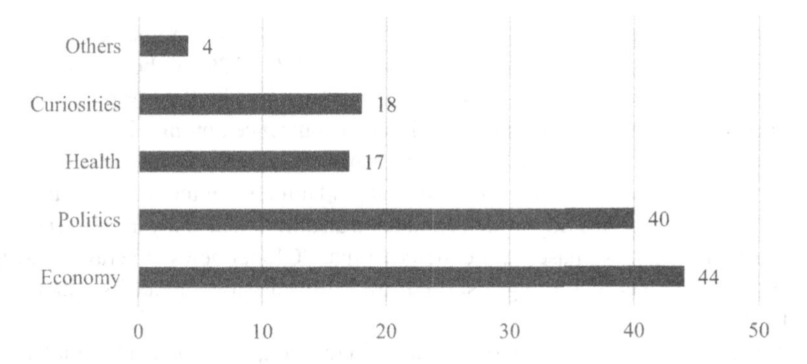

Fig. 5 Themes. Own elaboration

AI, in matters that speak of threats to democracy and privacy, regulation of the sector, and the plans and proposals of public authorities in the adoption of these technologies, including budgetary allocations. Although the discussions raised in these articles are in line with the social mission of journalism, what is noticeable is that the lack of sources and depth of debate hampers the full achievement of this role.

Figure 6 presents the main news peg for the publication of the stories. It should also be noted that the proportional participation of academia (15%) and civil society (7%) in guiding the theme in the Portuguese media is quite restricted.

In some cases, although the news peg for the publication was codified as 'Market', as in the text "'Artificial intelligence needs to be regulated', says the executive president of Google" (Público 2020), from a statement by the CEO of a large technology company, the subject was classified as political/public power, because it deals mostly with the regulatory issue—one of the subjects, by the way, more present in our sample, given the high temperature of the European news on the regulation of so-called big techs. Hence the disparity between the 'economy' theme (36%) and the 'market' news peg (46%) in the sample. Although below the index found in the United Kingdom, we consider it relevant to note that almost half of all texts collected come

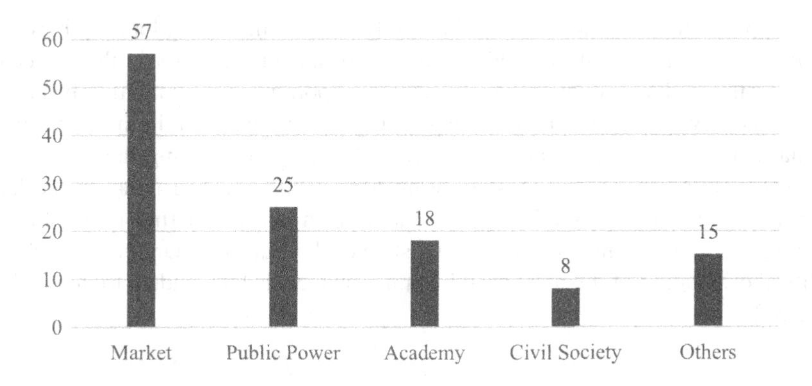

Fig. 6 News pegs. Own elaboration

from company announcements and technology executives. Another curiosity: the British media fetish by businessman Elon Musk, cited in 12% of the articles studied by Brennen and others [3], is not even by far accompanied by the Portuguese media, which only mentioned the businessman in one occurrence among 123 texts.

Regarding the discussion on the political aspect of AI in the Portuguese media, the results are in line with the Stanford University's Human-Centered Artificial Intelligence Institute's global study, which conducted an automated analysis of over 60,000 English-language news bases and concluded that "Global news coverage of Artificial Intelligence has increasingly shifted toward discussions about its ethical use" [26:150].

After classifying the most frequent topics in the sample, we sought to understand whether mentions of artificial intelligence were positive, negative, or neutral, and how they were distributed in relation to newspapers and subjects. The coverage in relation to AI in general is positive in 59% of the cases, neutral in 27% and negative in only 15%.

The comparison between the topics covered and the value given to AI reveals how the media frames the "intelligent" technologies because it is in the field of health news that we have the highest rate of texts favorable to AI: 94% of the stories are positive, and 6%, neutral. No article that mentions AI when dealing with health applications is critical, skeptical, or frankly negative.

In the sequence, 80% of the pieces with the subject economy are positive, 16% neutral and only 5% negative. When we make a second cut for the texts dealing specifically with entrepreneurship and startups, the most numerous subsets of the 'economy' category, we find 100% positive mentions. In other words, whenever the Portuguese press referred to a startup in the period, it was in a favorable manner.

The approach only changes in matters dealing with politics (a set subdivided into the categories 'government', 'democracy', 'regulation', 'human rights' (which includes 'privacy') and 'geopolitics'. In this group, 47% is neutral, 33% negative, and 19% positive. In other words, one third of the texts that frame AI as a political issue provide critical coverage, citing potential threats to privacy and human rights, to the future of work, and to democracy itself. These findings are summarized in Fig. 7.

In opposite to what was observed in a study in the United Kingdom [3], in which the political alignment of the media proved to be a decisive factor in the choice of topics by the media (newspapers on the right wing pointing out economic advantages of AI, and newspapers on the left wing seeing a threat to jobs), in the Portuguese media there was no significant difference in the way of covering the topic. This conclusion reinforces a previous study by Santana Pereira and Nina on the lack of clear identification of the Portuguese media with political parties or ideologies: "Portugal is, after Denmark, the member state of the European Union in which the political orientation of the most prominent newspapers is less evident to the public" [31:233].

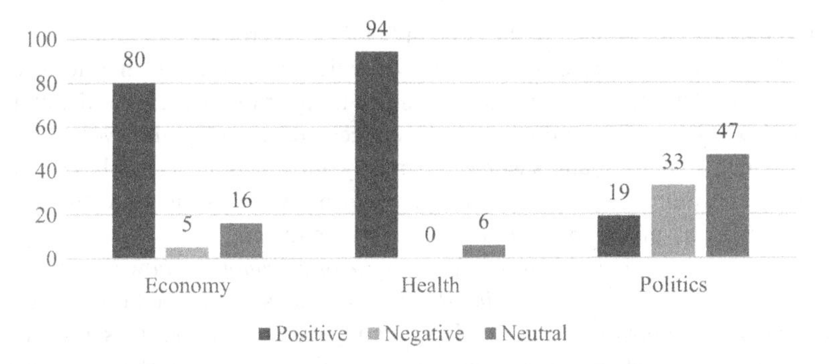

Fig. 7 Value attributed to AI, by theme. Own elaboration

4 Conclusion

Artificial intelligence has entered the daily life of societies through literature and cinema. Movies such as *Metropolis* (Fritz Lang), *2001: A Space Odissey* (Stanley Kubrick) or *Blade Runner* (Ridley Scott), to name a few references from different decades, are examples of this process. In the case of literature, authors such as Isaac Asimov or Arthur C. Clarke, have made artificial intelligence aware to the public opinion through the problems produced by their works.

Although this theme has been present in society for over 100 years, it was the technological development associated with the popularization of Internet and, more recently, smartphones, that put AI in our daily lives. The increase in processing and storage capacities, together with the immeasurable amount of personal data that circulates on the networks, began to be used in the development of products and services in all fields of human activity, with immediate impact on journalistic content. Therefore, in this work we seek to know what kind of approach the Portuguese media take on artificial intelligence.

Compared to other science-related topics, artificial intelligence has an important presence in the media, with an average of 61 news items per month in the five Portuguese newspapers analyzed. Most of the news (68%) was published in the two newspapers (*Público* and *Expresso*), which is not surprising given the nature of the topic.

Economic news predominates, mostly with a positive tone, as occurred in the United Kingdom [3]. The difference is that in Portugal no consistent association was identified between ideological approaches and the editorial orientation of the media. Next comes politics, an umbilical situation linked to discussions about privacy and democracy in its connection to misinformation processes.

The mostly positive approach has to do with the way the subject is approached, appearing represented as the solution to current and future problems, namely in themes related to ecology, health and well-being. The exception are the political themes, where the negative approach overcomes the positive one, but the neutral tone prevails, which annuls the weight of negativity.

In terms of sources, most of the content (71%) has zero or a single source, which is usually another media or a press release, which proves a tendency to resort to republication. The 'news' genre has the largest number of sources, but the plurality is small because, when the documentary sources are removed, the average is 0.61 source, which, to make matters worse, are mostly other media (51%). Market sources (25%) and academics (14%) are the other two privileged sources, with the three powers (7%), and members of civil society (3%) at the end.

Recovering the title of this work, *Apocalypse or redemption: how the Portuguese media cover the issue of artificial intelligence*, we can say that the Portuguese press clearly attributes a positive value to AI, presenting it as a salvation. This trend may be influenced by the existence of many news pegs coming from the market, as companies seek to sell products/services and therefore favor a positive approach. Using only one source (the market) and with ever smaller newsrooms where science journalists are a rarity, we can guess that this trend will continue for a long time.

Future studies should increase the observation period and look for news that accomplishes one of the journalism basic rules—the contrast of sources—situation only possible when there is more than one, something with little presence in this study.

References

1. Bellman, R.E.: An Introduction to Artificial Intelligence: Can computers think? Boyd & Fraser Publishing Company, San Francisco (1978)
2. Brennen, B.: The future of journalism. Journalism **10**(3), 300–302 (2009). https://doi.org/10.1177/1464884909102584
3. Brennen, S., Howard, P.N., Nielsen, R.K.: An Industry-Led Debate: How UK Media Cover Artificial Intelligence. University of Oxford, Oxford, RISJ Fact-Sheet (2018)
4. Bruns, A.: Gatewatching: Collaborative Online News Production. Peter Lang Publishing, New York (2005)
5. Chui, M., Malhotra, S.: AI Adoption Advances, but Foundational Barriers Remain. McKinsey & Company, San Francisco (2018)
6. Colson, V., Heinderyckx, F., Paterson, C.P., Domingo, D.: Do online journalists belong in the newsroom? A Belgian case of convergence. In: Making Online News: The Ethnography of New Media Production, pp. 143–154. Peter Lang Publishing, New York (2008)
7. Craig, C.: How Does Government Listen to Scientists? Palgrave Macmillan, London (2019)
8. Fenton, N.: New Media, Old News: Journalism and Democracy in the Digital Age. Sage Publications, London (2010)
9. Fernandes, J.M.: Tudo o que precisa de saber sobre o Observador. Observador (2014). Available at https://bit.ly/3cBliAp. Accessed 20 Jul 2020
10. Fígaro, R.: As mudanças no mundo do trabalho do jornalista. Editora Atlas, São Paulo (2013)
11. Dodd, M., Seruwagi, L., Grant, A.: Artificial Intelligence Through the Eyes of the Public. Worcester Polytechnic Institute, Worcester (2011)
12. Harari, Y.N.: Sapiens: A Brief History of Humankind. Penguin Random House, London (2014)
13. Havenstein, H.: Spring comes to AI winter. Computer World (2005). Available at https://bit.ly/39uTCLN. Accessed 24 Jul 2020
14. Hosch, W.L.: Machine Learning. Encyclopædia Britannica (2020). Available at http://www.rsc.org/dose/title of subordinate document. Accessed 09 Sept 2020

15. Kang, H., Bae, K., Zhang, S., Sundar, S.S.: Source cues in online news: is the proximate source more powerful than distal sources? Journal. Mass Commun. Q. **88**(4), 719–736 (2011). https://doi.org/10.1177/107769901108800403
16. Kischinhevsky, M., Chagas, L.: Diversidade não é igual a pluralidade – Proposta de categorização das fontes no radiojornalismo. Galáxia **36**, 111–124 (2017)
17. Manning, P.: News and News Sources: A Critical Introduction. Sage Publications, London (2001)
18. Manninen, V.J.E.: Sourcing practices in online journalism: an ethnographic study of the formation of trust in and the use of journalistic sources. J. Media Pract. **18**(2–3), 212–228 (2017). https://doi.org/10.1080/14682753.2017.1375252
19. Martínez-Plumed, F., Loe, B.S., Flach, P., ÓhÉigeartaigh, S., Vold, K., Hernández-Orallo, J.: The facets of artificial intelligence: A framework to track the evolution of AI. In: Proceedings of the Twenty-Seventh International Joint Conference on Artificial Intelligence, IJCAI 2018, Stockholm, Sweden, pp. 5180–5187 (2018)
20. McCarthy, J., Minsky, M.L., Rochester, N.: Shannon CE (2006) A proposal for the dartmouth summer research project on artificial intelligence, august 31. AI Mag. **27**(4), 12–12
21. McQuail, D.: Teoria da comunicação de massas. Fundação Calouste Gulbenkian, Lisbon (2003)
22. Metzger, M.J.: Making sense of credibility on the Web: models for evaluating online information and recommendations for future research. J. Am. Soc. Inform. Sci. Technol. **58**(13), 2078–2091 (2007)
23. Micó, J.L., Canavilhas, J., Masip, P., Ruiz, C.: La ética en el ejercicio del periodismo: Credibilidad y autorregulación en la era del periodismo en Internet. Estudos em Comunicação **4**, 15–39 (2008)
24. Néveu, E.: Jornalismo sem jornalistas: uma ameaça real ou uma história de terror? Braz. Journal. Res. **6**(10), 29–57 (2010)
25. Obozintsev, L.: From Skynet to Siri: An exploration of the nature and effects of media coverage of artificial intelligence (PhD Thesis). University of Delaware (2018)
26. Perrault, R., Shoham, Y., Brynjolfsson, E., Clark, J., Etchemendy, J., et al.: The AI Index 2019 Annual Report. AI Index Steering Committee, Human-Centered AI Institute, Stanford University, Stanford (2019)
27. Popescu, M., Toka, G., Gosselin, T., Pereira, J.S.: European media systems survey 2010: Results and documentation. European Media Systems Survey (2012). Available at http://www.mediasystemsineurope.org/results.htm. Accessed 02 Mar 2020
28. Prodanov, C.C., de Freitas, E.C.: Metodologia do trabalho científico: métodos e técnicas da pesquisa e do trabalho acadêmico. Editora Feevale, Novo Hamburgo, Brazil (2013)
29. Ribeiro, V.: Fontes sofisticadas de informação. Media XXI, Lisbon (2009)
30. Russell, S.J., Norvig, P.: Artificial Intelligence: A Modern Approach. Pearson Education, New York (2016)
31. Santana Pereira, J., Nina, S.R.: A democracia nos media portugueses: Pluralismo político-partidário na televisão e na imprensa. In: Bello, E., Moura Ribeiro, S.S. (eds.) Democracia nos meios de comunicação: pluralismo, liberdade de expressão e informação, pp. 225–247. Lumen Juris, Rio de Janeiro (2016)
32. Schudson, M.: The Sociology of News. Norton, New York (2003)
33. Shoemaker, P.J., Vos, T.P.: Teoria do gatekeeping: seleção e construção da notícia. Penso Editora, Porto Alegre, Brazil (2016)
34. Sundar, S.S., Nass, C.: Conceptualizing sources in online news. J. Commun. **51**(1), 52–72 (2001). https://doi.org/10.1111/j.1460-2466.2001.tb02872.x
35. Winston, P.H.: Artificial Intelligence. Addison-Wesley, Boston (1992)

João Canavilhas Ph.D. and DEA in Communication, Culture and Education (USAL, Spain) and Bachelor in Social Communication (UBI). Associate Professor at Universidade da Beira Interior (Covilhã) and researcher at Labcom—Communication and Arts (Evaluation FCT: Very Good). He is the author or co-author of 10 books, 38 chapter books and 50 papers in national and international scientific journals. His research work focuses on various aspects of journalism and new technologies.

Renato Essenfelder Journalist and Ph.D. in Communication Sciences at the University of São Paulo (ECA-USP), Master in Portuguese at the Catholic University of São Paulo and Bachelor in Journalism. Professor of Communication at the Fernando Pessoa University (Porto, Portugal). Develops research in the areas of journalistic narratives, storytelling and artificial intelligence. Also, a digital columnist for one of the major journalistic companies in Brazil, *O Estado de S. Paulo*, since 2014.

Horizon 2030 in Journalism: A Predictable Future Starring AI?

Bella Palomo, Bahareh Heravi, and Pere Masip

Abstract The future of journalism is more stable than we imagine. The search for innovative solutions always requires that we take a look back at the past, even if it is to address the opportunities and challenges posed by artificial intelligence in the realm of journalism. Computational algorithms are capable of streamlining such diverse tasks as content discovery, filtering, analysis, production, publishing or distribution, although the financial investment of implementation represents one of the main impediments and generates a visible divide in the present and future media system. Furthermore, the ethical dilemmas it incites will depend on the choices made by its creators, whose mixed profiles will be in high demand.

1 Introduction

The future of journalism raises contradictory opinions that have been discussed in industry reports, professional forums, specialized congresses and academia. Where some foresee one outcome, others trust in the arrival of a new form of journalism [23], yet despite the myriad desires and rational predictions, all point to a hybrid future [10] with no single clear-cut solutions, a larger cooperation network [27] and artificial intelligence (AI) occupying a central role [23].

The degree of development will depend on the culture of innovation fostered by each company or individual. Innovation is an omnipresent concept in reflections on the future of journalism. Yet these initiatives in more traditional newsrooms are essentially a survival strategy. It is a desperate form of innovation. The spread of

B. Palomo (✉)
University of Malaga, Malaga, Spain
e-mail: bellapalomo@uma.es

B. Heravi
School of Information and Communication Studies, University College Dublin, Dublin, Ireland
e-mail: bahareh.heravi@ucd.ie

P. Masip
Ramon Llull University, Barcelona, Spain
e-mail: peremm@blanquerna.url.edu

J. Vázquez-Herrero et al. (eds.), *Total Journalism*, Studies in Big Data 97,
https://doi.org/10.1007/978-3-030-88028-6_20

media labs over the past decade has paradoxically heightened concerns about the near future, about cloning safe actions and conforming to means of using technology that are often determined by the competition, designing immediate isomorphic innovations instead of cultivating and developing genuine novelties, alternative disruptors [24]. This trend is partly due to the fact that an increasingly high and more demanding level of originality is now required to surprise society, one that cannot be the result of improvisation.

The absence of a forward-thinking culture centred around planning long-term strategies for generating new products, processes or means of organization or marketing heightens the risk of perpetuating an interest in shiny things, instead of correcting, reforming or transforming the weaknesses of the information ecosystem [20]. This short-sighted conception of innovation makes it difficult to surmount a complex structural crisis in which new players and roles have emerged, the scope of information is reliant on social media, inattentive audiences disaffected with the media are becoming increasingly fragmented and financial difficulties are a permanent reality. How to ensure the long-term sustainability of journalism and reassert the value of media content are the biggest concerns for media managers at present. Are change or die really the only two options?

1.1 Solutions for a Journalism Industry in Crisis

The current situation does not paint an optimistic picture. Journalism is currently subject to profound stress, from both a political and economic standpoint. The instability of the profession is compounded by a resistance to change evidenced by those opposed to journalists having a profile in constant transition and engaging in adaptive journalism [26]. And the emergence of disruptive events such as the power currently wielded by technology companies or obstacles such as the global pandemic also destabilize predictions, lead to greater dislocation and fuel uncertainty.

Another factor to bear in mind is the divide that exists between local newsrooms and those of global media outlets, which proves that innovations such as virtual reality are unachievable luxuries for small companies, given the sizeable investment this type of content requires. This situation will only grow worse in the coming years, and no single 'one size fits all' measure will protect journalism in the future [9]. Paradoxically, while technology introduces into the media system threats and discrimination which make it difficult to outline general trends, it also provides opportunities, and this imbalance has no easy solution [1]. This issue, however, is not limited to the realm of communication. Most organizations that aspire to be innovative are forced to contend with VUCA (volatility, uncertainty, complexity, ambiguity).

Yet despite the problematic context, there have been several memorable initiatives. In recent years, certain media outlets have used drones to cover demonstrations or reach inaccessible areas; 360° videos have been made to promote immersive experiences and transport audiences to places they have never been before; documentaries have been created using virtual reality; cybermedia supply content in an infinite scroll; and robot journalism is automating corporate earnings and sports game results. Thus, bearing in mind the above-mentioned divide, if this is the present, what does the future hold?

An exploratory analysis of the scientific literature concerning the future of journalism reveals two trends: cautious, prudent studies, more concerned with explaining what is being done rather than what could be done and, secondly, those of a more alarmist nature, which view journalism as in decline or even augur its extinction [9].

Determining which technology and which new opportunities will ultimately benefit journalism and, therefore, society within a decade is a constant challenge for the industry and a responsibility that will depend on the choices made in the present, because to understand the future, we must look to the past [38].

Incremental innovations are the simplest to predict and imagine, those which encourage more multimedia solutions, facilitate consumption by enhancing usability and the quality of the news products, help to identify manipulation in real-time and reduce costs. By 2025, the constant expansion of mobiles is expected to coincide with the advent of the Internet of Things, at which point 5 billion connected citizens will coexist alongside 50 billion devices [5], and settings such as streets and buildings will become ‘smart’, in addition to new content distribution points. However, the scalable readings of the future are not limited to the realm of technology. According to Boczkowski and Mitchelstein [2], the public service mission of journalism, understood as a case of social activism, must be advanced in order to stimulate positive change and win back lost confidence. To favor this model, more ethical, more independent and more networked journalism is needed [33], hence the proposal to create national laboratories for journalism in which inter-media collaboration would help to identify new spaces for growth and, together, address urgent and pressing topics facing the community [5]. Over the next few years, the application of an inclusive, customized, individualized, problem-solving, useful, necessary and action-oriented model of journalism will be fundamental in achieving stable levels of audience engagement and reclaiming a central role in democratic societies.

At the other end of the spectrum from the incremental model are disruptive innovations, which are riskier to predict because they require forward thinking and more drastic changes to reinvent products and work methods and generate new audience needs. Furthermore, prototypes are often protected by trade secrets, and their visibility is limited until the success of their launch can be guaranteed.

One of the most original predictions forecasts the disappearance of social media in their current form and the development of social tools which will help to build bridges between polarized and fragmented communities. Cyber-insurance is also expected to become a reality, to protect the victims of cybercrimes. Thus, following the furore caused by the expansion of social media in the initial participatory phase, a new era of concern has settled in, in which there is increasing awareness that the model of addictive consumption, aggressive surveillance, arbitrary censorship and subtle manipulation pushed by big tech companies, and which has affected behavior, attitudes and beliefs, has had an adverse effect on society.

In a study by Pew Research Center involving 979 technology innovators, developers, business and policy leaders, researchers and activists, 49% of these experts believed that the current use of technology will ultimately erode central features of democracy and may even affect democratic representation in the next decade. However, two-thirds of those interviewed also trust that technology will provide a solution to some of the problems it has created [37].

One of the other main question marks has to do with training. How, in the classroom, can we prepare future journalists to assume tasks and roles that still do not exist? Training new generations to work in flexible settings will be key, and this requires a commitment to creating mixed profiles capable of combining structured and systemic thinking and willing to be up against a constant learning curve. At present, few developers and engineers choose to work for media companies, although the virtues of interdisciplinary teams are altering traditional structures. The inclusion of more technical profiles in newsrooms will advance the introduction of new routines linked to precision, while sociologists, linguists and psychologists are already helping and training engineers and machines to process the language better and analyze it rapidly.

According to several authors, the future of professional communication will be determined by non-human journalism actors [23] and, more specifically, by artificial intelligence, robotics, virtual reality, 3D printing and blockchain. These predictions come as no surprise, as algorithmic gatekeepers are already a reality and already form part of our daily routines. Everything from using facial recognition to unlock mobile devices to automatically loading content onto a user's Amazon, Facebook or LinkedIn page depends on software which, to run, requires sponsors, parameters and pre-established limits. A 2018 McKinsey Global Institute report estimates that, by 2030, European workers will dedicate 16% less time to manual and physical tasks, with similar developments also due to occur with basic cognitive tasks in this new era of automation. However, higher cognitive, social and emotional skills and technological requirements will increase over the next decade, particularly in tasks that demand empathy, creativity, critical thinking, decision making and complex information processing [3].

Whittaker [38] foresees a revolution in local information regarding healthcare, crime and government services, with algorithmic solutions that could transform open data into something more usable, as well as the increasing prominence of automated assistants similar to Siri, Alexa and Cortana. However, this author harbours doubts about the feasibility of algorithmic writers in creating complex and comprehensive stories that require analysis, context and reactions and constitute a truly premium service, content which is relevant to the audience.

Scientific literature focused on analyzing the relationship between journalism and artificial intelligence has been transdisciplinary and particularly fertile since 2015, with fundamentally descriptive and exploratory approaches, yet no prospective analyses. These studies revolve around five main issues: data journalism, big data, news production, application in social networks and fact-checking [4].

On a professional front, in 2018, the number of media outlets that had begun automating journalistic texts was close to thirty [36], and this trend, which will not be exempt from ethical dilemmas, is projected to grow in the coming years.

2 The Influence of AI on Professional Practice

Artificial intelligence (AI) was developed as an academic discipline in the 1950s. Since then, it has gone through waves of popularity and decline. The peak of its popularity was in the 1980s, which roughly coincides with when computers were becoming more accessible at universities, workplaces and homes. By the 1990s and early 2000s, AI was widely taught in universities all over the world.

The basic idea behind artificial intelligence is that the machine will have some level of intelligence that would allow it to mimic human cognition and/or behaviors. The core to this is learning (from data or environment) and decision making. Just like a kid learning about her surroundings as she grows up to be able to make intelligible decisions when she grows up, the machine would also need to learn to be able to successfully mimic near-human behavior, when and if possible.

Many tasks or functionalities that are now considered basic or simple machine, computer or software tasks, would have been categorized as AI in the past. Examples include more advanced machines such smartwatches and fitness trackers that tell us about our stress level, or about how much more water we should drink after our exercise, to voice-activated personal assistants such as Siri and Alexa, to some very basic and longstanding functions such as spell checkers and autocorrects.

Artificial intelligence draws on a series of disciplines including computer science, information science, information engineering, statistics, cognitive science, and linguistics. Over the years AI research has included many topics such as logic and machine reasoning, machine learning, expert systems, data mining, business intelligence, knowledge discovery and representation, knowledge engineering, natural language processing, and voice and image processing. Some of these terms the younger journalist would be very familiar with, e.g. machine learning. But they may or may not have heard of some other terms that would have encompassed AI or utilized AI methods in the past 70, or even 20 years.

AI is not new. A quick search on Google Books Ngram shows the popularity of the term "artificial intelligence" in books. Figure 1 shows that 'artificial intelligence' as a term started appearing in books in the mid-1950s, when it was coined, and reached its maximum popularity around the 1990s, after which its popularity started declining. In the past decade, the term has started to regain popularity, as for many other data-driven and centric disciples—presumably due to the greater access to data and computational power, and the opportunities that they present.

Google Ngram shows that unlike artificial intelligence, and many other established disciplines and topics such as information science or even machine learning if you would like to try, data science was a new term, making a presence in the books around 2010. This, however, did not mean that the technologies used in data science were

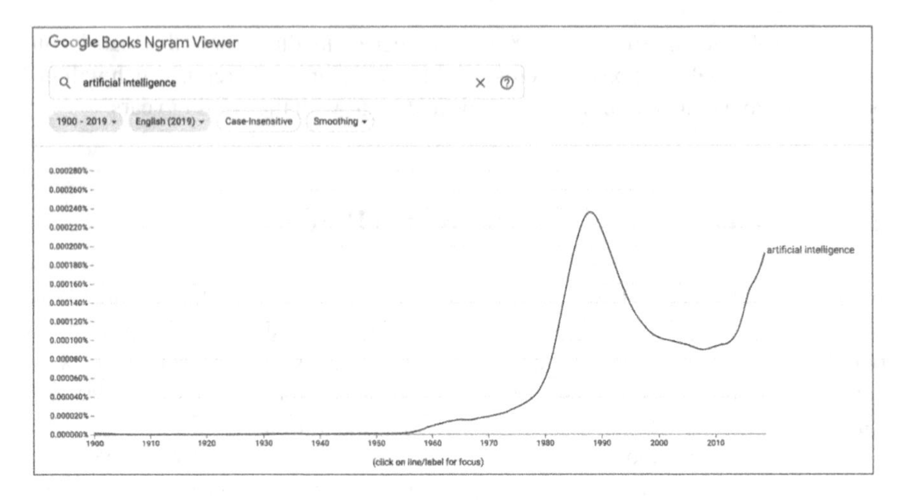

Fig. 1 Google Ngram viewer search result for 'artificial intelligence. Captured on 18 April 2021

also new. In fact, they were not. They were fundamentally based on statistics and computer science techniques, wrapped into a new, shiny term.

The term data science was in fact coined in 2008 to highlight the importance of 'data', which is now more than ever available to everyone, and the new values that could be derived from data to improve our day to day life and practices. But soon it reverted to its complex and specialized roots of statistics and computer science, something that was not the task of non-technical users, drawing the funding back to now 'data science' specialists, and pushing the salaries of already well-paid computer scientists, and the not as well statisticians, even higher. Making it less and less accessible for smaller media houses, and a commodity of large and rich organizations to be able to tap into the openly existing and in these cases non-openly existing highly profitable customer data.

At some level, many of the new terms, which in some cases such as the recent use of the term AI may be the recycling of old terms, are the new commodities of the richer than ever tech world. Same old concepts (or the natural progression of the same old topics) wrapped into a new shiny package, which will create a buzz, sell better, and/or attract more funding.

In academia, for example, it isn't hard to find professors who change their title every few years according to the latest buzzword. A professor who may sign as Professor of Artificial Intelligence today, might have called themself Professor of Machine Learning two years ago. Few years before that they may have called themselves Professor of Data Science, a couple of years before that Professor of Big Data, a little before that Professor of Data Analytics, and likely at some stage Professor of Computer Science. Along the way, they might have also called themselves Professor of Knowledge Engineering or Expert Systems. In many cases, the latest buzzwords have become paramount to whether or not an academic may receive funding for their latest research, or for the continuation of their existing and past research.

When it comes to the application of such computational models and tools to practical domains, such as journalism and news media, this sudden rise and fall of buzzwords often causes confusion, anxiety and disarray in newsrooms. This may partly be due to the fact that the practitioners may or may not have the technical knowledge to know that AI is not very different to what they called machine learning last year, and to what they called data science a couple of years before that. These technological advances and their natural progression over the years, while they may not reshape or transform journalism, can provide great assistance to the journalist and more importantly to the overall workflow of journalism.

2.1 What Can AI Do for Journalism?

AI methods, as we call them today, are essentially computational algorithms that use concepts in other fields to learn and act like we do as human beings, often in very specific and repeatable scenarios, with the aim of helping us and hopefully improving our lives one way or another. Like any other disciplines, journalism has been taking advantage of the computational advances for many years. Computers and computational tools and methods have been widely used in journalism since the 1960s with the emergence of Computer Assisted Reporting. Hence, there is no surprise that journalism will continue taking advantage of the computational tools and methods, as they evolve.

Journalism is to a great extent about finding new stories, whether it is in the form of daily and breaking news stories, or longer-term and more investigative pieces. They all involve sources, whether in the shape of human being or data, and methods to process what the sources have to offer. Access to data provides journalists with an enhanced set of sources like never before. The journalist will only be able to utilize these new sources of data if they are equipped to process and analyze this data. Computational advances, call it AI or any other buzzword it may be called tomorrow, could help journalism in processing and making sense of the data, and use computational methods in their day-to-day practice in numerous ways. In the following, we provide a non-exhaustive, yet ambitious framework of tasks that could potentially be improved, or achieved using the emerging and developing computational advances (Table 1).

The computational methods that may be used for the above tasks are expansive. They may include various types of data analysis, text analysis, named entity recognition, linking, and disambiguation, semantic annotation and linking, opinion mining, network analysis, and sentiment analysis, to name a few. We will not go into the details of each of the tasks listed above or the methods. Instead, in the following, we briefly discuss some of the advances so far, and the barriers to their implementation in newsrooms.

While we would have hoped, and some may claim that AI will magically facilitate and enable all the above tasks, many of these have indeed been problems and technological methods under study and development for years now. For example,

Table 1 How artificial intelligence and computational methods can help journalism

Content discovery	Personalized discovery Data newswires Newsworthiness assessment Verification
Filtering, analysis and contextualization	Data cleaning Data wrangling (Semi-) Automated exploratory data analysis for story lead construction Topic detection Pattern analysis/discovery Outlier detection Linking to relevant background stories and relevant data sources Network analysis
Production	Story lead construction/recommendation Automated story generation (Semi) Automated visualization
Publishing and distribution	(Semantic) Annotation (Semantic) Linking Audience and engagement analysis Personalization and recommendation
Preservation and archiving	Semantic annotation Entity linking Semantic discovery

in 2012 Heravi and colleagues proposed Social Semantic Journalism [14, 16, 17], which employed a set of computational methods for a framework of news discovery, verification, contextualization and publication from the social media sources [15]. The methods and technical problems explored for Social Semantic Journalism over a period of five years, and in collaboration with news organizations, included event detection [18, 21, 22], location detection and disambiguation [19], topic detection [34], named entity recognition, extraction and linking [35], sentiment analysis, semantic linking, social network analysis, bias analysis, verification [17], and semantic annotation and archiving [13, 29]. These technologies and methods have been in fact under study by many scholars in the past ten or fifteen years. However, nearly ten years after Social Semantic Journalism was first proposed, pretty much the same problems, in some cases in slightly different contexts, are still being discussed in journalism, and are unresolved, with almost identical technological methods—now under the name of AI—are being explored to solve them. Time has passed and the technologies have evolved, arguably not as fast as some would have hoped for. Yet, the problems are not fully solved yet to a practical degree, and in many cases, they are still within the scope of 'academic R&D'.

Part of the reason behind this slow progression is the complexity of the computational methods themselves, but when it comes to their implementation in news and journalism there exist specific practical barriers to their implementation and integration in journalism workflows that we will discuss in the following.

2.2 Barriers to Implementing AI in Newsrooms

To be able to use AI for specific tasks (or any tasks) the algorithms should be designed and trained for the specific problem at hand. This training process is intensive, time-consuming and costly, and sometimes results in unacceptable or questionable efficacy. Even if the final efficacy is reliably acceptable, as soon as the domain or problem at hand changes, the training would likely need to be done for the new domain or problem, and possibly the algorithm itself should be amended. This makes AI, and such computationally heavy methods very expensive to implement, and not economically feasible in many day-to-day, deadline-driven, journalistic production work. Especially when it comes to smaller newsrooms.

In the context of newsrooms, we can distinguish between the use of AI for story production, and story distribution and promotion. While AI methods are rarely used in story production, they are commonly used in other areas of news work, such as audience engagement analysis, personalization and recommendation [30]. While automated story production already happens in some newsrooms, particularly in more structured reporting such as sport, automation of investigative stories is not expected to happen any time soon [30]. Yet, we believe other tasks such as news lead generation and certain levels of exploratory data analysis and linking is within reach in the next foreseeable future.

Most journalistic stories are one-off and rarely repeated. Given how expensive designing algorithms and training them for a specific problem is, the time and cost spent are not justified for one single story, or even for a series of stories. In reality, one AI project could take months, or maybe even years in the case of many research projects if the technological method itself is still being developed further. In reality, there are very few projects ever in journalism, which would have used AI methods successfully for investigative and storytelling purposes [30]. Individual stories have vastly varied topics and potential underlying data, and one AI tool cannot do magic for all types of stories, in all news organizations across the world. If we look into one news organization, or one investigative team, we are essentially talking about custom-made software development for each story, or for a small number of stories. Something that has been done by software engineers since they have existed.

Another barrier is the cost of experts. Programmers, computer, data or information scientists or engineers, fair or not, are often paid considerably higher than journalists. This makes it not cost-effective for news organizations to invest in AI specialists, as they may be able to hire two or even three journalists for every AI/computational specialist. Additionally, in the case of newsrooms that have inhouse journalists who learn the computational skills, sometimes referred to as ‘nerd journalists’—journalists who learn coding, statistics and computational skills—the reality is that they will soon be able to find considerably higher paid jobs outside of the newsrooms, which would make their retention rather difficult.

While AI may still have a long way to go to be useful for day-to-day story production in newsrooms of all sizes, AI methods and established computational tools work very well for business problems associated with journalism, including audience analytics and consequent recommendations. The underlying data structure for such tasks is relatively unchanged over time, and the tasks are repeated on an ongoing basis. These are essentially business problems.

2.3 *Ethical Issues*

The introduction of technology into newsrooms has brought with it reflections on the ethical consequences of its implementation. The application of artificial intelligence in journalism is no exception, and, albeit timidly and hesitantly, the discussion surrounding the new ethical dilemmas that its emergence is generating has finally been put on the table.

Ethical reflections concerning the application of artificial intelligence in journalism involve a series of factors that make the issue even more complex. On the one hand, its application is not solely in the hands of journalists, as journalists have been relegated by developers, who are capable of designing and controlling the entire process. Furthermore, the media industry is currently in the midst of an economic crisis, exacerbated in certain countries by the COVID-19 pandemic and a loss of confidence [25], adding to the monopoly power wielded by large technology platforms and an increasingly polarized society, in which radical and authoritarian discourses now dot the landscape. Thus, despite the technological possibilities derived from the application of AI, a poor use—whether done consciously or not—could serve to aggravate some of the above-mentioned problems.

Due to its nature, ethical reflections on the application of AI in journalism should not be limited to journalists, but must take place within a much broader framework that also encompasses engineers, technologists and lawyers. The ramifications of the incorrect application of AI extend well beyond journalism, as has been observed on countless occasions in recent years [28, 40].

In discussions on AI and ethics, it is important to keep in mind that we are talking about ethics for the creators of robots and AI-based solutions. The debate should not focus on the need for an ethics for robots, but rather for their creators and users, in this case journalists and media organizations. Humans are more prone to error than machines.

It is therefore reasonable to ask ourselves whether the ethical principles of journalism are enough to face the challenges posed by the application of AI. The answer should be affirmative, in that the technology will not alter the moral principles upon which these principles rest. Nevertheless, regard must be had for the impact that the development of any technological innovation may have and which may complicate compliance with these moral principles or the need to update them prior to being applied to new scenarios. In this regard, for instance, algorithms are frequently considered black boxes due to their complexity and opacity, as well as their capacity for learning and self-improvement, which could escape their creators' control.

One of the fears associated with computational journalism is the risk of promoting the creation of filter bubbles. Algorithms, like humans, have ideological biases that are reflected in the final product, in the decisions that the algorithm makes based on the supplied data [11]. These biases have multiple origins, they may emanate from the quality of the data, as well as, for instance, the rating, association and filtering techniques adopted by programmers. Algorithms are not racist or sexist. They do not foster ideological extremism. However, the training given to those who use them and the quality of the data on which they are based may give rise to unethical or illegal practices.

Journalists must therefore ensure the credibility and accuracy of the data they supply to algorithms, as well as the methods used to prevent their unethical use, which may result in the creation of closed spaces in which people are only exposed to content related to their interests. They must strengthen their ethical responsibility with regards to data, which should be treated like sources of information, to which the principles of accuracy, independence, responsibility and privacy apply. Poor quality data yield poor quality results.

Personalization may fuel this problem, in that it determines what is offered to the reader—in accordance with a series of criteria that are not always transparent—and what is hidden. There are more and more media outlets that personalize content, decide what type of content should be distributed through social media or provide the audience recommendations based on their consumption patterns [39]. These practices, geared towards maximizing profits, may have an impact on the quality of the information, prioritizing clickbait or certain extreme content over other kinds of news stories, with the subsequent consequences this may hold for the quality of democracies [31, 32].

Although, as we have seen, AI influences numerous facets of journalism, ethical reflections should revolve around the news production process and its consequences. Traditionally, the media has verified journalistic products and processes based on a series of ethical precepts; although those precepts remain equally valid today, their application requires greater commitment and complexity. This, however, does not justify the lack of accountability on the part of producers, who, conversely, must guarantee the quality of the information and the ethical values on which it is based.

Despite the continued development of AI-based initiatives, very few ethical codes and guidelines have integrated reflections on the consequences of practices such as automated news production and personalization. While some initiatives do exist, they have a tendency to generalize. A recent study by Lauri Haapanen [12] confirms that little attention is being paid by European media councils to automation technology and the use of algorithms. Finland's Council for Mass Media is probably one of the most active in this regard, having published a series of recommendations [6] that highlight the need for editorial offices to retain journalistic decision-making power and for citizens to have the right to know whether information has been developed in automated form or whether user data has been used to personalize the content.

Instead of appealing to the classical ethical principles atop which journalism rests, several authors [7, 8] are in favor of promoting transparency by reducing the challenges stemming from the use of computational journalism. Transparency would provide further insight into how news is produced and potentially recover part of the confidence that has been lost. Yet despite this, transparency is by no means an effective measure. Media outlets have economic interests, and revealing the inner workings of their algorithms would put them at a disadvantage with respect to their competitors. Furthermore, self-regulatory transparency systems implemented in other fields—e.g. The Trust Project—have proven to be ineffective and remain largely unknown to the public.

The emergence of computational journalism has added new elements on which to reflect, such as recovering the postulates of Kantian positivism or historical materialism, in efforts to reduce the complexity of human behavior to mere mathematical calculations and objective conditions. In short, to algorithms which govern the way we interact with our most immediate surroundings.

Journalism should embrace the possibilities offered by AI, yet not with a view to replacing the labor of journalism, but, in line with Whittacker [38], to complement and streamline human capabilities through augmented journalism. The key to this relationship will lie in a reasonable dependency on automation, crafting products through hybrid authorship as part of a collaboration between journalists and machines, to help human intelligence understand and transform the world. An adaptation that will be take into account the singular nature of each journalist and the perfect comprehension of natural language; one which, at least in the short-term, will generate further scepticism.

3 Is AI the Future of Journalism?

The future of journalism is journalism. The future of the field of computer science is all the new tools, technologies, methods and models that will be developed. These tools and methods are built to improve tasks in varying domains, and one of those domains is journalism.

AI is not a shiny new thing, it can help newsrooms in many ways, some of which outlined in this paper. But no, it won't transform journalism.

Technology changes, and while the pace of change may seem fast, it is in fact changing incrementally, and surprisingly quite slowly. In many cases, the big change you might see today, seemingly arriving with a bang, may have been the problem under study and development during the 10 or even 20 years ago. After all this is a gradual change.

Artificial intelligence, or any other techy buzzword, will not be the future of journalism. But, journalism will also not freeze in time when it comes to technology. To go forward, like any other domain, e.g. business, journalism should also remain on board with the moving train of technological development to grow with the times. In other words, there will be no sudden new technology coming in and changing everything. There will not be a single AI tool, or a suite of AI tools, that would magically solve all the problems in journalism, improve the workflows, find new stories, and reach new audiences.

The technology changes, both slowly and fast at the same time, and it will continue changing. Like any other domain, journalism will need to pay attention to the new developments, integrate them, help them flourish and grow with them. To take the most out of technological development, journalism cannot remain only a passive user. Unless journalism makes itself part of this change, as an agent for change and a contributor, it will not be able to take the best advantage out of the new developments. It is perhaps time for journalism to roll up its sleeves and 'contribute' to the development of AI and any other computational models and tools that would enhance the domain and benefit the industry as a whole. That would make a difference in the future of journalism in relation to new computational technologies, whatever name they may have at any given time.

Funding This work was supported by the projects *News consumption, social networks and pluralism in the hybrid media system* (RTI2018-095775-B-C44) and *The impact of disinformation on Journalism: Contents, professional routines and audiences* (PID2019–108956RB-I00).

References

1. Berger, G.: Is there a future for Journalism? Journal. Pract. **12**(8), 939–953 (2018). https://doi.org/10.1080/17512786.2018.1516117
2. Boczkowski, P.J., Mitchelstein, E.: Scholarship in online journalism: Roads traveled and pathways ahead. In: Boczkowski, P.J., Anderson, C.W. (eds.) Remaking the news, pp. 15–26. The MIT Press, Cambridge, Essays on the Future of Journalism Scholarship in the Digital Age (2017)
3. Bughin, J., Hazan, E., Lund, S., Dahlström, P., Wiesinger, A., Subramaniam, A.: Skill Shift: Automation and the Future of the Workforce. McKinsey & Company, New York (2018)
4. Calvo, L.M., Ufarte, M.J.: Artificial intelligence and journalism: Systematic review of scientific production in Web of Science and Scopus (2008–2019). Commun. Soc. **34**(2):159–176 (2021). https://doi.org/10.15581/003.34.2.159-176
5. Chan, S.P.: The future of journalism. The Aspen Institute (2017). Available at https://bit.ly/3uGKzzO. Accessed 16 Jan 2021
6. Council for Mass Media.: Statement on marking news automation and personalization. Julkisen sanan neuvosto (2020). Available at https://bit.ly/2Q6Y0dl. Accessed 14 March 2021
7. Diakopoulos, N.: Algorithmic accountability: journalistic investigation of computational power structures. Digit. Journal. **3**(3), 398–415 (2015). https://doi.org/10.1080/21670811.2014.976411
8. Diakopoulos, N., Koliska, M.: Algorithmic transparency in the news media. Digit. Journal. **5**(7), 809–828 (2016). https://doi.org/10.1080/21670811.2016.1208053
9. Dimitrov, R.: Do social media spell the end of journalism as a profession? Glob. Media J. **8**(1), 1–16 (2014)
10. Doherti, S.: Journalism Design. Interactive Technologies and the Future of Storytelling. Routledge Focus, London (2018)
11. Fleischmann, K., Wallace, W.: Value conflicts in computational modeling. Computer **43**(7), 57–63 (2010). https://doi.org/10.1109/MC.2010.120

12. Haapanen, L.: Media councils and self-regulation in the emerging era of news automation. Julkisen sanan neuvosto (2020). Available at https://bit.ly/3tEHGON. Accessed 14 Mar 2021
13. Harrower, N., Heravi, B.: How to archive an event: reflections on the social repository of Ireland. New Rev. Inf. Netw. **20**(1–2), 104–116 (2015). https://doi.org/10.1080/13614576.2015.111632
14. Heravi, B., Boran, M., Breslin, J.: Towards Social Semantic Journalism. In: Proceedings of the International AAAI Conference on Web and Social Media, vol. 6(1) (2012)
15. Heravi, B., Harige, R., Stasiewicz, A., Torres-Tramon, P.: Social Semantic Journalism in action: RTÉ News360. In: 11th International Conference on Social Media & Society, 11–13 July, London (2016)
16. Heravi, B.R., McGinnis, J.: A framework for Social Semantic Journalism. In: Proceedings of the 7th International AAAI Conference on Web and Social Media. AAAI, Cambridge, pp. 13–18 (2013)
17. Heravi, B.R., McGinnis, J.: Introducing social semantic journalism. J. Media Innov. **1**(2), 131–140 (2015). https://doi.org/10.5617/jmi.v2i1.868
18. Heravi, B.R., Morrison, D., Khare, P., Marchand-Maillet, S.: Where is the news breaking? Towards a location-based event detection framework for journalists. In: Gurrin, C., et al. (eds.) MultiMedia Modeling. MMM 2014. Lecture Notes in Computer Science 8326. Springer, Cham (2014). https://doi.org/10.1007/978-3-319-04117-9_18
19. Heravi, B., Salawdeh, I.: Tweet location detection. In: Proceedings of Computational + Journalism Symposium (2015). Available at https://bit.ly/3uEVERY. Accessed 20 Apr 2021
20. Hermida, A., Young, M.L.: Journalism innovation in a time of survival. In: Luengo, M., Herrera, S. (eds.) News Media Innovation Reconsidered. Wiley Blackwell, Hoboken, pp. 40–52 (2021)
21. Khare, P., Heravi, B.: Towards social event detection and contextualisation for journalists. In: Proceedings of the AHA! Workshop on Information Discovery in Text. ACL, Dublin, pp. 54–59 (2014)
22. Khare, P., Torres-Tramón, P., Heravi, B.: What just happened? A framework for social event detection and contextualisation. In: 48th Hawaii International Conference on System Sciences. IEEE, Kauai, pp. 1565–1574 (2015). https://doi.org/10.1109/HICSS.2015.190
23. Marconi, F.: Newsmakers: Artificial Intelligence and the Future of Journalism. Columbia University Press, New York (2020)
24. Mills, J., Wagemans, A.: Media labs: constructing journalism laboratories, innovating the future. Convergence, Online First. (2021). https://doi.org/10.1177/1354856521994453
25. Newman, N., Fletcher, R., Schulz, A., Andı, S., Nielsen, R.K.: Digital News Report 2020. Reuters Institute for the Study of Journalism, Oxford (2020)
26. Palomo, B., Palau, D.: El periodista adaptativo. Consultores y directores de innovación analizan las cualidades del profesional de la comunicación. El Profesional de la Información **25**(2), 188–195 (2016). https://doi.org/10.3145/epi.2016.mar.05
27. Quackenbush, C.: Collaboration is the future of journalism. Nieman Reports (2020). Available at https://bit.ly/33vyLol. Accessed 10 Mar 2020
28. Rosenberg, M., Confessore, N., Cadwalladr, C.: How Trump consultants exploited the Facebook data of millions. The New York Times (2018). Available at https://nyti.ms/2R1MLnf. Accessed 22 Apr 2021
29. Salawdeh, I., Heravi, B., Harrower, N.: Social repository of Ireland: Collecting & archiving event-based social media. In: 11th International Conference on Social Media & Society, 11–13 Jul, London (2016)
30. Stray, J.: Making artificial intelligence work for investigative journalism. Digit. Journal. **7**(8), 1076–1097 (2019). https://doi.org/10.1080/21670811.2019.1630289
31. Sunstein, C.R.: Republic.com. Princeton University Press, Princeton (2001)
32. Sunstein, C.R.: #Republic: Divided Democracy in the Age of Social Media. Princeton University Press, Princeton (2017)
33. Taylor, A., Heinonen, S., Ruotsalaien, J., Parkkinen, M.: Highlighting Media & Journalism Futures 2030. Finland Futures Research Centre, Turku (2015)
34. Torres-Tramón, P., Hromic, H., Heravi, B.R.: Topic detection in Twitter using topology data analysis. In: Daniel, F., Diaz, O. (eds.) Current Trends in Web Engineering. ICWE 2015. Lecture

Notes in Computer Science 9396. Springer, Cham (2015). https://doi.org/10.1007/978-3-319-24800-4_16
35. Torres-Tramón, P., Hromic, H., Walsh, B., Heravi, B., Hayes, C.: Kanopy4Tweets: Entity extraction and linking for Twitter. In: Dadzie, A.-S., et al. (eds) #Microposts2016 6th Workshop on Making Sense of Microposts @WWW2016. CEUR, Montréal, pp. 64–66 (2016)
36. Túñez, J.M., Toural, C., Cacheiro, S.: Uso de bots y algoritmos para automatizar la redacción de noticias: percepción y actitudes de los periodistas en España. El Profesional de la Información **27**(4), 750–758 (2018). https://doi.org/10.3145/epi.2018.jul.04
37. Vogels, E.A., Rainie, L., Anderson, J.: Experts predict more digital innovation by 2030 aimed at enhancing democracy. Pew Research Center (2020). Available at https://pewrsr.ch/2RdYqiy. Accessed 15 Mar 2021
38. Whittaker, J.: Tech Giants, Artificial Intelligence, and the Future of Journalism. Routledge, New York (2019)
39. Wang, S.: The New York times built a Slack bot to help decide which stories to post to social media. NiemanLab (2015). Available at https://bit.ly/3hizKjO. Accessed 15 Mar 2021
40. Wong, J.C.: The Cambridge Analytica scandal changed the world—but it didn't change Facebook. The Guardian (2019). Available at https://bit.ly/3f3WcKI. Accessed 16 Mar 2021

Bella Palomo Full Professor of Journalism Department at the University of Malaga. She has focused her line of research on digital journalism during the last two decades. She has been visiting scholar at the Universities of Washington, Rutgers, Miami (US), and Federal de Bahia (Brazil). She is leading the project "The impact of disinformation in Journalism: Contents, professional routines and audiences".

Bahareh Heravi Assistant Professor of Information and Communication Studies at University College Dublin. She researches the application of new technologies in journalism, news and media industry. She is particularly focused on research in the areas of data and computational journalism and data storytelling.

Pere Masip Professor of Communication at Ramon Llull University. His main research interests are digital journalism, media pluralism and the impact of technology on journalistic and communication practices.

Printed in the United States
by Baker & Taylor Publisher Services